Geospatial and Hydrological Modelling for Sustainable Land and Water Management

This book presents a holistic scientific approach to addressing current and future challenges in the sustainable management of land and water resources, supported by geospatial and hydrological modelling. It explores remote sensing integrated with hydrological modelling for watershed management, land and water resource planning, soil erosion and sediment transport studies using satellite data, and climate change impact assessments on land and water systems all illustrated through case studies.

- Focuses on both geospatial and hydrological modelling for effective land and water resource management
- Covers topics including Google Earth Engine, forest land management, coastal land issues, geo-hydrology, and salinity problems
- Discusses Intergovernmental Panel on Climate Change (IPCC) guidelines for water resource management
- Incorporates big data and machine learning applications in water resource management
- Reviews digital image processing techniques applied to satellite data

This book is intended for researchers and professionals in hydrological engineering, geospatial technologies, and environmental engineering.

Urban Remote Sensing for Energy and Environmental Sustainability
Series Editor: *Deepak Kumar and Sulochana Shekhar*

This series aims to provide compelling new insights into how urban remote sensing and environmental sustainability are attempting to address sustainability challenges via major applications of geospatial technology in an urban area. It will elucidate the role of remote sensing and geospatial techniques for urban resource management, to cover the theory and practice of urban sustainability transitions. It will provide the case studies and contextualized tools for the governance of urban transitions to present various applications of geospatial techniques in an urban environment.

Geospatial Technologies: Environmental and Climate Science Applications and Challenges
Edited by Pradeep Kumar, Rajeev Bhatla, Sudhir Kumar Singh and Arti Choudhary

Geospatial and Hydrological Modelling for Sustainable Land and Water Management
Edited by Deepak Kumar, Anil Kumar, and Shekhar Singh

For more information about this series, please visit: www.routledge.com/Urban-Remote-Sensing-for-Energy-and-Environmental-Sustainability/book-series/CRCURSEES

Geospatial and Hydrological Modelling for Sustainable Land and Water Management

Edited by
Deepak Kumar, Anil Kumar, and Shekhar Singh

CRC Press
Taylor & Francis Group
Boca Raton London New York

CRC Press is an imprint of the
Taylor & Francis Group, an **informa** business

First edition published 2026
by CRC Press
2385 NW Executive Center Drive, Suite 320, Boca Raton, FL 33431

and by CRC Press
4 Park Square, Milton Park, Abingdon, Oxon, OX14 4RN

CRC Press is an imprint of Taylor & Francis Group, LLC

ISBN: 9781032974231 (hbk)
ISBN: 9781032979144 (pbk)
ISBN: 9781003596080 (ebk)

DOI: 10.1201/9781003596080

Typeset in Times
by Newgen Publishing UK

Contents

Foreword

It gives me immense pleasure—and, indeed, a sense of intellectual satisfaction—to pen this foreword for the meticulously curated volume titled *Geospatial and Hydrological Modelling for Sustainable Land and Water Management.* This scholarly work could not be more apropos, appearing at a time when our planet is besieged by multifarious ecological perturbations, chief among them the insidious, often imperceptible, yet inexorable transformations in our natural resources. These spatiotemporal vicissitudes, if left unaddressed, threaten to destabilize not merely ecosystems but the very fabric of life as we know it.

In an era increasingly defined by the confluence of climate cataclysms, burgeoning demographics, and an almost rapacious appetite for finite natural assets, it is patently clear that we must transcend the silos of conventional thinking. What is now required is a paradigmatic shift—an epistemological reorientation—toward holistic, science-anchored methodologies that seamlessly amalgamate environmental consciousness with technological ingenuity. In this intellectual endeavor, hydrological modelling has assumed a paramount role. It is no longer merely a scientific tool, but a veritable compass, guiding us through the turbulent waters of planetary change, allowing us to assess, predict, and judiciously manage our water resources. This book stands as a testament to such integrated thinking. It offers a perspicacious exploration of modelling frameworks, deftly weaving together the analytical precision of hydrological science with the spatial acuity of remote sensing and geospatial analytics. More significantly, it ventures beyond the confines of technical abstraction to incorporate socioeconomic variables, thereby rendering its assessments not only scientifically robust but also pragmatically relevant.

The authors—each an erudite in their own right—have coalesced their collective intellectual prowess to craft a narrative that is both illuminating and accessible. This volume caters to a diverse constellation of readers: from seasoned researchers and environmental practitioners to curious students and discerning policymakers. It invites them to engage with its contents not as passive recipients of information, but as co-thinkers in the larger mission of environmental stewardship.

I extend my unequivocal commendation to the editors and contributing authors for this admirable contribution to the ever-evolving corpus of environmental literature. Their work epitomizes the quintessence of interdisciplinary—where science, technology, and societal concerns converge in a harmonious symphony of insight. May this book inspire its readers to move beyond intellectual inertia and embrace actionable, evidence-based strategies for soil and water conservation. For it is only through such concerted and enlightened efforts that we may aspire to bequeath a water-secure, ecologically balanced, and genuinely sustainable world to posterity.

Dr. Sudheer Chintalapati
Additional Director (S)/ Scientist E
Ministry of Environment and Climate Change

Preface

Land and water resources are dynamic natural resources which are changing its coverage both in terms of space and time. With development of industrial, agricultural, and urban areas, land and water resources have suffered in terms of quality and quantity. We have to see these resources as most valuable resources and develop plans to conserve them for future generations. A sustainable approach of land and water management is necessary for keeping water resources and land resource safe for future generations. Thus, planning, designing, and implementation are three important tools for sustainable development of natural resources. A coupled approach of geospatial science and mathematical models can be used to study change detection and to develop optimal management plans for land and water resource management.

Sustainable land and water management involves a system of complex processes which include soil erosion, groundwater flow, crop growth, water allocation, and climate change impact assessment, among others. Geospatial science and mathematical modelling offer a powerful framework to analyze these interrelated factors, enabling policymakers, researchers, and practitioners to make informed decisions. This book aims to bridge the gap between theory and practice, showcasing real-world applications of mathematical models in land and water management.

The book covers a wide range of themes which include remote sensing coupled hydrological modelling for watershed management, remote sensing for land and water management, soil erosion using satellite data, climate change impact assessment on land and water resources, emerging sensor technologies, and trends in mathematical modelling for sustainability.

Geospatial and Hydrological Modelling for Sustainable Land and Water Management is intended for a diverse audience, including researchers, academics, professionals in the fields of hydrology, agriculture, environmental science, and mathematics. This book will be equally helpful for environmental policymakers, government officials, and Non-Governmental Organizations (NGOs) working on land and water resource management.

Deepak Kumar
Anil Kumar
Shekhar Singh
Editors

About the Editors

Deepak Kumar is currently working as Assistant Professor in the Department of Soil and Water Conservation Engineering, G.B. Pant University of Agriculture and Technology (GBPUA&T), Pantnagar, India. His specific research interests focus on surface and groundwater remediation methods, soil and water conservation technologies, artificial intelligence, and remote sensing technologies in agriculture and allied sectors. He holds a PhD in Civil Engineering with emphasis on Water Resource Engineering from IIT Delhi; MTech in Agricultural Systems and Management from IIT Kharagpur, India. Dr. Kumar also has post-doctoral research and teaching experience at the Department of Hydrology, IIT Roorkee. He has worked as a guest faculty member at School of Technology, Assam University, Silchar (a Central University). He has also worked as an Adjunct Fellow at the School of Engineering, Design and Built Environment, Western Sydney University, Australia. He has published in international journals including the *Journal of Contaminant Hydrology*, *Neurocomputing*, *Journal of Irrigation and Drainage Engineering*, *Water Resource Management*, *Journal of Water and Land Development*, and *Journal of the Geological Society of India*. He has also served as a reviewer of research articles for Taylor & Francis and several other internationally acclaimed publishing houses. He has guided several MTech, and BTech students. He has delivered invited lectures in various engineering, research institutions, and technical associations like Western Sydney University, Australia; IIT Roorkee; Graphic Era Deemed to be University and Uttarakhand; and Assam University, Silchar. He has conducted several workshops, training programs, and international programs, and is an active coordinator of IIRS and ISRO programs.

Anil Kumar is Professor of Soil and Water Conservation Engineering at G.B. Pant University of Agriculture and Technology, Pantnagar, Uttarakhand, India. He has been head of the department for six years. He is an active member of various national and international professional societies. He has over 40 years of teaching, research, and extension experience. His work is particularly focused on teaching and research in the field of soil and water conservation; hydrology and hydrologic modelling; watershed planning and management; rejuvenation of natural water springs; and drought assessment and mitigation for livelihood sustainability. He has conducted various field studies on the design and construction of cost-effective harvesting of surface runoff; rejuvenation of natural water springs using biological and mechanical measures; vegetative measures to check landslides along hill roads; and use of soil amendments to check the loss of runoff, sediment and nutrients from sloped lands in Himalayan region. He has taught various courses to undergraduate, post-graduate and PhD students. He has more than 140 publications to his credit, including over 90 research articles in peer-reviewed international and national journals, 19 book chapters, and 45 conference proceedings.

Shekhar Singh holds a PhD in Soil and Water Conservation Engineering from the Department of Soil and Water Conservation Engineering, GBPUA&T, Pantnagar,

India. His core research areas include climate change, climate extremes, hydrological modelling, and land and water management. Dr. Singh has worked as a Post-Doctoral Research Associate at the Indian Institute of Technology Indore, Madhya Pradesh, India, where he was actively involved in research related to water policy research. His work at IIT Indore focused on addressing pressing issues related to water governance in the Indian context. He has authored and co-authored various research papers and book chapters published in reputed Science Citation Index (SCI) and Scopus-indexed journals. In addition to academic publishing, he has presented his research at a wide range of national and international conferences, receiving recognition and awards for his scholarly excellence. He also serves as a reviewer for prestigious journals including the *Journal of Hydrology*, *Discover Water*, *Theoretical and Applied Climatology*, and *Water Resources Management* among others.

Contributors

K. Anoop Krishnan
National Centre for Earth Science Studies (NCESS)
Thiruvananthapuram
Ministry of Earth Sciences
Kerala, India

Ankit Bajaj
City Bank
Toronto, Canada

Ravi Kant Bhardwaj
Texas A&M University
Texas, USA

Kamal Srogy Darwish
Department of Geography
Faculty of Arts
Minia University, Minia
Egypt

Akash Deep
Department of Civil Engineering
Graphic Era, Deemed to be University
Dehradun, Uttarakhand, India

Shubham Dixit
Indian Institute of Technology, (BHU)
Varanasi, Uttar Pradesh, India

Divyansh Ghildiyal
College of Technology
GBPUA&T, Pantnagar, India

Abhilash Gogineni
Department of Civil Engineering
National Institute of Technology
Jamshedpur, Jharkhand, India

Rohit Kumar Gupta
Department of Agricultural and Food Engineering
IIT Kharagpur, India

Alaa Hassan
Department of Social Sciences
Mansoura University, Mansoura City
Egypt

Gauri Jayaswal
Institute of Himalayan Risk Reduction (IHRR)
Lalitpur, Nepal

Jeenu Jose
Cochin University of Science and Technology
Kochi, Kerala, India

Harish Khali
Department of Civil Engineering
Graphic Era, Deemed to be University
Dehradun, Uttarakhand, India

Shubhi Khare
Department of Soil and Water Conservation Engineering
GBPUA&T, Pantnagar, India

A. Krishnakumar
National Centre for Earth Science Studies (NCESS)
Thiruvananthapuram
Ministry of Earth Sciences
Kerala, India

Aekesh Kumar
Department of Soil and Water Engineering
Punjab Agricultural University
Ludhiana, Punjab, India

Anil Kumar
Department of Soil and Water Conservation Engineering
GBPUA&T, Pantnagar, India

Deepak Kumar
Department of Soil and Water Conservation Engineering
GBPUA&T, Pantnagar, India

Manish Kumar
Dr. Rajendra Prasad Central Agricultural University
Pusa, Bihar, India

Sanju Kumari
Agha Khan Rural Support Programme
Patna, India

Sandeep Kunwar
Department of Civil Engineering
Graphic Era, Deemed to be University
Dehradun, Uttarakhand, India

A. R. Malavika
Gandhigram Rural Institute
Dindigul, Tamil Nadu, India

Priyadarshi Maurya
Department of Science, Technology and Society
University School for Advanced Studies IUSS Pavia
Italy

Ashutosh Mishra
College of Fisheries
GBPUA&T, Pantnagar, India

Nitin Mishra
Department of Civil Engineering
Graphic Era, Deemed to be University
Dehradun, Uttarakhand, India

Mahabir Singh Negi
Department of Geography
Hemwati Nandan Bahuguna Garhwal University
Srinagar, Uttarakhand, India

S. K. Nirala
Dr. Rajendra Prasad Central Agricultural University
Pusa, Bihar, India

Krishnakant Nivesh
College of Technology
GBPUA&T, Pantnagar, India

Victor Nyabuti Ongéra
University School for Advanced Studies
IUSS Pavia, Italy

Vishnu Prasad
Water Technology Centre
ICAR-Indian Agricultural Research Institute
New Delhi, India

Akash Raj
Dr. Rajendra Prasad Central Agricultural University
Pusa, Bihar, India

Vikas Rawat
Department of Geography
Hemwati Nandan Bahuguna Garhwal University
Srinagar, Uttarakhand, India

Srija Roy
Department of Civil Engineering
National Institute of Technology
Jamshedpur, Jharkhand, India

Amneesh Singh Saini
Department of Agricultural and Food Engineering
IIT Kharagpur, India

V. Santhosh
Gandhigram Rural Institute
Dindigul, Tamil Nadu, India

A. P. Sharma
College of Fisheries
GBPUA&T, Pantnagar, India

Amit Kumar Sharma
Department of Civil Engineering
Graphic Era, Deemed to be University
Dehradun, Uttarakhand, India

I. J. Singh
College of Fisheries
GBPUA&T, Pantnagar, India

Kritika Singh
College of Technology
GBPUA&T, Pantnagar, India

Mahesh Chand Singh
Department of Soil and Water Engineering
Punjab Agricultural University
Ludhiana, Punjab, India

Shekhar Singh
Department of Civil Engineering,
IIT Indore, India; Department of Soil and Water Conservation Engineering, GBPUA&T, Pantnagar, India

Shiva Tiwari
Department of Soil and Water
Conservation Engineering
GBPUA&T, Pantnagar, India

Benson Turyasingura
Department of Environment and Natural Resources
Kabale University, Kabale
Uganda

U. Vishnu
University of Kerala
Palayam, Thiruvananthapuram, Kerala, India

Jayantpal Yadav
Farm Equipment Manager
Reliance Bio-Energy Limited, Sudhadihikateni, Odisha, India

Acknowledgments

It gives me immense pleasure to present this edited book, and I take this opportunity to express my heartfelt gratitude to all those who have contributed to its completion. First and foremost, I am highly grateful to the contributing authors whose valuable chapters form the core of this book. Further, I extend my sincere thanks to all reviewers and editorial advisory members for their thoughtful suggestions which have significantly improved the quality and coherence of the book.

I also acknowledge the support and encouragement received from my home institution, G.B. Pant University of Agriculture and Technology, for providing the academic environment and resources essential for the success of this project. My special thanks to Taylor & Francis for their professionalism, timely coordination, and assistance throughout the publication process.

Last but not the least, I am thankful to my family, friends, and research scholars for their constant motivation, patience, and unwavering support during the compilation of this volume. This book is a collective effort, and I remain indebted to all who contributed, directly or indirectly, to its realization.

1 Impact of Land Use Land Cover (LULC) Change on Watershed Hydrology

A Review

Nitin Mishra, Sandeep Kunwar, and Akash Deep

1.1 INTRODUCTION

Many developing countries are now facing environmental issues mainly caused by land use land cover (LULC) changes (Matlhodi et al., 2019). The conversion of forest and woodland into many types of land uses is widely seen in the changes in LULC, which are led by increased agricultural activities and expansion in residential, urban, and industrial lands. This conversion demonstrates significant transformations in the natural landscape with important implications for hydrology and water resources within a watershed (Mariye et al., 2022; Seyam et al., 2022). For instance, since 1990, worldwide deforestation has led to the loss of around 420 million hectares of forest (FAO, 2020). Ibrahim et al. (2022) highlighted the environmental sustainability risks linked to the conversion of forest-rich and natural ecosystems into agricultural land, particularly in the less developed areas of Nigeria.

A similar scenario is also noted in Ethiopia's Central Rift Valley, a region with high biodiversity noted for its multiple ecosystem services (Elias et al., 2019). The area is experiencing immense pressures resulting from several factors, such as rapid population increase, uncontrolled urban growth, intense agricultural expansion, and related LULC changes. This will, in fact, give a profound transformation of the region's landscape and therefore challenge the hydrology and water resources of the region. Thus, having these changes in a sustainable and meticulously planned manner is integral to environmental preservation. Hydrological processes within a catchment are dominated by a sophisticated interaction between several factors, which cover LULC, climatic fluctuations, topography, and soil properties. The LULC changes mainly influence these processes, which naturally change both spatially and temporally.

An important review of 200 peer-reviewed articles reveals a wide range of aspects on how land use and climate change impact river hydrology. Recent studies have demonstrated that changes in land use related to deforestation or urbanization have big effects on hydrological processes. For example, modeling in the upper Mara River Basin in Kenya demonstrated that both land use and climate change have substantial effects on hydrology, and enhanced resource management strategies should be put in place (Mango et al., 2011). The same conclusions came from the Yellow River Basin

DOI: 10.1201/9781003596080-1

in China, where change in land cover was demonstrated to change the dynamics of hydrology; hence, integrated management should be embraced (Cuo et al., 2013). Interactions between climate change and land use also have implications for biodiversity and ecosystem functions. Oliver and Morecroft (2014) further emphasize that while complexities of attribution problems make it difficult to understand the effect in question, the combined action of climate and land-use change poses risks to biodiversity, yet it also creates opportunities for conservation strategies. Peters et al. (2019) further explain that the interaction of climate and the land use is essential in shaping tropical mountain biodiversity, thus underlining a need to consider these factors in ecological assessments.

Besides, the effect of hydrology on aquatic species has been expressed, especially as regards the impact of climate change and wildfires on habitats for salmonid species (Isaak et al., 2010). These combine to cause a significant shift in temperatures in streams, which directly alters the fish populations and the general aquatic ecosystem health. There are many studies that promote the significant influence of LULC change on catchment hydrology. For instance, in Ethiopia, at the Didessa River basin in the Blue Nile River basin, average monthly flow increased over a variety of time spans as a result of LULC changes (Chimdessa et al. 2018). This is corroborated by surface runoff and evapotranspiration (ET) increase based on land-use change, as seen with Belihu et al. (2020). These findings, along with those from other investigations, have highlighted the need for an understanding of the effects of LULC changes on catchment hydrology for effective watershed management and conservation planning (Legesse et al., 2003; Rientjes et al., 2011; Wolde Yohannes et al., 2018).

The Bilate Catchment is an agriculturally significant and populous area within the Rift Valley Lakes Basin, Ethiopia and undergoes a series of critical environmental challenges like deforestation, overgrazing, soil erosion, and agricultural land expansion (Elias et al., 2019; Garedew et al., 2009; Godebo et al., 2018). The separate impacts of climate change and the LULC on hydrological processes in the Heihe River Basin prove that both factors significantly affect runoff and ET. The study uses the Soil and Water Assessment Tool (SWAT) model to quantify this impact, which calls for the integration of climate and land use considerations into effective water resource management across tropical regions. Along the same lines, Zhang et al. (2010) examine the impact of climate changes on the water resources in arid areas, always emphasizing the need for grasping hydrological dynamics in reaction to climate variability. These issues are compounded by the sensitivity of subsistence farming, typical to the region, to patterns of rainfall variability, such as the crop failure in the short rainy season of 2021.

1.2 LITERATURE REVIEW

Escalating irrigation through increased groundwater extraction and expanding urbanization are widespread land-use activities affecting the water balance in the Upper-Kharun Catchment region, Chhattisgarh, India. Intensive pumping of groundwater for irrigation has significantly reduced contributions from groundwater to streamflow, causing a decline in discharge and the water yield. The rise in annual surface runoff through urbanization occurred over the decades. Apparently, at the catchment scale, the change in land use seems to have a less significant effect on water balance, as several effects that happen to be opposing, such as intensified cropping versus urban expansion, occur simultaneously. However, at the subcatchment level, the effects are more evident and noticeable. In the case of urban land, it has lower penetration limit because of impervious areas, expulsion of vegetation, and in this manner transpiration. Consequently, urbanization widens flood hazard both due to changing flood recurrence allocations and due to loss of the economy (Hurkmans et al., 2009). The use of geographical information (GIS) analysis is aimed at evaluating the spatial distribution of all land cover types and sub-watersheds that contribute runoff to sampling sites along the Njoro River in Kenya. Downstream, nutrient levels have dramatically declined due to efficient nutrient removal as the stream passes through areas of large-scale farming, protected riparian zones, and in-stream vegetation. Mixed agricultural and bare lands are responsible for more than 55% of the phosphorus load in the upper and middle sections of the river. Furthermore, sub-watershed size explains about 53% of the variation in soluble phosphorus levels, meaning that land use is a critical parameter that most influences the quality of water within the watershed. The study is significant in that it stresses that both upstream and in-stream land uses have significant effects on water quality. High phosphorus loss from the watershed is also depicted to be lowered by intensive farming, plantation forests, and indigenous forest cover (Shivoga et al., 2007). The spatial extent of LULC changes plays a vital role in the development of these impacts (Petchprayoon et al., 2010). Assessing the impact of LULC change on hydrology is fundamental to building strategies that ensure sustainable water resource management.

Understanding how changes in each LULC class affect hydrological components is important for improving the robustness of hydrological analyses against changes in LULC. In the Upper San Pedro watershed, hydrologic modeling combined with multiple regression analysis was employed to understand the effects of changing different LULC classes on hydrological dynamics. For the period 1973–1997, urban development was found to be an important contributor to increased surface runoff and total water yield (0.4–2.2%). At the same time, the replacement of grassland and desert scrub with mesquite results in reduced percolation and base flow along with increased ET. Urbanization was determined to be an important land cover that affects runoff and water yield, placing considerable environmental stress on the hydrological component of the Upper San Pedro Basin (Nie et al., 2011). Similarly, LULC changes within the Upper-Brantas watershed in East Java, Indonesia, were analyzed over a 20-year period (1995–2015) using the SWAT model. This analysis revealed an increase in the settlement areas and dry-land agriculture, accompanied by a decline in forested regions, sugar plantations, and rice fields. Simulations reflected an increase in runoff 8% and water yield 0.28% coupled with a decline in groundwater flow (–1.8%) and ET (–1.15%) (Astuti et al., 2019). Multi-temporal land-use arrangements of three river catchments, analyzed for 1989/1990, 2000/2001, and 2009/2010, further highlighted the drastic impact of land-use change on hydrological patterns.

The principal changes in land use within the Mula and Mutha rivers catchment, Pune, India, involved an expansion of urban areas from 5.1% to 10.1% and an expansion of cropland areas from 9.7% to 13.5% of the catchment zone. At the catchment scale, land-use change impacts tend to cancel each other out. At the sub-basin scale, expansion of urban areas resulted in a rise in water yield by as much as 7.6% while decreasing ET. Conversely, expansion of cropland increased ET by as much as 5.9% (Wagner et al., 2013). Such extensive studies of LULC change in the Lake Victoria Basin (LVB) using diverse models and methodologies have, to a great extent, demonstrated impacts. It is reported by Liu et al. (2022) that, in the LVB, forest and grassland degradation along with agricultural expansion have drastically altered basin hydrological patterns. This trend is similar to that mentioned in the paper by Owuor et al. (2016), who observed that LULC change in semi-arid regions has strong impacts on dynamics in groundwater recharge and surface runoff. Bhatta et al. (2019) used the SWAT model to study the hydrological impact of climate variability in the Himalayan River Basin. This approach offers valuable insights for understanding the interplay between climate and LULC dynamics in the LVB. In fact, Pokhrel et al. (2018) underscored the need for multi-factorial analyses to better examine the added effects of climate changes, land-use changes, and damming in light of hydrological alterations in the Mekong River Basin. Such analysis would lead to a disentangling of complex mechanisms in the LVB.

In the Pennar Basin, Andhra Pradesh, ET, runoff potential, and base flow were simulated for LULC scenarios to estimate their impacts on the basin's hydrological system. Between 1985 and 2005, built-up areas increased by 13.94%, whereas deciduous forest areas diminished by 2.44%. The extensive grid-level spatial analyses showed significant changes in hydrologic parameters. Overflow generation between 1985 and 2005 increased up to an extreme value of 265 mm and decreased by 48 mm between 1985 and 1995. ET estimated increases up to 400 mm and a reduction of 570 mm under different LULC conditions. The base flow parameters showed increments of 70 mm and decrements of 100 mm. Therefore, LULC variations can substantially influence a river basin's hydrology, and hence there is a demand for more detailed spatial-scale studies (Garg et al., 2017).

The significant reasons for urban floods involve an increment in precipitation because of environmental change impact, exceptional change in LULC, and related hydrological effects. In the Poisar River Basin, Mumbai, India, expansion in built-up zones increased from 16.64% to 44.08%, and there was a decrease in open area from 43.09% to 7.38% with regard to the entire catchment area between 1966 and 2009. Due to the effect of urbanization in 1966–2009, there is an increase in peak discharge of about 2.61–20.9% for return times of 200 to 2 years. As the infrastructural development of the region is rapid, natural areas that were initially used as detention or retention ponds are now reclaimed for urban dwellers' needs. Results show that during low return period precipitation events, hydrological impact is greater because of geographical characteristics of the area. Detention pond systems reduce peak flow as well (Zope et al., 2017).

The conceptual HBV rainfall-runoff model was used to evaluate change in LULC on water balance in catchment, including the soil moisture, ET, runoff, and groundwater recharge. The forest and grassland areas in 1986 were put to be approximately 11% and 18%, respectively, while in 2015, they were reduced to 5% and 10%.

Meanwhile, the cultivated land area increased at different degrees, covering 70% in 1986, while in 2015 it covered 82%. Despite this extensive LULC change, the HBV model indicated only minor variations of ±5% in the water balance components, with runoff and the others being stable elements. This outcome highlights that the simplified precipitation-runoff models such as HBV have flaws as they always represent LULC changes in a generalized and aggregated manner. In this case, the study proposes integrating statistical and physically based modeling approaches with high spatial and process-based LULC representations to fully capture the effect of LULC on the catchment hydrology (Birhanu et al., 2019).

The impacts of land cover changes on the relation of rainfall-runoff in the case of extreme precipitation were estimated for the period 1989–2009 in the Yarkon–Ayalon watershed, central Israel. The hydrological model Kinematic Runoff and Erosion (KINEROS) alongside the rainfall data of an extreme precipitation occasion has been used in the study. The outcomes reveal an increment in volume of runoff and peak discharge because of land cover changes. A strong correlation was established between the expansion of vegetation and the quantity of runoffs. Urbanization and clearing of the area of vegetation had the greatest effects on the volume of runoffs (Levi et al., 2015). The Njoro River, being a small watershed, is highly responsive to short, intense precipitation events that provoke its response by overland flow processes. The present study assesses the LULC changes in terms of the extent, timing, and direction of this watershed with the Automated Geospatial Watershed Assessment (AGWA-SWAT) model. Simulations were conducted based on all LULC change activities, indicating increased surface runoff while reduced groundwater recharge was observed. Hydrological changes in the Njoro watershed between 1986 and 2003 were primarily caused by LULC change in the upland forested areas, which lowered the surface roughness and litter cover, therefore reducing infiltration and average annual ET. This underscores the role that land use or vegetation change plays in modifying hydrologic processes in small watersheds (Baker and Miller, 2013).

Land-use change profoundly influences the energy balance of watersheds with effects on ET rates as well as fluxes of water. For example, Spera et al. (2016) indicated how the changes in land use at agricultural frontiers in Brazil modify the recycling of water such that less water recurs. Similarly, Wagner et al. (2013) utilized the SWAT model to assess the hydrological balance in the Pune region, revealing that urbanization increases water yield while decreasing ET, echoing the concerns noted in the original research by Talib and Randhir regarding altered basin hydrology. Another important dimension is land-use changes and climate variability interaction. Zhang et al. (2016) indicate that for the Heihe River Basin, changes in land use interact with climatic trends to drive hydrological processes. Of note, it therefore seems that these two variables are interconnected and need to be combined in policies designed for water resource management. This actually resonates with a larger theme in the literature review efforts in which complex land-use change interactions with climate impacts modulate watershed hydrology.

This study employed a holistic methodology that combined the SWAT and Partial Least Squares Regression (PLSR) models to assess the hydrological elements' responses in the Northwest Benin Owena River Basin (NWBORB) in Nigeria to land degradation. The results obtained included an 18.56% reduction in forest cover area from 1986 to 2015, of which 16.19% of the said area was converted to plantations. This land cover change translated to an 18.28% rise in annual water yield and a 17.26% increase in runoff coefficient, while base flow and actual ET declined by 22.58% and 21.72%, respectively. A higher negative correlation exists between forest area loss and plantation land gain (–0.833), runoff coefficient (–0.723), and water yield (–0.532). The results in these findings underscore the need to consider land-use changes in river basin water resource management structures (Aladejana et al., 2018). The study looked at the historical urbanization patterns and impervious surface coverage that occurred between the 1960s and the 2010s by digitizing topographic maps into a hydrological model to understand the current storm runoff responses in comparison to past conditions. The findings from the peri-urban catchment showed an increase in impervious surfaces from 11% in the 1960s to 44% in the 2010s. Coupled with the installation of a large storm drainage system in the early 2000s, this resulted in a 50% reduction in the Muskingum routing parameter 'k' and reduced the natural flood duration by half and the peak flow more than fourfold. When viewed comparatively with more urbanized areas, the largest increases in peak flows and flood duration reductions occurred at lower levels of urbanization. The stormwater management systems accelerate storm runoff velocity highly compared to the effect of impervious surfaces alone in the study (Miller et al., 2014).

Hydrological processes in the Tons River Basin, located in Madhya Pradesh, India, were observed at topographical changes over 50 years from 1985 to 2035. In this area, striking changes in mixed-crop land usage were recorded from 72.68% to 78.27%, deciduous forest land usage reduced to 9.77% from 15.57%, and residential land usage increased to 2.81% from 0.29%. The SWAT simulation demonstrated that these changes in land use did not have a significant effect on the water processes. Surface runoff changed slightly from 62.29% to 62.14%, whereas groundwater changed between 37.42% and 37.62%. Total water yield varied between 774.61 mm and 776.74 mm. These slight alterations were due to low alterations in LULC alteration together with very limited anthropogenic disturbance of the area. In addition, farming in TRB for wheat, millets, soybeans, and pulses does not require massive water resources (Kumar et al., 2018). Rainfall-runoff models are traditionally used to explore hydrological techniques at different scales ranging from a small catchment to a global model. However, most of these models are associated with limitations such as complexity, the requirement of high data, and obscurity over limitations (Devi et al., 2015).

This study innovated on the application of rainfall-runoff models in arid catchments. This is by incorporating the GSMaP_MVK+ satellite-derived precipitation dataset, whose recalibrated source of rainfall data allowed for accurate identification of rainstorms. The study made use of the Hydrologic Engineering Center-Hydrologic Modeling System (HEC-HMS) model, along with its extension, HEC-GeoHMS, in ArcView 3.3. Another model applied in this research was Java-based Identification of unit Hydrographs and Component flows from Rainfall, Evaporation and Streamflow data (IHACRES). Although the IHACRES model had some drawbacks, the HEC-HMS model succeeded as the calibrated streamflow values came out closely comparable to observed streamflow. One major strength of the IHACRES model is that it has a minimal number of data inputs. As concluded in the study by Abushandi and Merkel (2013), calibrating both the HEC-HMS and IHACRES models will minimize any possible prediction errors resulting from an individual storm event. The effects of changes in climate and land-use patterns on discharge behavior have been simulated here, resulting in significant mismatches due to differences in model structure and calibration methods. Consistent with the outcomes of the land-use scenarios from the West African Catchment from 1998 to 2005 are results using the SWAT, Water balance Simulation Model (WaSiM), and Unit Hydrograph Parameter-Hydrologic Response Unit (UHP-HRU) models. The model results show that, based on land-use changes, the surface runoff is going to increase. However, climate change scenarios introduce significant variability among models, indicating the kind of uncertainty also present in these scenarios. The findings here can be used to understand the plausible range of future developments (Cornelissen et al., 2013). The Variable Infiltration Capacity (VIC) model is applicable in wetland areas. It can be used to optimize the management of water resources for agricultural purposes. However, MIKE SHE is a model that suits small catchments and requires high data input as well as specific physical parameters. The SWAT model needs minimal calibration to yield reliable hydrologic predictions, offering it as an asset in hydrological studies (Devi et al., 2015).

1.3 IMPACT ON SURFACE RUNOFF AND STREAM FLOW

The Integrated Biosphere Simulator (IBIS) and the Terrestrial Hydrology Model with Biogeochemistry (THMB) were used to simulate idealized forest clearing and overgrazing scenarios in the Niger and Lake Chad basins of West Africa. Complete deforestation increased the runoff ratio from 0.15 to 0.44, thereby enhancing annual streamflow by 35–65%. On the other hand, if all the grassland and the savanna were to be cleared, modeled streamflows would increase by 33–91%. However, when deforestation rates were below 50% and overgrazing rates were below 70% for savannas and 80% for grasslands, water yield and discharge did not experience significant increases. On the contrary, once these thresholds were exceeded, considerable changes in water yield and discharge had been experienced (Li et al., 2007).

For the Banha watershed in Jharkhand, the effect of LULC on sediment yield and runoff was analyzed using the SWAT model. Lower runoff and sediment yield values have been obtained for sub-watersheds having more extents of forest cover compared to those with lesser amounts of forest coverage. This is because forests can regulate surface runoff; consequently, there will be soil erosion reduction. The sub-watersheds with a large area of cultivated regions experienced higher runoff and sediment yield. LULC has the capability to be detected through analyzing temporal and spatial variations in watershed attributes, as pointed out by the study of Mishra et al. (2007).

A deterministic, physically based hydrological model, MIKE SHE, was used in assessing hydrological changes due to land-use change in the Gyeongancheon watershed in Korea. Validation was performed by comparing simulated and observed data on streamflow from 1988 to 1991. The analysis estimated a considerable growth of total watershed runoff and overland flow for the period 1980–2000, largely contributed by urban expansion. The increase in impervious surfaces decreased the infiltration and ET, thus increasing total runoff by 5.5% and overland flow by 24.8%. However, in regions where paddy fields are flooded, this method is not applicable (Im et al., 2008). Combining ground-truth data, the remote sensing technique, and GIS tools with the GeoSFM hydrological model, LULC changes in Mara River Basin across Kenya and Tanzania from 1973 to 2000 were assessed. The results showed a 32% loss of forest area, an increase of open forest/tea area by 214%, as well as an increase in agricultural fields by 203%. These modifications led to a 7% increase in high flood flows and a greater number of peak flow events (Mati et al., 2008). The study also considered the effects of atmospheric variability and land-use changes on streamflow variability in the Nyangores River, Kenya.

The existence and breakpoint of trends in discharge data from 1965 to 2007 were assessed using the Mann–Kendall (MK) test and sequential MK test. Land-use changes were found to be the main cause of streamflow change at about 97.5% and climate variability only about 2.5%. The impacts of increased precipitation and potential ET neutralized each other's individual impact, thereby bringing out minimal net change in runoff. The main cause of alterations in discharge was deforestation in the watershed headwaters (Mwangi et al., 2016). For effective management of watersheds and ecological restoration, there should be an in-depth awareness of the hydrological process in the area. Applying the SWAT model, this study analyzed the hydrological impact of land-use changes from 1967 to 2007 in the Zanjanrood basin, Iran. The results showed that a 12% increase in rainfed agriculture translated into an increase of 43% in average yearly surface runoff. Similarly, a 14% increase in shrubland resulted in a 34% increase in surface runoff. The study also showed how sensitive surface runoff was to changes in grassland, bare ground, and irrigated agriculture. An increase of 5% in bare ground and a 2% rise in irrigated agriculture led to a rise in average annual surface runoff by 31 and 21%, respectively. Land-use changes also resulted in a 22% alteration in groundwater. It therefore emphasizes that the impacts of land cover change on hydrological responses are nonlinear with a limit to these impacts (Ghaffari et al., 2010).

Variations in surface runoff and soil erosion due to LULC changes were analyzed using the ArcSWAT model and GIS data in the Modjo watershed of Ethiopia. The results indicated a 14% increase in surface runoff and a 37% increase in soil loss with the reduction of forest, grassland, and shrub areas and an expansion of bare and cultivated lands (Gessesse et al., 2015). A study on the Xinjiang River basin of Poyang Lake explored the impacts of climatic variations and LULC changes on hydrological responses and streamflow using the SWAT model. The results showed the large impacts of land use and vegetation changes on seasonal hydrology and water balance but with minimal change in annual discharge. Forestation, through the reversion of cultivated lands to forests, decreased runoff during the rainy season while enhancing it during the dry season, which reduced the severity of flooding and drought. Deforestation, on the other hand, enhanced the possibility of flooding and the manifestation of drought (Guo et al., 2008).

A hybrid model combining Artificial Neural Networks (ANN) with the Soil Conservation Service-Curve Number (SCS-CN) hydrology method was, therefore, developed to simulate daily streamflow and predict the type of hydrological changes due to land use/cover. Results of this model were successfully tested on 10 watersheds in western Georgia, USA, using input parameters like infiltration depth, precipitation, ET, and surface runoff. The created ANN model anticipated increments in mean stream and flashiness in the case of the urban and pasture prevailing situations. In the case of forest-commanded situations, the model anticipated progressively stable hydrology with lower normal streams (Isik et al., 2012).

This assessment presents the evaluation of the effects of LULC changes on the extreme flood flow design using the HEC modeling and upgraded Transient Rainfall Infiltration and Grid-based Regional Slope-Stability (TRIGRS) modeling under the precipitation events during 30 years (from 1984 to 2013) earlier at the Kelantan basin. During the period of 30 years, the land cover enhancement in zones covered with oil palm and rubber and secondary forest and a reduction in scrub and forest zones. The results presented that excessive water flow increases due to deforestation and increments in urban areas, especially due to the decrement in the canopy coverage of forest areas (Saadatkhah et al., 2016).

Urbanization is the most dramatic of all the land-use changes affecting the hydrology of a region. As forests are cleared and soils are packed, stripped, and covered with buildings, roads, and other impervious surfaces, precipitation that formerly was absorbed by vegetation or that flowed slowly into and through the layer of soil as subsurface stream is now converted to overland flow. The combination of reduced retention and increased conveyance results in more runoff volumes, more storm flow events, lower stream base flows and higher peak stream flows that quickly rise and recede (Figure 1.1) (Glasoe and Christy, 2004).

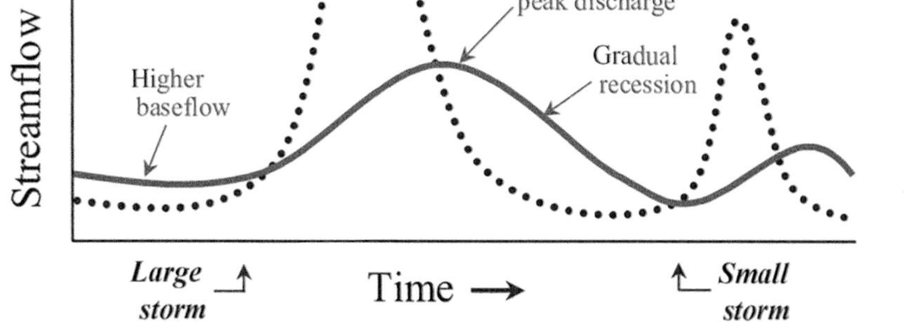

FIGURE 1.1 Stream hydrograph in response to urbanization.

Hydrological modeling and PLSR were used to assess the impacts of changes in land use from 1978 to 2007 on sediment yield and stream flow for the Du watershed, China. The principal or major land-use conversions that affected the streamflow were conversion to the urban region, forest, and farmland. Forest and farmland were the dominant factors for sediment yield. The results indicated that changes in grassland resulted in no significant impact on streamflow or sediment yield (Yan et al., 2013). Extensive observation was performed to investigate the response runoff on the day-to-day scale along with yearly and month-wise time scales. SWAT was used over a period of 9 years, from 2002 to 2010, in Jinjiang, which is a coastal catchment in China. Runoff was simulated for two different land-use scenarios, namely, 1985 and 2006, with decreased forest and increased areas of cropland and urbanized regions. With an increase in urban area and plantation area and a diminishing in forest regions for the interval 1985–2006, changes in runoff differed for different time scales and at three different catchments. Yearly overflow depicted the minor variation; monthly overflow showed medium changes, expanding entirely in months but October–November, and daily runoff depicted maximum change with expansion seen in flood peaks but reducing in dry season streams as a function of variable effect on ET/interception loss, percolation to depth, and antecedent soil water storage (Lin et al., 2015).

Effects of LULC changes were evaluated for discharge volume, runoff, and low flows in 37 catchments in East Africa. It has been found that despite the loss of forest cover, 63% of the total watersheds represent show non-critical variations in yearly releases while 31% represent expanding patterns. Loss of forest cover estimates that increases yearly releases and runoff by 16 ± 5.5% and 45 ± 14%, separately. Increase in peak flows by a mean of 10 ± 2.8% while low streams diminished by a mean of 7 ± 5.3%. With an increment in forest cover, yearly discharges and runoff decrease by 13 ± 1.9% and 25 ± 5%, respectively. There exist heterogeneities in sizes of the progressions in hydrological fluxes with changes in LULCC. Investigations showed that as LULCC changes influence discharge, runoff, and low flows, other controlling factors were also identified in the catchment (Guzha et al., 2018).

Water resources, which are significantly affected by LULC changes and atmospheric shifts, are a primary limiting factor for fragile ecosystems in arid and semi-arid areas. Simulations done over the Jinghe River Basin (JRB) in northwestern China showed that LULC combined with atmospheric shifts enhanced the surface runoff in the 1970s and 1980s by 29.75 m^3/s. However, in the period of the 1980s to 2000s, both the LULC and atmospheric shift contributed to reducing runoff by an amount of 28.24 m^3/s. It was concluded that LULC transitions consistently influenced the cycle of hydrology, and LULC transition caused a proportion of runoff change greater than environmental change, which contributed to 70.67% of the streamflow decrease since the end of the 1990s. This research underlines the importance of the role of LULC at the transition between a semi-humid area and a semi-arid region. LULC significant changes that took place during this period include grass spread, showing an amount of 16% reduction between the 1970s and 1990s followed by a 6% increase from the 1990s to the 2000s. The change is dominated by the successful Grain for Green Program implemented by the government (Yin et al., 2017).

Creating a storm runoff always requires mimicking the effects of LULC on storm runoff by taking into account various factors such as land cover properties, soil infiltration characteristics, temporal and spatial variations in precipitation, runoff generation processes, and antecedent moisture conditions in the catchment. However, modeling these effects is often limited by the lack of reliable data, spatial and temporal heterogeneity, and the limited understanding of the physical and stochastic processes, especially during extreme precipitation events (Niehoff et al., 2002). A conceptual rainfall-runoff model applied to 95 catchments in the Rhine basin demonstrated the impact of land-use changes on runoff. Results indicated that urbanization increased lower peak runoff during summer storms, whereas impact on higher peaks from winter precipitation was negligible. Highly significant reductions were found in both peak runoff and volume of total runoff as a result of increased afforestation (Hundecha and Bardossy, 2004).

A study on the impact of deforestation on regional ET and runoff in the Amazon River system showed considerable modifications of large river systems due to deforestation, even without significant changes in precipitation. It was observed, for example, that river flow increased by 25% in rivers such as the Tocantins and Araguaia, without any change in rainfall. However, beyond 30% deforestation of the Amazon basin, atmospheric feedbacks are significant enough to alter the water balance and cause an increase in river flow. These land cover change-related feedbacks reflect the regional land surface fluxes' effects but have opposite hydrological responses. It is, therefore, deduced that ongoing deforestation will keep on having considerable hydrological impacts on the watersheds of the Amazon basin (Coe et al., 2009).

It has been shown that the impacts of land use and management changes in terms of catchment discharge are sensitive to the spatial distribution of the changes, as well as the size and timing of storm events. For instance, clear-cutting in a Norwegian catchment area resulted in doubling peak discharge during a 50-year return period storm. Reafforestation was the most effective therapy when used on a catchment-wide scale to reduce both peak flow and total runoff, especially for smaller storms. However, the effectiveness of reforestation depended on its location within the catchment and was more effective near the outlet of the catchment than farther upstream. That means spatial distribution of change in land use and interventions are to be incorporated to effectively manage the watersheds and mitigate hydrological impacts (Kalantari et al., 2014).

Beijing, China experienced rapid urbanization between 1992 and 2009. By 2009, the urban areas expanded to 12.78% from 4.28% in 1992, while the residential areas expanded from 3.21% to 6.66%. Agricultural lands instead declined from 27.20% in 1992 to 11.44% in 2009. Along with the expansion of the urban areas, the long-term surface runoff also increased by 30% in the region, with modeled runoff rising by 35% in urban areas (Sun et al., 2013). Global and local land-use datasets were utilized to model surface runoff in the Manali and Kurumali watersheds using SWAT. In these watersheds, the forest zone decreased by 60% and 32%, respectively. However, changes in surface runoff did not directly correlate with variations in the forest zones, remaining within a 20% range. The highest runoff increase was only 15%, a lesser effect than expected. These might include changes from forest to agricultural land, especially plantations, which have several characteristics in common with forests, except for ET (Sajikumar and Remya, 2014).

In the densely urbanized North Moat area of Beijing, improvements in Percent Pervious Cover (PPC) spatially responded to runoff; improvements in pervious cover were closely linked to reductions in runoff, regardless of the spatial patterns or levels of precipitation. Increases in PPC to 30–40% improved the spatial continuity of pervious patches and significantly reduced runoff. When PPC was <30%, runoff was largely eliminated by routing water from impervious areas over pervious zones to the outlet. When PPC fell within the 30–60% interval, the potential for runoff reduction was limited only to light and moderate rain, while stream routing had little impact on controlling extreme rainstorm floods. The study concluded that the intensity and spatial distribution of urban pervious and impervious surfaces significantly influence runoff (Zhang et al., 2018).

1.4 IMPACT ON GROUNDWATER QUALITY AND RECHARGE

The changes in land cover affect water balance through alteration of the level of groundwater. The effects due to changes in LULC on groundwater recharge were assessed for the Haramaya watershed, Ethiopia. Land-use changes for future scenarios were simulated utilizing CLUE-S model. The outcomes showed an expansion in Chat creation and decrease in cultivated area. The WetSpass water balance model revealed that LULC change severely affects groundwater recharge. The yearly range of groundwater recharge varies from 0-90 mm in 2011 to 0-83 and 0-87 mm in 2028. The model depicts a lowering rate of groundwater recharge due to the continuous decrement in forest area (Gebere et al., 2016).

Urbanization and successive variations in LULC have extreme unfavorable consequences for the local ecology (Figure 1.2).

This study aims to look at how urban development changes over time and how it affects weather and water conditions in the HMC region of West Bengal. The research uses a mix of changes in LULC, the Normalized Difference Built-up Index (NDBI), GIS, and remote sensing methods. To study the distribution of rainfall, temperature, and groundwater levels, the Inverse Distance Weighting (IDW) method was used. The findings show that urban growth increased by 1,813 hectares from 1975 to 2000 and by 1,193 hectares from 2000 to 2015, leading to a big loss of plants, wetlands, water bodies, and lowland areas. The fast growth of urban areas, like concrete and asphalt, which soak up and keep solar heat, has raised both the average highest and lowest temperatures while also decreasing rainfall in the area. The combined effects of higher temperatures and less rain, along with urban growth, have harmed the groundwater level before and after the monsoon (Patra et al., 2018).

In this study looking at how urbanization affects groundwater quality, researchers used multiple linear regression to study the Entropy Weighted Water Quality Index (EWQI) and seven types of LULC in the Curved Streamline Searchlight Model (CS-SLM) of wells in Xi'an City for the years 2005, 2010, and 2015. The results show that urban land in Xi'an increased by more than 160%, while forest and agricultural lands dropped by 52.54% and 83.08%, respectively, from 2005 to 2015. The regression analysis indicates that agricultural land, urban land, and industrial areas harmed groundwater quality, while forests improved groundwater quality during the same time period (He et al., 2020).

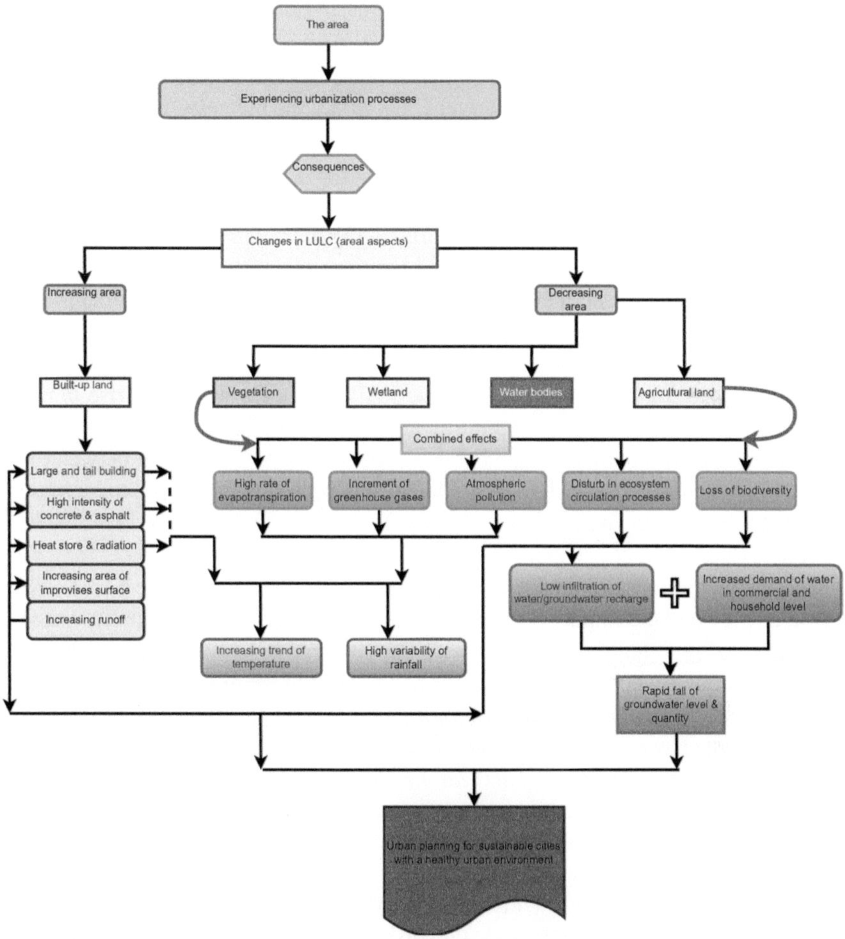

FIGURE 1.2 Hydrological and environmental impacts of urbanization process.

The investigation presents an analysis to differentiate the potential causes of groundwater fluctuation from 1995 to 2014 and values of water quality with the help of LULC change location techniques from 1975 to 2014. For WQI, eight parameters of water quality were taken into account. Image differencing and discriminant function techniques, unsupervised classification implemented for the detection of LULC change. Examinations show a noteworthy increment in the built-up region when compared with increment in the agricultural land. It was observed that the agricultural activities were the principal factors for the fluctuation in the groundwater table for the period 1995 to 2014 (Sahoo et al., 2016).

Effects of a change in the pattern of land use on the quality of groundwater were assessed in Maheshwaram watershed, Hyderabad, India. Specific parameters of water quality were included in Groundwater Quality Index (GQI) to derive a final value that could be utilized for various spatio-temporal investigations. Multispectral satellite images were employed to check the changes in land use for the time period 2003–2008. The investigation exposes a rapid declination in the quality of groundwater especially due to increment in built-up area with unsewered sanitation and poultry farms. It was noticed that for the interval 2003–2008, the mean GQI diminished from 84.16 to 83.26 (Khan et al., 2010).

Clearing of trees and farming more land is said to greatly affect the quality of the water in the aquifer. A new study in Northern Kelantan, Malaysia, uncovers ways through which human activity has been influencing the quality of groundwater due to changes in land use. The research applied time-series analysis and geospatial modeling that the levels of nitrates (NO_3-N) rose by about 8.1% annually in agriculture wells and 3.89% yearly in home wells between the years 1989 and 2014. By 1989, only 1% of the area had nitrate levels above 10 mg/L but by 2014 this number shot to 48%. Huge increases were observed at one shallow aquifer where the rise in nitrate levels was 3.74% yearly. Both agriculture and home wells have risen nitrogen levels; however, areas covered by forest and grass declined by 4.5% every year between 1989 and 2014. According to the ARIMA model, nitrate pollution is foretold to increase by 2.64–3.9% every year until the year 2030 within the agriculture and the home areas as well as (Narany et al., 2017).

This further elaborates that remote sensing can be used to analyze LULC over a period of time and the impacts of that changeover on the river water quality based on the sources of pollutants. The Analysis of Variance (ANOVA) for datasets of LULC versus water quality points out that developed regions are significantly contaminating water with contaminants such as *Escherichia coli*, biological oxygen demand (BOD), total coliforms, electrical conductivity (EC), total suspended solids (TSS), chemical oxygen demand (COD), Hg, Fe, and Zn. Farming activities cause TSS, salinity, EC, total coliforms, *E. coli*, arsenic, and iron contamination, while open areas result in salinity, EC, turbidity, and TSS contamination (Hua, 2017).

People's changes to water resources, like using too much water and dumping untreated waste, are important factors that affect groundwater quality. A study in the lower Shiwalik hills, Rupnagar, Punjab, India, looked at LULC changes from 1989 to 2005 with a focus on groundwater quality and quantity. The results showed that the amount of groundwater increased because of natural and artificial recharge, which improved the LULC trend. Enriched fertilizer added to the soil decreases its quality but increases the critical quality of groundwater. River zone area decreased by 17% from 1989 to 2005, the land used for cropland decreased by 31%, and dense forest areas decreased by 22%. However, the settlement area increased by 534%, and salt-affected land increased by 94% during the same time due to over-extraction of groundwater, fertilizer, and pesticide use (Singh et al., 2010).

Land use has impacts on groundwater resources due to the modification of recharge and alteration in the demand for water. Unsound land use, particularly mismanagement of the land increases inherent groundwater quality problems at all times. Severe groundwater quality problems arise from unsound land use and management very much through uses of harmful chemicals. Groundwater is highly susceptible to anthropogenic activities at the land surface, primarily those which involved filtration of contamination whether point or diffuse. There are three major services associated with groundwater that provides the base flow, supply of drinking water, and health of aquatic ecosystems. Present land-use instruments have only been designed to solve quality issues in groundwater and never take into account recharge (Lerner and Harris, 2009).

Reductions in groundwater recharge beneath forested areas can exacerbate water supply deficits. This study employed a groundwater stream model Modular Three-Dimensional Finite-Difference Ground-Water Flow Model (MODFLOW) alongside a soil moisture balance model to predict the effects of land-use changes in forested areas on groundwater recharge and levels by 2025. The analysis focused on two open supply borehole sites located on the unconfined Sherwood Sandstone aquifer in Nottinghamshire. The results from the simulations at both sites revealed a decrease in groundwater resources in response to the conversion of land use to forest. Notably, the recharge at both borehole sites was more significantly reduced in the winter than in the summer when land use shifted to forest. However, the groundwater tables were constrained due to the maximum storage capacity of the sandstone aquifer (Zhang and Hiscock, 2010).

The impact of LULC changes on groundwater is increasingly recognized, particularly at the watershed scale. A study of LULC changes in the Dang River Watershed from 1987 to 2007 used satellite images to assess land cover changes, while the FAO Penman–Monteith method evaluated water use linked to different land cover types. The results showed a decrease in food crop areas, while cash crops, especially cotton, expanded significantly. Agricultural irrigation accounted for the majority of water use. The total water consumption by cash crops increased from 14.1% in 1987 to 71.6% in 2007, making them the largest consumers of water (Zhang et al., 2014).

In another study, the nitrate-nitrogen (NO_3-N) status of shallow groundwater (<30 m below the surface) in areas with mixed agricultural land uses was examined. Groundwater NO_3-N concentrations were sampled at two intervals per year over 3 years at 88 sites in the Waikato Region, New Zealand. The 'DRASTIC' model was employed to assess the risk of NO_3-N leaching to groundwater. None of the three methodologies used—land use, topsoil properties, or hazard evaluation models—was highly effective in predicting groundwater nitrate concentrations. However, it was found that within a region, NO_3-N concentrations in groundwater increased as the proportion of land used for dairy farming increased (McLay et al., 2001).

Groundwater recharge was lowered and groundwater utilization was elevated from 1985 to 2010, which resulted in decreasing groundwater depth in Linze County. The area with a groundwater depth above 5 m decreased by 187 km^2 between 1985 and 2010 due to industrial growth, farming and economic development. The regions that employ inappropriate land use—such as farmlands, forests, buildings, and grasslands—had extreme increases in the levels of groundwater. Between 1985 and 2010, groundwater depth increased slightly (0–1.2 m) at the periphery of the desert (24.21 km^2), mainly due to irrigation from neighboring farmland (Wang et al., 2015).

An increase in hard surfaces concluded to be a major cause for fewer groundwater inlets in the Indira Sagar Command area, Madhya Pradesh, India. It has land-use changes, formulas for groundwater storage, recharge, preparation of a digital elevation model, and diverse drainage situations to monitor how these affect the water balance and the groundwater flow with the help of GIS. In some areas, canal construction (a land-use change) enhanced the groundwater capacity of villages because the needed water was provided through canals. Second, better plants implied higher infiltration of water in many areas (Prabhakar and Tiwari, 2015). LULC studies further revealed that the leveling of runoff and groundwater recharge is relatively lower in forests than in other types of lands and environments, particularly in the tropical and subtropical semi-arid regions. Rehabilitation of land without cover significantly reduces groundwater recharge from 42% of rainfall to between 6% and 12%, contingent upon how the land is used. Forest clearing for rangelands increased groundwater recharge by 7.8 ± 12.6%, while land change to cropland or pasture increased it by 3.4 ± 2.5% and 4.4 ± 3.3%, respectively (Owuor et al., 2016).

Changes in land use that lower groundwater recharge can be studied by looking at how chloride is fixed in soil water. This involves comparing land uses before and after the change using chloride mass balance. In the Guyuan terrace area, groundwater recharge under natural sparse grass was 100 mm/year, but changing to winter wheat about 100 years ago reduced it to 55 mm/year. Recommendations on nature-based water management in the Loess Plateau of China reveal that local tree planting and other forms of land-use changes that increase demand on water result in soil water depletion with a higher concentration of solutes. These effects undermine the process of groundwater recharge and ecological restoration (Huang and Pang, 2011). From 1992 to 2011, the change rate of LULC in the Upper Ganga Canal area grew hard surfaces from 10.60% to 35.97%. This has a resultant effect of hastening the collection of rainwater leading to more runoffs. According to the researchers, the growth and development of housing areas had decreased the rates of recharging the groundwater (Mishra et al., 2015). The seasonal water balance model WetSpass was used to estimate the recharge to groundwater in the Guishui River Basin. According to the calculation by the model, 21.16% of the recharge was attributed to precipitation and 72.54% loss to ET. Cases were positive for each and every kind of land use type, which ranged from 122.74 mm/year for cropland to 76.14 mm/year in the forests. Between 1980 and 2005, the annual volume of recharge of groundwater declined by 4×10^6 m^3. This has mainly been ascribed to having fewer areas under cropland and grassland, with more urbanized areas and rural settlements (Yun et al., 2011).

1.5 CONCLUSIONS

A comprehensive review of the research works done on the impact of LULC changes on the hydrology of a watershed has been conducted in this study. The impact of LULC change is of relatively minor significance at the catchment scale because there is a compensation of different effects and intensities (e.g., intensification of cropping vs. urbanization) but very pronounced and critical at the subcatchment scale (increase: surface runoff, actual ET; decrease: percolation and groundwater contribution to the stream flow). The effects of LULC changes can be identified by the temporal and spatial changes in the attributes of a watershed. Hydrological response to land cover change is nonlinear and has a threshold effect, below which there is no dramatic effect on the water yield and stream discharge, but beyond it the changes are drastic. Urbanization is the most important of all the land-use changes impacting the hydrology of a zone. It was revealed that the intensities and spatial distributions of urban pervious and impervious surfaces had strong effects on runoff. Results indicated an increment in runoff and streamflow that results due to deforestation and increase in urban regions, especially due to canopy cover decrease of the forest region. Replanting proved most effective in reducing peak flow and total runoff, especially during smaller storms. Replanting close to the outlet proved more effective than higher upstream. Hydrological modeling, as well as long-term ecological monitoring, will play important roles in developing sustainable water resource management strategies under changing environmental conditions. Agricultural land, urban land and industrial land have negative impacts on quality of groundwater, while the forests have positive effect in a similar period. The expanded impervious zone is the most significant factor diminishing groundwater infiltration. The abatement of cropland and grassland together with the expansion of urban zones and rural settlements are the significant reasons for declining groundwater recharge.

REFERENCES

Abushandi, E. and Merkel, B. (2013). Modelling rainfall runoff relations using HEC-HMS and IHACRES for a single rain event in an Arid Region of Jordan. *Water Resources Management*, 27, 2391–2409.

Aladejana O. O., Salami A. T. and Adetoro O. O. (2018). Hydrological responses to land degradation in the Northwest Benin Owena River Basin, Nigeria. *Journal of Environmental Management,* 225, 300–312.

Astuti I. S., Sahoo K., Milewski A. and Mishra D. R. (2019). Impact of land use land cover (LULC) change on surface runoff in an increasingly urbanized tropical watershed. *Water Resources Management*, 33(12), 4087–4103.

Baker T. J. and Miller S. N. (2013). Using the Soil and Water Assessment Tool (SWAT) to assess landuse impact on water resources in an East African watershed. *Journal of Hydrology*, 486, 100–111.

Belihu M., Tekleab C., Abate B. and Bewket W. (2020). Hydrologic response to landuse land cover change in the upper Gidabo Watershed, rift valley lakes basin, Ethiopia. *HydroResearch*, 3, 85–94.

Bhatta, B., Shrestha, S., Shrestha, P. K. and Talchabhadel, R. (2019). Evaluation and application of a SWAT model to assess the climate change impact on the hydrology of the Himalayan River Basin. *Catena*, 181, 104082.

Birhanu A., Masih I., Zaag P. V. D., Nyssen J. and Cai X. (2019). Impacts of landuse and land cover changes on hydrology of the Gumara catchment, Ethiopia. *Physics and Chemistry of the Earth*, 112, 165–174.

Chimdessa K., Quraishi S., Kebede A. and Alamirew T. (2018). Effect of landuse land cover and climate change on river flow and soil loss in Didessa River Basin, South West Blue Nile, Ethiopia. *Hydrology*, 6, 1–20.

Coe M. T., Costa M. H. and Filho B. S. S. (2009). The influence of historical and potential future deforestation on the stream flow of the Amazon River – Land surface processes and atmospheric feedbacks. *Journal of Hydrology*, 369, 165–174.

Cornelissen T., Diekkrüger B. and Giertz, S. (2013). A comparison of hydrological models for assessing the impact of landuse and climate change on discharge in a tropical catchment. *Journal of Hydrology*, 498, 221–236.

Cuo, L., Zhang, Y., Gao, Y., Hao, Z. and Cairang, L. (2013). The impacts of climate change and land cover/use transition on the hydrology in the upper Yellow River Basin, China. *Journal of Hydrology*, 502, 37–52.

Devi G. K., Ganasri B. P. and Dwarakish G. S. (2015). A review on hydrological models. International Conference on Water Resources, Coastal and Ocean Engineering. *Aquatic Procedia* 4, 1001–1007.

Elias E., Seifu W., Tesfaye B., Girmay W. and Tejada Moral M. (2019). Impact of landuse/cover changes on lake ecosystem of Ethiopia Central Rift Valley. *Cogent Food and Agriculture*, 5(1), 1–20.

FAO. (2020). *Global forest assessment resources 2020 (No. 163)*. Forestry Paper.

Garedew E., Sanderwall M., Sanderberg U. and Campbell B. M. (2009). Landuse and landcover dynamics in the Central Rift Valley of Ethiopia. *Environmental Management*, 44(4), 683–694.

Garg V., Aggarwal S. P., Gupta P. K., Nikam B. R., Thakur P. K., Srivastav S. K. and Kumar A. S. (2017). Assessment of landuse land cover change impact on hydrological regime of a basin. *Environmental Earth Sciences*, 76, 1–17.

Gebere S. B., Alamirew T., Merkel, B. J. and Melesse A. M. (2016). Land use and land cover change impact on groundwater recharge: The Case of Lake Haramaya watershed, Ethiopia. In Landscape Dynamics, Soils and Hydrological Processes in Varied Climates, 93–110. Cham: Springer International Publishing.

Gessesse B., Bewket W. and Bräuning A. (2015). Model-based characterization and monitoring of runoff and soil erosion in response to land use/land cover changes in the Modjo watershed, Ethiopia. *Land Degradation & Development*, 26(7), 711–724.

Ghaffari G., Keesstra S., Ghodousi J. and Ahmadi H. (2010). SWAT-simulated hydrological impact of landuse change in the Zanjanrood Basin, Northwest Iran. *Hydrological Processes*, 24, 892–903.

Glasoe S. and Christy A. (2004). *Coastal Urbanization and Microbial Contamination of Shellfish Growing Areas*. Puget Sound Action Team Publication (PSAT), Olympia, WA, 04-09.

Godebo, M. M., Ulsido, M. D., Jijo, T. E. and Geleto, G. M. (2018). Influence of landuse and land cover changes on ecosystem services in the BilateAlabasub-watershed, Southern Ethiopia. *Journal of Ecology and the Natural Environment*, 10(9), 228–238.

Guo H., Hu Q. and Jiang T. (2008). Annual and seasonal streamflow responses to climate and land-cover changes in the Poyang Lake basin, China. *Journal of Hydrology*, 355, 106–122.

Guzha A. C., Rufino M. C., Okoth S., Jacobs S. and Nobrega R. L. B. (2018). Impacts of landuse and land cover change on surface runoff, discharge and low flows: Evidence from East Africa. *Journal of Hydrology: Regional Studies*, 15, 49–67.

He S., Li P., Wu J., Elumalai V. and Adimalla N. (2020). Groundwater quality under land use/land cover changes: A temporal study from 2005 to 2015 in Xi'an, Northwest China. *Human and Ecological Risk Assessment: An International Journal*, 26(10), 2771–2797.

Hua A. K. (2017). Landuse land cover changes in detection of water quality: A study based on remote sensing and multivariate statistics. *Journal of Environmental and Public Health*, 2017(1), 7515130.

Huang T. and Pang Z. (2011). Estimating groundwater recharge following landuse change using chloride mass balance of soil profiles: A case study at Guyuan and Xifeng in the Loess Plateau of China. *Hydrogeology Journal*, 19, 177–186.

Hundecha Y. and Bardossy A. (2004). Modeling of the effect of landuse changes on the runoff generation of a river basin through parameter regionalization of a watershed model. *Journal of Hydrology*, 292, 281–295.

Hurkmans R. T. W. L., Terink W., Uijlenhoet R., Moors E. J., Troch P. A. and Verburg P. H. (2009). Effects of land use changes on streamflow generation in the Rhine basin. *Water Resources Research*, 45(6), W06405.

Ibrahim F., Osikabor B., Olatunji B. T. and Olatunji T. (2022). Understanding forest land conversion for agriculture in a developing country context: An application of the theory of planned behavior among a cohort of Nigerian farmers. *Folia Forestalia Polonica. Series A. Forestry*, 64(3), 117–130.

Im S., Kim H., Kim C. and Jang C. (2008). Assessing the impacts of landuse changes on watershed hydrology using MIKE SHE. *Environmental Geology*, 57, 231–239.

Isaak, D. J., Luce, C. H., Rieman, B. E., Nagel, D. E., Peterson, E. E., Horan, D. L., Parkes, S. and Chandler, G. L. (2010). Effects of climate change and wildfire on stream temperatures and salmonid thermal habitat in a mountain river network. *Ecological Applications*, 20(5), 1350–1371.

Isik S., Kalin L., Schoonover J. E., Srivastava P. and Lockaby B. G. (2012). Modeling effects of changing landuse/cover on daily streamflow: An Artificial Neural Network and curve number-based hybrid approach. *Journal of Hydrology,* 485, 103–112.

Kalantari Z., Lyon S. W., Folkeson L., French H. K., Stolte J., Jansson P. E. and Sassner M. (2014). Quantifying the hydrological impact of simulated changes in landuse on peak discharge in a small catchment. *Science of the Total Environment*, 466–467, 741–754.

Khan H. H., Khan A., Ahmed S. and Perrin J. (2010). GIS-based impact assessment of landuse changes on groundwater quality: Study from a rapidly urbanizing region of South India. *Environmental Earth Sciences*, 63, 1289–1302.

Kumar N., Singh S. K., Singh V. G. and Dzwairo B. (2018). Investigation of impacts of landuse/land cover change on water availability of Tons River Basin, Madhya Pradesh, India. *Modeling Earth Systems and Environment*, 4, 295–310.

Legesse D., Vallet-Coulomb C. and Gasse F. (2003). Hydrological response of a catchment to climate and landuse changes in tropical Africa: Case study South Central Ethiopia. *Journal of Hydrology*, 275(1–2), 67–85.

Lerner D. N. and Harris B. (2009). The relationship between landuse and groundwater resources and quality. *Landuse Policy*, 26S, S265–S273.

Levi N. O., Karnieli A., Egozi R., Givati A. and Peeters A. (2015). Modeling the effects of land-cover change on rainfall-runoff relationships in a Semiarid, Eastern Mediterranean watershed. *Advances in Meteorology*, 2015(1), 838070.

Li K. Y., Coe M. T., Ramankutty N. and Jong R. D. (2007). Modeling the hydrological impact of landuse change in West Africa. *Journal of Hydrology,* 337, 258–268.

Lin B., Chen X., Yao H., Chen Y., Liu M., Gao L. and James A. (2015). Analyses of landuse change impacts on catchment runoff using different time indicators based on SWAT model. *Ecological Indicators,* 58, 55–63.

Liu, Y., Wu, G., Fan, X., Gan, G., Wang, W. and Liu, Y. (2022). Hydrological impacts of land use/cover changes in the Lake Victoria basin. *Ecological Indicators*, 145, 109580.

Mango, L. M., Melesse, A. M., McClain, M. E., Gann, D. and Setegn, S. (2011). Land use and climate change impacts on the hydrology of the upper Mara River Basin, Kenya: results of a modeling study to support better resource management. *Hydrology and Earth System Sciences*, 15(7), 2245–2258.

Mariye M., Jianhua L. L. and Maryo M. (2022). Landuse land cover change analysis and detection of its drivers using geospatial techniques: A case of south-central Ethiopia. *All Earth*, 34(1), 309–332.

Mati B. M., Mutie S., Gadain H., Home P. and Mtalo F. (2008). Impacts of landuse/cover changes on the hydrology of the transboundary Mara River, Kenya/Tanzania. *Lakes and Reservoirs: Research and Management*, 13, 169–177.

Matlhodi B., Kenabatho P. K., Parida B. P. and Maphanyane J. G. (2019). Evaluating landuse and land cover change in the Gaborone dam catchment, Botswana, from 1984–2015 using GIS and remote sensing. *Sustainability*, 11(19), 5174.

McLay C. D. A., Dragten R., Sparling G. and Selvarajah N. (2001). Predicting groundwater nitrate concentrations in a region of mixed agricultural landuse: A comparison of three approaches. *Environmental Pollution*, 115, 191–204.

Miller J. D., Kim H., Kjeldsen T. R., Packman J., Grebby S. and Dearden R. (2014). Assessing the impact of urbanization on storm runoff in a peri-urban catchment using historical change in impervious cover. *Journal of Hydrology*, 515, 59–70.

Mishra A., Kar S. and Singh V. P. (2007). Prioritizing structural management by quantifying the effect of landuse and land cover on watershed runoff and sediment yield. *Water Resources Management*, 21, 1899–1913.

Mishra N., Khare D. and Kumar S. (2015). Impact of landuse change on groundwater recharge in upper Ganga Canal Command. *International Journal of Applied Engineering Research*, 10(33), 24284–24288.

Mwangi H. M., Julich S., Patil S. D., McDonald M. A. and Feger K. H. (2016). Relative contribution of landuse change and climate variability on discharge of upper Mara River, Kenya. *Journal of Hydrology: Regional Studies*, 5, 244–260.

Narany T. S., Aris A. Z., Sefie A. and Keesstra S. (2017). Detecting and predicting the impact of landuse changes on groundwater quality, a case study in Northern Kelantan, Malaysia. *Science of the Total Environment,* 599, 844–853.

Nie W., Yuan Y., Kepner W., Nash M. S., Jackson M. and Erickson C. (2011). Assessing impacts of landuse and landcover changes on hydrology for the upper San Pedro watershed. *Journal of Hydrology*, 407, 105–114.

Niehoff D., Fritsch U. and Bronstert A. (2002). Landuse impacts on storm-runoff generation: Scenarios of landuse change and simulation of hydrological response in a mesoscale catchment in SW-Germany. *Journal of Hydrology*, 267, 80–93.

Oliver, T. H. and Morecroft, M. D. (2014). Interactions between climate change and land use change on biodiversity: Attribution problems, risks, and opportunities. *Wiley Interdisciplinary Reviews: Climate Change*, 5(3), 317–335.

Owuor S. O., Bahl K. B., Guzha A. C., Rufino M. C., Pelster D. E., PinEs E. D. and Breuer L. (2016). Groundwater recharge rates and surface runoff response to landuse and land cover changes in semi-arid environments. *Ecological Processes*, 5, 16.

Patra S., Sahoo S., Mishra P. and Mahapatra S. C. (2018). Impacts of urbanization on landuse/cover changes and its probable implications on local climate and groundwater level. *Journal of Urban Management*, 7(2), 70–84.

Petchprayoon P., Blanken P. D., Ekkawatpanit C. and Hussein K. (2010). Hydrological impacts of landuse/land cover change in a large river basin in central–northern Thailand. *International Journal of Climatology*, 30, 1917–1930.

Peters, M. K., Hemp, A., Appelhans, T., Becker, J. N., Behler, C., Classen, A., Detsch, F., Ensslin, A., Ferger, S.W., Frederiksen, S.B. and Gebert, F. and Steffan-Dewenter, I. (2019). Climate–land-use interactions shape tropical mountain biodiversity and ecosystem functions. *Nature*, 568(7750), 88–92.

Pokhrel, Y., Burbano, M., Roush, J., Kang, H., Sridhar, V. and Hyndman, D. W. (2018). A review of the integrated effects of changing climate, land use, and dams on Mekong river hydrology. *Water*, 10(3), 266.

Prabhakar A. and Tiwari, H. (2015). Landuse and land cover effect on groundwater storage. *Modeling Earth Systems and Environment*, 1, 1–10.

Rientjes T. H. M., Haile A. T., Kebede E., Mannaerts C. M. M., Habib E. and Steenhuis T. S. (2011). Changes in land cover, rainfall and stream flow in upper Gilgel Abbay catchment, Blue Nile basin – Ethiopia. *Hydrology and Earth System Sciences*, 15(6), 1979–1989.

Saadatkhah N., Tehrani M. H., Mansor S., Khuzaimah Z., Kassim A. and Saadatkhah R. (2016). Impact assessment of land cover changes on the runoff changes on the extreme flood events in the Kelantan River basin. *Arabian Journal of Geosciences*, 9, 1–17.

Sahoo S., Dhar A., Kayet N. and Kar A. (2016). Detecting water stress scenario by landuse/land cover changes in an agricultural command area. *Spatial Information Research*, 25, 11–21.

Sajikumar N. and Remya R. S. (2014). Impact of land cover and landuse change on runoff characteristics. *Journal of Environmental Management*, 161, 460–468.

Seyam, M. M. H., Haque, M. R. and Rahman, M. M. (2022). Identifying the landuse land cover (LULC) changes using remote sensing and GIS approach: A case study at Bhaluka in Mymensingh, Bangladesh. *Case Studies in Chemical and Environmental Engineering*, 7, 100293.

Shivoga W. A., Muchiri M., Kibichi S., Odanga J., Miller S. N., Baldyga T. J., Enanga E. M. and Gichaba M. C. (2007). Influences of landuse/cover on water quality in the upper and middle reaches of River Njoro, Kenya. *Lakes and Reservoirs: Research and Management*, 12, 97–105.

Singh S. K., Singh C. K. and Mukherjee S. (2010). Impact of landuse and land-cover change on groundwater quality in the Lower Shiwalik hills: A remote sensing and GIS based approach. *Open Geosciences*, 2(2), 124–131.

Spera, S. A., Galford, G. L., Coe, M. T., Macedo, M. N. and Mustard, J. F. (2016). Land-use change affects water recycling in Brazil's last agricultural frontier. *Global Change Biology*, 22(10), 3405–3413.

Sun Z., Li X., Fu W., Li Y. and Tang D. (2013). Long-term effects of landuse/land cover change on surface runoff in urban areas of Beijing, China. *Journal of Applied Remote Sensing*, 8(1), 084596–084596.

Wagner P. D., Kumar S. and Schneider K. (2013). An assessment of landuse change impacts on the water resources of the Mula and Mutha Rivers catchment upstream of Pune, India. *Hydrology and Earth System Sciences*, 17, 2233–2246.

Wang J., Gao Y. and Wang S. (2015). Landuse/cover change impacts on water table change over 25 years in a desert-oasis transition zone of the Heihe River Basin, China. *Water*, 8(1), 11.

WoldeYohannes A., Cotter M., Kelboro G. and Dessalegn W. (2018). Landuse and land cover changes and their effects on the landscape of Abaya-Chamo basin, Southern Ethiopia. *Land*, 7(1), 1–17.

Yan B., Fang N. F., Zhang P. C. and Shi Z. H. (2013). Impacts of landuse change on watershed streamflow and sediment yield: An assessment using hydrologic modelling and partial least squares regression. *Journal of Hydrology,* 484, 26–37.

Yin J., He F., Xiong Y. J. and Qiu G. Y. (2017). Effects of landuse/land cover and climate changes on surface runoff in a semi-humid and semi-arid transition zone in northwest China. *Hydrology and Earth System Sciences*, 21, 183–196.

Yun P., Huili G., Demin Z., Xiaojuan L. and Nobukazu N. (2011). Impact of landuse change on groundwater recharge in Guishui River Basin, China. *Chinese Geographical Science*, 21(6), 734–743.

Zhang H. and Hiscock K. M. (2010). Modelling the impact of forest cover on groundwater resources: A case study of the Sherwood Sandstone aquifer in the East Midlands, UK. *Journal of Hydrology*, 392, 136–149.

Zhang L., Nan Z., Xu Y. and Li S. (2016). Hydrological impacts of land use change and climate variability in the headwater region of the Heihe River Basin, Northwest China. *PloS One*, 11(6), e0158394.

Zhang N., Luo Y. J., Chen X. Y., Li Q., Jing Y. C., Wang X. and Feng C. H. (2018). Understanding the effects of composition and configuration of land covers on surface runoff in a highly urbanized area. *Ecological Engineering*, 125, 11–25.

Zhang Q., Xu C. Y., Tao H., Jiang T. and Chen Y. D. (2010). Climate changes and their impacts on water resources in the arid regions: A case study of the Tarim River basin, China. *Stochastic Environmental Research and Risk Assessment*, 24(3), 349–358.

Zhang X., Zhang L., He C., Li J., Jiang Y. and Ma L. (2014). Quantifying the impacts of landuse/land cover change on groundwater depletion in Northwestern China – a case study of the Dunhuang oasis. *Agricultural Water Management*, 146, 270–279.

Zope P. E., Eldho T. I. and Jothiprakash V. (2017). Hydrological impacts of landuse–land cover change and detention basins on urban flood hazard: A case study of Poisar River basin, Mumbai, India. *Natural Hazards*, 87, 1267–1283.

2 Urban Land Use Dynamics

A Remote Sensing and GIS Approach

Divyansh Ghildiyal, Jayantpal Yadav, Shubhi Khare, Gauri Jayaswal, and Deepak Kumar

2.1 INTRODUCTION

The land use and land cover (LULC) is a concept of obtaining different aspects of an area by gathering data about change in vegetation, agricultural land, urban infrastructure, and other forms of information, which can be gathered through satellite images. The terms "land use" and "land cover" are often used against one another, but have different meanings. Land use means the utility of a land for a purpose it is being used, which could be either a construction purpose or a commercial purpose, whereas the land cover is defined by the cover, that is, vegetation, barren land, wetland, snow mountains, etc. Observing the featural changes in Earth's surface is used to determine the management of resources between natural phenomena and human activities (Ganaie et al., 2021). LULC changes are challenging in certain countries like Ethiopia, where a significant part of the economy comes from the agricultural sector (Berihun et al., 2019). Rainfall has a direct impact on agricultural productivity; for example, regular rainfall supports healthy crop production, while irregular patterns can lead to extreme events such as floods and droughts (Singh and Kumar, 2021; Singh et al., 2023a; Singh et al., 2023b; Singh et al., 2024). Land-use intensity is increasing due to an increase in landscape modification, and the splitting of vegetation is affected by the modified area operations (Mugiraneza et al., 2019).

Land cover is used to define an area's terrain. Depending on the features present in the study area, its land cover could be determined. There are certain criteria by which different land cover classes could be classified. Considering the class of land cover as per the United States Geological Survey (USGS), there are eight classes in which land cover could be differentiated, as shown in Figure 2.1 and described below.

1. *Developed:* Area related to built-up activities includes developed areas like high-density residential areas, commercially used areas, industries, etc. The population in the expanded area contributes economically to society, which is necessary for growth, employment, etc.
2. *Cropland:* It covers the area used for food and fiber production. The area under the cropland includes orchards, hay lands, cultivated croplands, and

DOI: 10.1201/9781003596080-2

uncultivated croplands. These areas provide food and fodder to the population and livestock.

3. *Grass/Shrub:* Land covered with grasses, shrubs, or other forms of herbaceous vegetation with little or no tree cover. These grasslands are called by different names on different continents.
4. *Tree Cover:* Area where the tree cover is more than 10% of the total area will be classified as tree cover. The forest includes a major portion of tree cover due to the high density of trees.
5. *Water:* Area covered by water bodies, including all fresh water and saline water bodies like lakes, streams, oceans, seas, etc. The Earth consists of 71% water, of which only 3% is freshwater, the rest being saline.
6. *Wetland:* Places where land is flooded with water or where water saturation is high are called wetlands. The wetlands play a crucial role in biodiversity by being home to different terrestrial and aquatic animals. The Sundarban delta of India is one of the wetlands, being the largest mangrove forest.
7. *Ice/Snow:* Area where ice and snow are present all year round, that is, they are entirely not melted. Areas near the poles where the amount of incident sunlight is less and has low temperature throughout the year are referred to it.
8. *Barren:* An area where less than 10% of the total land area is vegetated is barren land. The reason for sparse vegetation could be human factors, environmental restrictions, overgrazing, a greater number of soil rocks, temperature, etc.

As per the Ministry of Statistics and Programme Implementation of India, out of a geographical area of 329 million hectares, statistics are available only for 305 million hectares, which makes 7% of the area still not covered or classifiable under the ninefold classification shown in Figure 2.1.

As shown in Figure 2.1, forests include all lands classed as forest under any legal enactment dealing with forests or administered as forests, whether state-owned or private, and whether wooded or maintained as potential forest land. The area of crops raised in the forest and grazing lands, or areas open for grazing within the forests, should remain included under the forest area. Area under nonagricultural uses includes the land under nonagricultural use, which is predominant, where settlements, along with other infrastructural facilities such as roads, buildings, and railways, as well as areas under water bodies, are more prevalent. Barren and unculturable land includes the barren land exposed to different rock types, which occur as massive rocks or stony wastes in mountains and desert areas. Land that cannot be brought under cultivation except at a high cost should be considered unculturable—permanent pastures and other grazing lands. Pasture is land used for grazing, which is grazed by livestock such as cattle, sheep, goats, or horses. Pastures are mainly grazed in the rainy season, during which climatic conditions are favorable for pasture growth. It includes all permanent pastures or meadows, but does not include the village common grazing land under this head. Land under miscellaneous tree crops, including all cultivable land not included in the net area sown, is put to some agricultural use. Land under casuarina trees, thatching grasses, bamboo bushes, and other groves for fuel, etc., which are not included under "orchards," are classified under this category. Culturable waste

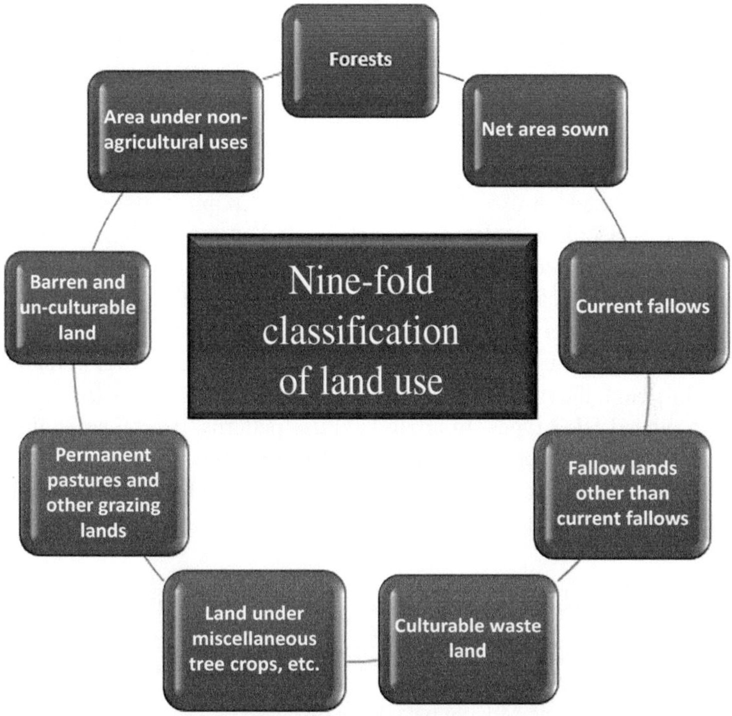

FIGURE 2.1 Types of nine-fold classification of land use.

land includes lands available for cultivation, whether not completely taken up for cultivation or taken up for cultivation once but not cultivated during the current year and the last 5 years or more in succession for some reason or other. Such lands may either fallow or be covered with shrubs and jungles, which are not put to any use. They may be assessed or unassessed and may lie in isolated blocks or within cultivated holdings. Fallow lands other than current fallows include all land which is taken up for cultivation but is temporarily out of cultivation for a period of not less than 1 year and not more than 5 years. Current fallows include cropped areas, which are kept fallow during the current year. If any seeding area is not cropped in the same year due to some reasons, such as low rainfall in the current year or unavailability of cultivation resources, it may be treated as current fallow. Net area sown includes all land that is taken for cultivation in continuation without any major period of gap. The pattern of distribution of net cultivated area depicts variations among villages and blocks.

Urbanization is the process of shifting a large population from rural areas into cities or urban areas. The population moves to occupy the vacant, agricultural, and fallow land for their requirements. In the scenario of two categories, as developed nations and developing nations, urban growth is higher in developing nations than in developed nations. The process of urban expansion in developing nations is started early, but the rate of urban expansion is increased rapidly at a faster rate. Several studies (e.g., Bhatta, 2009; Thapa and Murayama, 2010) attributed unprecedented population growth and socioeconomic development as the major drivers of such unplanned urban expansion. In developing nations, urban expansion mainly provides more opportunities for employment, production, and goods and services (Cohen, 2006). This results in rural–urban migration being carried out faster over a shorter span. The process of urbanization has been characterized not only by population growth, but also by industrial expansion, increasing economic and social activities, and intensified use of land resources (Mundia and Aniya, 2005). The increased population exerts pressure on the landscape as people need resources such as food, water, shelter, and fuel to fulfil other necessities. The demand for resources and socioeconomic factors results in the area being used for developmental and infrastructural activities. Based on definitions of urban land, estimates of the global urban area vary from less than 1–3% of the world's land surface (Liu et al., 2014). Even though urban areas cover a small fraction of the world's land, the rapid urban expansion in different parts has led to significant changes in other LULC types. In urban land-use patterns, nearby agricultural land, barren land, forest land, and fallow land are changed into built-up areas consisting of buildings, roads, and drainage systems. It may cause a reduction in the agricultural cover of land use, which impacts the ecosystem's biodiversity.

2.2 URBAN LAND-USE STUDY USING REMOTE SENSING

Remote sensing is a powerful tool for gathering information about urban expansion and its development. In this section, a few landmark research studies carried out for urban LULC using remote sensing have been discussed.

Landsat Thematic Mapper (TM) images were used in a study by Xu et al. (2000) to analyze land-cover types and changes in Fuqing City and its surrounding areas in Fujian Province, China, over a 5-year period (1991–1996). Three images from different years were processed to create land-cover maps, with the aid of geographic information systems (GIS). The analysis revealed rapid urban expansion, significant loss of agricultural land, and deforestation in the region during this time. The driving factors behind urbanization were found to be fast economic growth and improvements in transportation infrastructure. The study also provided recommendations for sustainable development in the region.

The need for reliable and timely information that identifies the composition and distribution of land resources and the nature and extent of their change over time is becoming increasingly important, especially in metropolitan areas that are experiencing rapid growth. A study by Yuan (2010) developed a methodology for mapping and monitoring land cover change using relevant multitemporal Landsat TM data in the seven-county Twin Cities Metropolitan Area of Minnesota for the years 1986, 1991, 1998, and 2002. Overall accuracies ranged from 93% to 95% for the classes to data activities, and of the mapped land-cover change maps produced using post-classification change detection methodology and accuracy evaluation using several approaches, ranging between 80% and 90%. Results showed an increase in urban or developed land extent from 23.7% to 32.8% from 1986 to 2002, when several rural cover approach types of agriculture, forest, and wetlands decreased from 69.6% to 60.5% as well. The results provided a quantification of land cover change in the metropolitan area and demonstrated the value of multitemporal Landsat data to supply an accurate and economical method to map land cover changes over time to be used as input to land management and policy decisions.

Urbanization has been an alarming problem all over the world. Unplanned construction, exploitation of natural resources have created a disbalance in the ecosystem that not only affects us humans but also the other animals. Vinod Kumar et al. (2007) conducted a study on Indore city of India to determine the urban growth by spatiotemporal analysis. The study was performed with Shannon Entropy, which is a measure of considering the compactness or dispersion of built-up land growth in the urban areas. The growth pattern was divided into four zones. The data was considered along with the growth pattern of roads to understand the impact of the built-up area. The study showed that Shannon is a useful method in determining the concentration or dispersion of built-up areas in a region. The result showed that the population has increased from 5.75 lakh in 1971 to 15.97 lakh in 2001, making a growth of nearly 300%.

A study by Kesgin and Nurlu (2009) has been conducted in the coastal zone of Candarli Bay, Turkey, over the past 15 years (1990–2005), using a Landsat TM image of 19 May 1990 and an ASTER image of 23 July 2005 to analyze LULC transition rates for this coastal zone of Turkey. The images were analyzed using supervised classification, and land cover changes were detected through a post-classification comparison, revealing a notable expansion of urban areas, while semi-natural and agricultural lands had declined.

Urban growth patterns have a direct impact on the urban development processes, not only within the city but also in the neighborhood areas. The growth of urban areas depends on several factors such as land and housing policy, current build-up area, available developable area, change in household count, population, economy, and many others. A study by Bhatta (2009) was carried out to analyze the land-use pattern in the Indian City of Kolkata, which is also one of the most populous cities in the country. Statistical methods were used for analysis to recognize and analyze this pattern in the city. The development of the city was dispersed with a declining population growth rate.

A study by Fichera et al. (2011) was conducted in an area of Avellino, situated in Southern Italy, which characterized the dynamics or changes of land cover for a period of the last 50 years from 1954 to 2004. The results observed during this period of last five decades, the urban land cover type had been expanded to five time from 893.45 ha to 5,104.44 ha (from the 1.6% to the 9.1% of the total area of study), mostly at the expense of the cultivated areas, suffered the effect of expansion of built up area, woodland, and grassland/pasture LC types has relatively lower increment, due to this specific zoning laws and urban planning programs are effectively implemented.

LULC configurations in China have been revolutionized with speeded-up urbanization due to rapid economic development under its economic reform policies since 1978. Urbanization has been accelerated, and severe environmental stress has ensued. This has been the case in the study by Weng and Lu (2009) conducted in the coastal area, like the Zhujiang Delta, where gigantic agricultural land is being lost every year, transitioning to urban or allied uses. Assessing the scale and trend of China's urban expansion is an immediate requirement. Additionally, due to the absence of proper land-use planning and the provisions for sustainable development, uncontrolled urban expansion has been causing serious environmental impacts. Therefore, it is a requirement to evaluate the environmental effects of the fast-paced urban expansion.

Using remote sensing and GIS, Landsat TM/ETM+ images from 1979 to 2008 were analyzed to measure land-use changes and urban expansion in rapidly growing Shanghai by Zhang et al. (2011). The results indicated that, from 1979 to 2008, Shanghai's urban area expanded by 4,242.06 ha/year, mainly replacing cropland (67.35%), forests and shrubs (9.12%), water (4.80%), and tidal land (2.19%). Meanwhile, cropland, forests, water bodies, and tidal land declined annually by 3,286.26 ha, 1,331.33 ha, 903.43 ha, and 315.72 ha, respectively. Shanghai's industrial transformation increased demand for investment and skilled labor, reducing cropland and lowering the farming population by 13.84%. However, with limited land and high population pressure, sacrificing cropland and water for economic growth is unsustainable; future policies must balance development with sustainable land use.

A study of LULC by remote sensing and GIS technique at Bahir Dar, Ethiopia, was performed by Haregeweyn et al. (2012) to explore the past and present LULC changes and urban expansion pattern. The data of 52 years (1957–2009) were taken for this study. Aerial photos for the years 1957, 1984, and 1994, as well as field mapping using GPS for the year 2009, were employed and analyzed using GIS. The urban area had expanded by 12%, 14%, and 5% during the time period of 1957–1984, 1984–1994, and 1994–2009, respectively. The results showed that area showed increase of 31% in the urban area, from 279 ha in 1957 to 4830 ha in 2009. Built-up areas increased from 80 ha in 1957 to 848 ha in 1994. A total of 242.2 ha of farmland was dispossessed of ownership from 271 households between 2004 and 2009, and 96% of those beneficiaries claimed that the compensation was insufficient, as the economic assessment was influenced by the government's land ownership system. Observing the data as per different assessments the current urban area would be doubled by 2024.

Urbanization is an evolving and multifaceted process that involves large-scale alterations in existing land use, primarily at the local scale. The principal intention of the study by Ramachandra et al. (2012) was to quantify the urban dynamics using temporal remote sensing data with the support of recognized landscape metrics. Bangalore, as an example of a rapidly urbanizing landscape in India, was selected for this research. The nuanced process of urban sprawl was assessed by means of spatio-temporal analysis. Land-use analysis reveals that 584% development in built-up area over the last four decades, alongside a decrease in vegetation by 66% and water bodies by 74%. Temporal data analyses indicate an increase in urban built-up area of 342.83% (1973–1992); 129.56% (1992–1999); 106.7% (1999–2002); 114.51% (2002–2006); 126.19% (2006–2010). The area was divided into four zones, with each zone subdivided into 17 concentric circles with a 1 km expanding radius to assess the patterns and level of urbanization at the local scale. Urban density gradient showed a radial pattern of urban land-use change.

A study by Rawat et al. (2013) used remote sensing and GIS techniques to analyze land-use changes in Ramnagar, Uttarakhand, over two decades (1990–2010), Landsat Images using Landsat TM of 1990 and 2010 were acquired by USGS Earth Explorer, and for supervised classification using maximum likelihood technique in ERDAS 9.3

was used in study. The result indicated that, over the past two decades, Ramnagar's built-up area and sandbars had increased by 8.88% (2.83 km^2) and 3.98% (1.27 km^2), respectively, while vegetation, agricultural land, and water bodies have decreased by 9.41% (3.00 km^2), 0.69% (0.22 km^2), and 2.76% (0.88 km^2), respectively. The study showed that Ramnagar town is expanding mainly southward along National Highway-121. It also emphasized the value of digital change detection techniques in analyzing the nature and location of these changes.

LULC changes were examined by Mallupattu and Sreenivasula Reddy (2013) in urban lands of Tirupati, over the years 1976–2003 with the GISs and remote sensing technology. These studies were done using the Survey of India topographic map 57 O/6 and IRS ID 2003 remote sensing data, collection-LISS III and PAN. The study area was categorized into eight types of classification based on field study, geographical conditions, and remote sensing images. The change comparison of LU/LC from 1976 to 2003 from the toposheet and remote sensing images showed a significant increase in built-up area, open forest, plantation, and other lands. In addition, the built-up area increased from 5.91 km^2 to 18.34 km^2. In addition to the above land uses, the study area showed substantial disappearance of agricultural land, water spread area, and dense forest area, which is likely due to the urbanization of the study area. In 1976, no mining activities were found in the study area; however, in 2003, there was a small addition of mining land.

Hyderabad, one of India's fastest-growing megacities, is experiencing significant economic, social, and environmental challenges due to rapid urban expansion. Effective resource planning and the provision of basic amenities require a comprehensive understanding of urban growth patterns. Monitoring land-use changes over time is crucial for detecting growth trends and making informed planning decisions. Remote sensing and GIS tools offer accurate and efficient methods for this analysis. The study by Wakode et al. (2014) focuses on mapping land-use changes in Hyderabad's urban area across three time periods: 1989–2000, 2000–2005, and 2005–2011, using Landsat TM and Enhanced Thematic Mapper (ETM) data. The study examined urban expansion and its effects on other land-use categories. Results indicated that high-density urban areas have significantly expanded, encroaching upon agricultural land, open spaces, and water bodies.

A study of LULC by remote sensing and GIS technique at Hawalbagh block of Almora District, Uttarakhand, India was performed to observe change in land dynamics. The study was made by Rawat and Kumar (2015) in order to detect the land consumption rate and the changes that had taken place during the last two decades using geospatial techniques. The data from 20 years of Landsat TM of 1990 and 2010 has been used with ERDAS 9.3 software had been used for supervised image processing. The landscape was divided into five different classes, namely, vegetation, agriculture, barren, built-up, and water body. The study concluded that the vegetation was increased by 3.51% (9.39 km^2), built-up land was increased by 3.55% (9.48 km^2), agriculture was decreased by 1.52% (4.06 km^2), barren land was decreased by 1.52% (4.06 km^2), and water bodies were decreased by 0.08% (0.22 km^2).

LULC changes result from both natural processes and human activities. In developing nations, rapid industrial growth and urban expansion pose significant risks to the natural environment. A study by Samal and Gedam (2015) was conducted in the Upper Bhima River Basin, which originates from the Western Ghats, to examine historical land-use changes and predict future impacts within a natural boundary. Multitemporal satellite images from Landsat and Indian remote sensing (IRS) satellites (1992–2009), along with a digital elevation model, were used to analyze past and present LULC patterns. Using an object-based image analysis technique and GIS-based change detection, the study found a significant increase of 288 km^2 in built-up areas between 1992 and 2009, primarily at the expense of wastelands and agricultural land. The rapid urbanization trend, particularly noticeable along major transportation networks after 2000, highlighted the ongoing expansion of urban areas within the river basin.

In the past thirty years, Malegaon district has experienced rapid urban growth, leading to significant ecological and environmental challenges in both urban and peri-urban contexts. The study by Poyil and Misra (2015) provided a quantifiable approach for land-use change studies within the Malegaon district by measuring urban agglomeration, population increase, and LULC change. The association with water and environmental degradation challenges surrounding the population has increased due to urbanization. Satellite image processing has been used as the primary tool, and morphological growth was based on an analysis of a series of satellite images (three different dates with time periods between them) with improvements to allow for the atmospheric effects and geometric corrections, and to classify based on prevalent land use such as urban, agriculture, forest, and water. The study determined that there was an increase in settlements by 78% from 1989 to 2006 and from a 26% increase between 2006 and 2008, along with an increase in population and rural migration due to economic gains and technological gains related to urbanization.

A study by Subasinghe et al. (2016) for the Colombo Metropolitan Area (CMA), Sri Lanka, was performed to observe changes in the spatiotemporal pattern of urban growth of the CMA. The data from 22 years (1992–2024) were observed using remote sensing data and GIS techniques. The LULC was divided into three time periods, that is, 1992, 2001, and 2014. Then, urban land change was observed from the intervals 1992–2001 and 2001–2014. The urban land cover of various driver variables (distance to roads) was observed, and a prediction for future growth was made from 2014 to 2050. The built-up land has increased by 24,711 ha (221%) over the past 22 years. By 2030, built-up land will increase to 42,500 ha, and in 2050, built-up land will be 56,000 ha.

A study of LULC by remote sensing and GIS technique at Barrackpore subdivision area, India was performed to observe change in LULC dynamics by Das and Angadi (2022). The data from 45 years (1972 and 2016) were observed. The landscape was divided into six different classes: vegetation cover, agricultural/cropland, wetland, water bodies, built-up, and fallow land. The study suggested that vegetation cover was decreased by 23%, agriculture/cropland was decreased by 7%, wetland was decreased by 6%, water bodies were decreased by 1.7%, built-up land was increased by 32.2%, and fallow land was increased by 6.3%.

In a developing nation, analysis of LULC changes, urban growth dynamics of the city of Chittagong, which is situated in Bangladesh, have suffered from rapid urban expansion, resulting in the conversion of rural or agricultural peripheral area of the city into built-up area this has been noted in one of the studies conducted by Hassan and Nazem (2016). A series of comprehensive geospatial datasets and Landsat images were collected for a period of 36 years from 1977 to 2013, geospatial data were collected from diverse sources, including the Survey of Bangladesh (SoB), the Chittagong Development Authority (CDA), the Chittagong City Corporation (CCC), Urban Development Directorate (UDD), and the Centre for Urban Studies (CUS). Landsat data from MSS 1977 (60 m spatial resolution), TM 1989 (30 m spatial resolution), ETM 1999 (30 m spatial resolution), and Landsat-8 (OLI TIRS, 30 m spatial resolution) 2013 were collected. During the study period, 56% of the land cover has undergone change, mainly because of the expansion of built-up areas by 618%, with an average annual rate of increase of 17.5%. As a result of this rapid urban expansion, 2178 ha of vegetated hills area were affected during this study period of 36 years.

LULC changes, due to rapid urbanization, are one of the noteworthy parameters for analyzing environmental changes. In this regard, a multilayer perception-Markov Chain Analysis (MLP-MCA) integrated method has been used to monitor and predict the future LULC changes in Patna District, Bihar has been done by Mishra and Rai (2016). A supervised maximum likelihood classification (MLC) method was applied to derive LULC maps in 1988, 2001, and 2013 using Landsat TM/ETM Plus (ETM+)/Operational Land Imagery (OLI) images, respectively. LULC maps of 1988 and 2001 were the basis of predicting the LULC scenario in 2013 using the MLP-MCA method. The results were further compared to the observed map of LULC map of 2013 for validation of the method using the kappa index statistics. Considering the results of the same, prediction for the future LULC changes in the years 2038 and 2050 was performed. The outcomes indicated a continuous decrease in the agricultural land area as a result of rapid growth in the built-up area of the city.

In concern of sustainable development and management of natural resources, it is important to analyze the trend of LULC changes, due to excessive human activities, effects on human livelihoods, as well as on the environment. The study was carried out at a place of Beressa Watershed, Northern Central Highland of Ethiopia, during the period of 1984–2015 with the help of a geographical information system and remote sensing by Meshesha et al. (2016). For analysis of LULC dynamics, satellite images of Landsat5 TM 1984, Landsat5 TM 1999, and Landsat8 TM 2015 with 30 m spatial resolution, ArcGIS10.2.2, and ERDAS Imagine14 have been used for image processing to produce 6 LULC classes. As a result of this land classification, during the study period farm land increased by 71.6 ha/year, settlement area increased by 16.8 ha/year, initially between 1984 and 1999, forest cover decreased by 5 ha/year, water body decreased 0.03 ha/year, but later between 1999 and 2015, forest cover increased by 15.6 ha/year, water body increased 7.1 ha/year.

Unregulated, but partial urban sprawl has become a threat to city development. Addressing this accelerated process of urban expansion requires identifying the driving forces behind urban growth, particularly when its threats to habitability are recognized. The study by Mondal et al. (2017) evaluated the issue of uncontrolled urban sprawl confronted by Kolkata using cellular automata—Markov chain. The central focus of this study was urban growth patterns, LULC changes, and spatial allocation alignment with the planning approach. This study found that, depending upon the driving forces, the outcome indicated a bidirectional urban development potential surface, possibly as a result of the skewed planning process and middle-class residential demand. These simulation results provided tangible evidence to the planning authorities on evaluating the efficiency of spatial allocation and, to the extent possible, changes in current allocation expectations. The Kappa values for 1990, 2001, and 2011 are 0.82, 0.85, and 0.87. The values were satisfactory for the urbanization changes.

Hassan et al. (2016) made a study to understand the dynamics of LULC of Islamabad, Pakistan, using GIS techniques. Remote sensing and GIS have improved the selection of areas for specific purposes, that is, commercial agriculture or build-up area, depending on the type of land. Two multispectral satellite images from Landsat 5 TM imagery, 1992, and Spot 5 imagery, 2012, of the study area were obtained for the month of October from the USGS and SUPERCO. The study was first conducted from 1992 to 2012. The study aims to determine different LULC categories and changes in land pattern. To identify major driving forces and their extent of effect on the LULC change. The result showed that the increase in build-up area was 213.5%, the water body was 11.5%, and the agricultural area was 180.6%, respectively. The decrease in forest area was 49.42% and the Barren land was 96.63% between 1992 and 2012.

Wubie et al. (2016) studied the causes, factors, and consequences of LULC in the Gumara watershed of Lake Tana basin in North Western Ethiopia. The study was observed over a time period of 48 years. The data sources were two sets of Ariel photographs from the years 1957 and 1985. For the year 2005, a multispectral SPOT 5 image was used as an input to generate 3GS-based land cover maps. The purpose of the study was to determine the strategies of managing open access resources and accurate land-use planning by identifying proper land for a specific purpose so that unnecessary changes in LULC types could be reduced. Therefore, the study was performed by observing historical patterns of LULC between 1957 and 2005 in the given watershed. The result showed that the urban settlement expanded by 21.99%. Forest land reduced by 85.3%. Shrub land reduced by 91.39%. The grassland was reduced by 76.15% and the wetland was reduced by 72.54% over the research period.

The study by Hassan (2017) examines urban growth and land cover changes in five rapidly expanding cities in Bangladesh (Rajshahi, Rangpur, Sylhet, Khulna, and Barisal) using Landsat imagery, socioeconomic data, and ecological analysis. Findings showed a 468% increase in built-up areas (from 2,356 ha in 1973 to 13,435 ha in 2014), replacing agricultural land, vegetation, and wetlands. This unplanned expansion had increased landscape complexity, posing challenges for planners and policymakers.

A study of LULC in remote sensing and GIS techniques at Gelda catchment, Lake Tana watershed, Ethiopia was performed to analyze the rate and patterns of LULC dynamics, and identify major driving forces in the Gelda catchment. This study was carried out by Hassen and Assen (2018). The data from 41 years were observed, from aerial photographs and Landsat images of 1990 and 2010 were used, and comparisons were made. The built-up area was expanded by 57.7% while grasslands/shrubs/forests decreased by 53.5%, 18.6%, and 83.3%. The study was useful in observing the transition of land use and land pattern. The remote sensing and GIS were helpful in understanding spatiotemporal patterns, based on the change detection matrix analysis technique.

A study of LULC by remote sensing and GIS technique for Kathmandu, Nepal was performed by Rimal et al. (2018) to explore the past and present LULC changes and urban expansion pattern of the cities of the Kathmandu valley and their surroundings. The data from 28 years (1988–2016) were observed using Landsat satellite images. The LULC was divided into seven time periods that is, 1988, 1992, 1996, 2000, 2004, 2008, 2013. Then urban land change was observed from the interval 1988–1992, 1992–1996, 1996–2000, 2000–2004, 2004–2008, 2008–2013, and 2013–2016. Markov-Chain and Cellular Automata (MC-CA) were used to determine the future urban sprawl in 2024 and 2032. The total urban area was 40.53 km^2 in 1988, which increased to 144.35 km^2 in 2016. The urban area was found to be expanded by 200 km^2 and 238 $km^{2,}$ and cultivated land will decline to 587 km^2 and 555 km^2 by 2024 and 2032.

Research on LULC change has been used in the fields of landslides, erosion, land-use planning, and global change by Liping et al. (2018). This research uses the CA-Markov model to predict land-use spatial patterns for 2025 and 2036. The study utilizes remote sensing and geographic information systems to obtain land-use pattern dynamic change. CA-Markov model integrated the strengths of cellular automata and Markov chain analysis to predict future land-use trends based on past land-use change studies. Using Landsat 5 TM images from 1992 and 2003, as well as Landsat 8 OLI images from 2014, the study created a land-use classification map for each year. The genetic transition probability from 1992 to 2003 was generated by IDRISI software. Based on the CA-Markov model, a predicted land-use map for 2014 was created, and compared to actual land-use results of 2014 with a Kappa index of 0.8128 for validation. Researchers then established land-use patterns for Jiangle Country in the years 2025 and 2036. The rate of change of construction land area was 12.01 km^2/a. The transfer rate and gain rate of construction land area are 5.23 km^2/a and 6.78 km^2/a, respectively, and the dynamic rate of construction land is 5.23 km^2/a.

Assessing the changing environmental conditions is necessary for planning the efficient and sustainable use of natural resources wisely. The objectives of the paper by Shawul and Chakma (2019) were to analyze the historical and future LULC changes using multitemporal Landsat images in Upper Awash Basin, Ethiopia. To determine the past LULC changes, the supervised image classification method was employed based on Landsat 1 MSS 1972, Landsat 5 TM 1984, Landsat 7 ETM + 2000, and Landsat 8 OLI TIRS 2014. To predict the future LULC changes, machine learning approaches of Land Change Modeler (LCM) were applied. A significant increment in the aerial extent of the cropland and urban areas and decreasing trends in pasture, forests, and shrubland coverage were observed during the study. LULC matrices indicated a majority of areas under shrubland converted to crop land areas. The urban area increased by 606.2% from 1972 to 2014, and an increment of 47.3% in cropland has been obtained. Further, a decrease in the forest area by 25.1%, pasture by 87.4%, shrubland by 28.8%, and water bodies by 21% was recorded in the same period. The modelled future LULC predicts a significant expansion of cropland and urban areas at the expense of forest, pasture, and shrubland areas.

The study by Islam (2021) examined land cover changes and shoreline shifts in Nijhum Dwip, Bangladesh, using supervised classification of Landsat images from 1998, 2008, and 2018. Results showed a loss of 285 ha of mangrove forest, a decline in agricultural land from 23.13% to 14.89%, and a rise in settlement areas from 3.73% to 11.93%. Water cover increased due to sea level rise, while land accretion exceeded erosion, shoreline expansion was observed in the northwest and southeast.

The rapid urban expansion in small cities of India is a major challenge for local government and urban planners in planning the urbanization in a systematic manner to efficiently utilize the natural resources. A study by Das and Angadi (2021) has been conducted to evaluate the LULC changes for an area of Barrackpore subdivision area, India, for a duration from 1972 to 2016 using remote sensing. For LULC maps, multitemporal Landsat images with the MLC method were applied, and spatio-temporal changes were performed by the change matrix identifier to study land transformation. Results stated that the area under vegetation cover, agricultural/cropland, wetland, and water bodies had decreased by 23%, 7%, 6%, and 1.7%, while the area under built-up and fallow lands had increased by 32.2% and 6.3%. The relationship between population growth and an increase in built-up areas indicated that the pressure of population growth enhanced the built-up area.

A study was conducted in Prayagraj, a city located in Uttar Pradesh, which has undergone significant transformation over the past three decades as a result of the urbanization process, conducted by Sarif and Gupta (2022) from 1998 to 2018. This study utilized multispectral imagery from the Landsat TM and the Operational Land Imager/Thermal Infrared Sensor, analyzing data for four different time periods spaced approximately a decade apart: 1988, 1997, 2008, and 2018. These images had been preprocessed and then classified using the maximum likelihood classifier in ERDAS IMAGINE version 2014. Results of the change matrix analysis from 1988 to 2018 revealed that built-up land experienced the most significant expansion, increasing by 64.52%, primarily at the expense of agricultural land (which declined by 24.41%), forest land (decreased by 10.56%), and barren land (reduced by 6.01%). These findings on LULC dynamics can provide valuable insights for urban planning and policy-making, aiding efforts to restore land capacity and promote environmental sustainability within the urban landscape.

Understanding how land is organized and changes over time is essential for effective city planning and development. A study by Zhai et al.(2021) was conducted for Wuhan, China, from 2000 to 2019, based on continuous time series mapping using Landsat observations with a support vector machine. The results indicate that rapid urbanization led to an increase in built-up area by 982.66 km^2 (228%), reducing cropland area by 717.14 km^2 (12%) and shrinking natural habitat, with reductions of 182.52 km^2, 23.92 km^2, and 64.95 km^2 for water, forest, and grassland, respectively. The patch-generating land-use simulation (PLUS) model predicts Wuhan's land use in 2029, showing growth in the northern suburbs. This insight helps urban planners enhance sustainable development.

Studying LULC changes helps analyze changes and regulate environmental sustainability. The study by Abijith and Saravanan (2022) focused on the Northern TN coast, for a duration of 2009–2019, using Google Earth Engine (GEE), TerrSet, and GIS tools. LULC changes from 2009 to 2019 were analyzed, and future projections for 2019–2030 were made. Landsat images were classified in GEE using the random forest (RF) algorithm, and future changes were predicted using the CA-Markov model. The study followed four steps: change analysis, transition potential, change prediction, and model validation. The region was divided into two zones, both showing a decline in water bodies and an increase in built-up areas, with barren land and vegetation converting into urban areas. These findings supported urban planning efforts, particularly given the region's vulnerability to coastal flooding.

Another study predicting future LULC changes in Mumbai and its surrounding areas to support urban development planning was carried out by Vinayak et al. (2021). Historical LULC dynamics were analyzed using a supervised classification of Landsat images from 1992, 2002, and 2011. The Multiple Perceptron Neural Network (MLPNN)-based Markov Chain Model (MCM) was used to simulate the 2011 LULC based on 1992 and 2002 data, with validation through kappa statistics, and the model was applied to predict LULC changes for 2050 using 2002 and 2011 data. The result of study suggested that by 2050, urban growth in Mumbai's suburban areas is expected to expand significantly, with urban land covering 46.87% (1328.77 km²), and compared to 2011, urban and forest areas will grow by 14.31% and 2.05%, respectively, while agricultural and barren lands will shrink by 16.87% or water and coastal areas will see minimal changes (<1%).

LULC change analysis helps understand human–environment interactions and supports sustainability efforts. The study by Muhammad et al. (2022) has been conducted in Linyi, China, over a period of 30 years (1990–2020), using multitemporal remote sensing data (1990–2020) and the CA-ANN model in QGIS to analyze changes and predict future trends (2030–2050). Results showed a rise in impervious surfaces (10.48%–26.91%) and a slight increase in water bodies, while forest cover, green space, and barren land declined. Predictions indicated continued urban expansion taking place at the cost of forests and green areas.

The paper by Hind et al. (2022) analyzed the spatiotemporal changes using a methodology based on GIS and remote sensing for the assessment of LULC in Algiers, the capital city of Algeria. Satellite images from 1987 to 2018 were used to categorize different types of LULC. The study was performed to predict changes in LULC by 2040. The spatial modelling of LULC in the future is determined by the CA-Markov simulation model. Results show that poor management has been a driving factor for the increase in urban areas. The calculated model revealed 69% urbanization of the study area till 2040 and savage consumption of forest area and agricultural land.

The urbanization in cities has led to changes in land-use cover. The land use has changed from agricultural to nonagricultural over time. A study using remote sensing and GIS techniques has been applied by Abebe et al. (2023) to analyze the LULC dynamics in Dukem town from 2003 to 2019. Initial data sources were collected using field observations, followed by a collection of data using remotely sensed satellite images. The results indicated a significant change from 698.06 ha (16.45%) in 2003 to 3091.67 ha (72.81%) in 2019. The agricultural land has decreased drastically from 2764.37 ha (65.12%) in 2003 to 999.05 ha (23.53%) in 2019.

2.3 CHANGE IN LULC IN RUDRAPUR CITY

To see the spatiotemporal variation of the urban area, a case study of Rudrapur city of Uttarakhand has been selected. The spatial locality of the study area, as shown in

Figure 2.2, Rudrapur, located at the foothills of the Himalayan Mountain range, is one of the fastest-growing cities in the Kumaon region of Uttarakhand. It has also become the second most populous city in the region. The city plays a vital role as a commercial hub within the Udham Singh Nagar district, mainly due to establishing and growing the State Infrastructure and Industrial Development Corporation of Uttarakhand Limited (SIIDCUL). The rapid growth can largely be attributed to the industrial development brought by SIIDCUL, which has attracted a significant number of people migrating from the hills in search of better job opportunities and an improved standard of living. Consequently, the city has expanded rapidly, pushing its boundaries outward and incorporating nearby rural areas into its urban framework. In 2023, Rudrapur officially included 11 rural areas within its municipal limits, and the number of wards was 40. These developments provide a strong foundation for choosing Rudrapur as the focus of this research study.

The bracket of the imagery is a critical step in LULC analysis. It can be conducted using either supervised or unsupervised bracketing methods, here supervised method has been used. Training samples representing colorful land cover classes were collected in a supervised bracket using tools like the Semi-Automatic Classification Plugin (SCP) in QGIS. These samples were also used to train a classifier to find areas with similar features to classify the entire image. To study the urban land-use change in Rudrapur, remote sensing data from 2015 to 2024 were taken from the Landsat 8 Satellite. The details of the data resolutions are mentioned in Table 2.1.

In Figure 2.3, the method to process the Landsat 8 data for LULC has been shown. This study used GEE and QGIS to process data and classifications. For the classification of land use, GEE used the random forest algorithm.

Covering the period from 2015 to 2024, a study was undertaken to ascertain the increase in built-up areas in Rudrapur, a city in Uttarakhand. The land was classified into three categories: built-up, water bodies, and vegetation. The results showed that the built-up area was the dominant land-use type over the 10 years. Water bodies were reduced from 4.02% in 2015 to 2.13% in 2024, while forest cover and cropland cover declined from 82.65% to just 75.89%. Meanwhile, the built-up area grew from 13.33% to 21.97% in the same timeframe. As shown in Figures 2.4 and 2.5, there is a relation between the increase in built-up areas by converting vegetation areas and water bodies. The expansion of urbanized areas due to fast-paced urbanization and population expansion poses many dangers to the environment, public health, and overall sustainability. Loss of natural habitats is one of the most notable impacts on the environment, with the destruction of forests, wetlands, and other ecosystems due to urban sprawl, causing loss of biodiversity and disrupting important ecological processes. This growth also intensifies soil erosion, water pollution, and the urban heat island, where cities get hotter because impervious surfaces such as asphalt and concrete are widespread. Additionally, urban areas create water scarcity and flood hazards by raising runoff and lowering groundwater recharge, which results in potential water shortages and enhanced exposure to extreme weather conditions. Socially, urban expansion tends to increase inequality as the poor communities tend to be excluded from basic services and reside in slums with poor infrastructure.

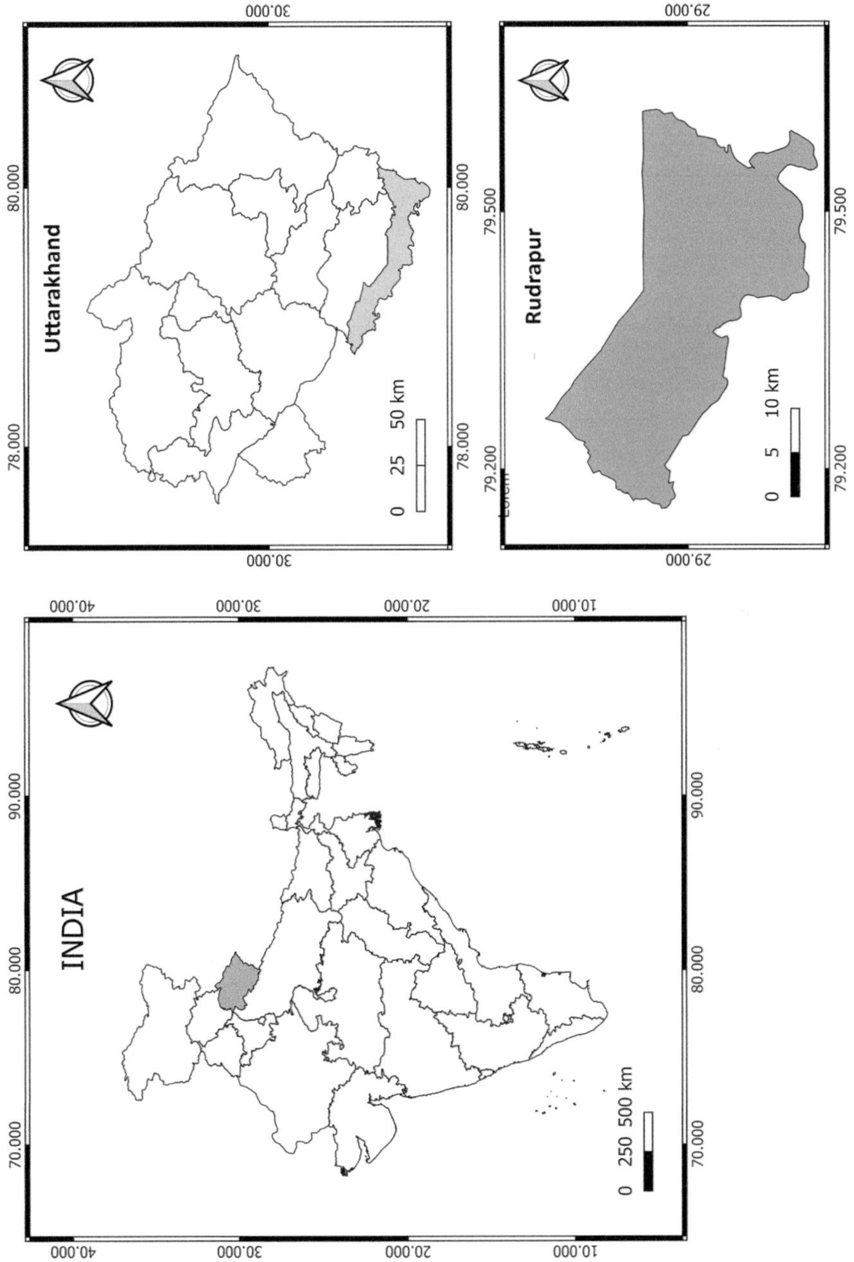

FIGURE 2.2 Study area map.

TABLE 2.1
Details of Satellite Data Used in the Study

1	Image	Landsat 8	Landsat 8
2	Path	146	146
3	Row	41	41
4	Sensor	Landsat 8 OLI/ TIRS C2 L2	Landsat 8 OLI/ TIRS C2 L2
5	Resolution	30 × 30	30 × 30
6	Bands	11	11
7	Acquisition date	4 October 2015	10 November 2024
8	Source	USGS Earth Explorer	USGS Earth Explorer

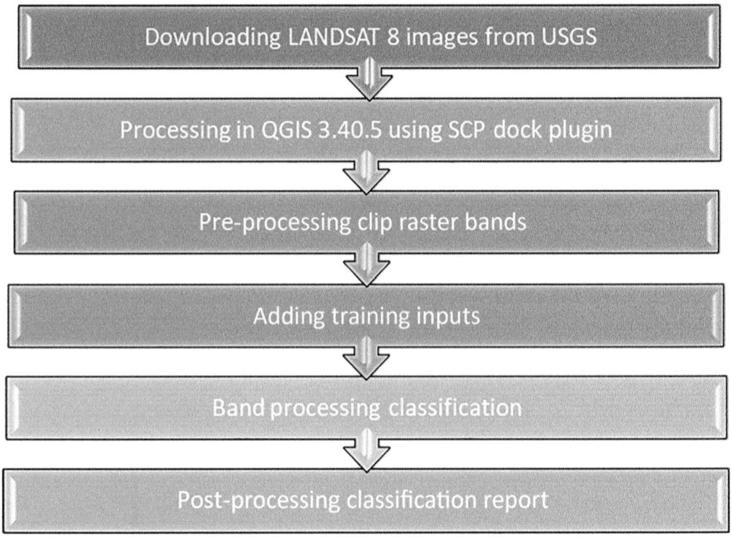

FIGURE 2.3 Process of data acquisition for LULC.

2.5 CONCLUSION

A rise in urban LULC in Rudrapur happens due to demographic, economic, and infrastructure factors. Population growth, both natural growth and rural-to-urban migration, is one of the main drivers, as it increases the requirement for housing, transportation, education, and other facilities. Economic growth also plays an important role, as increasing industries, firms, and business centers need to occupy more area, thereby turning agricultural or forest land into urban land. Developing infrastructure like roads, highways, and utilities helps and encourages increased urbanization. The cities that grow outwards toward low-density urban sprawl are acquiring more land, often made easier by lower-priced land on the outskirts of the city. Government policies, planning choices, and loose land-use regulations can further hasten urban growth. Advances in transportation and construction further simplify the land development process, while real estate investment and speculation heighten the pressure to

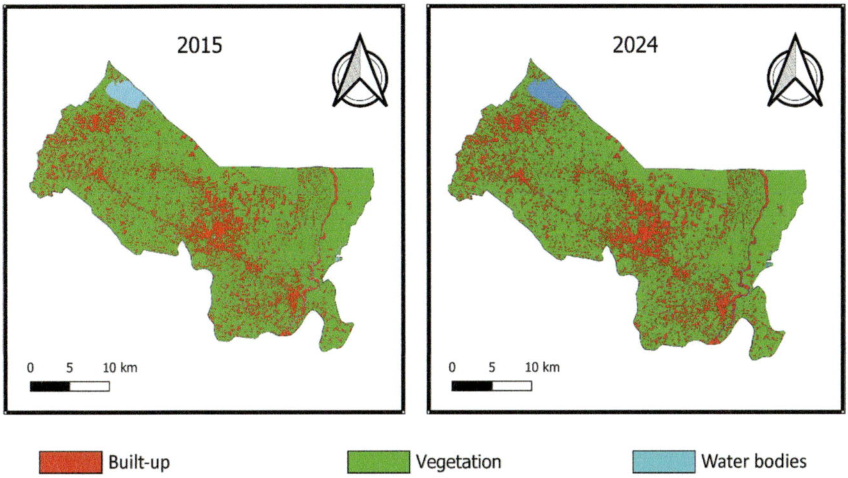

FIGURE 2.4 LULC analyzed during 2015 and 2024 of Rudrapur City.

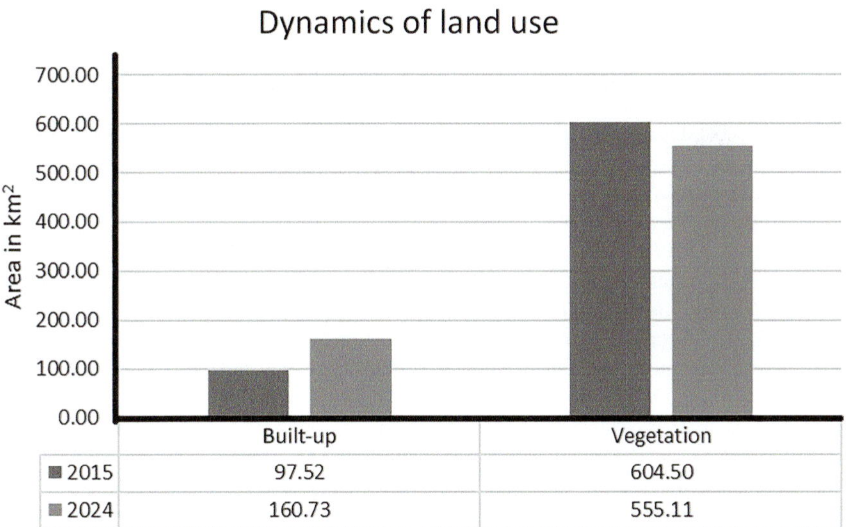

FIGURE 2.5 Land use change (km^2) for the study area.

convert land into urban land use. These factors combined result in a visible surge in urban LULC in expanding areas.

REFERENCES

Abebe B. G., Gemeda, B. S. and Sisay, G. (2023). Analysing land use land cover (LULC) dynamics by using remote sensing and GIS techniques: The case of Dukem town, Oromia special zone. *Geocarto International*, 38(1), 2210523.

Abijith, D., and Saravanan, S. (2022). Assessment of land use and land cover change detection and prediction using remote sensing and CA Markov in the northern coastal districts of Tamil Nadu, India. *Environmental Science and Pollution Research*, 29(57), 86055–86067.

Berihun, M. L., Tsunekawa, A., Haregeweyn, N., Meshesha, D. T., Adgo, E., Tsubo, M., Masunaga, T., Fenta, A.A., Sultan, D. and Yibeltal, M. (2019). Exploring land use/land cover changes, drivers and their implications in contrasting agro-ecological environments of Ethiopia. *Land use policy*, 87, 104052.

Bhatta, B. (2009) Analysis of urban growth pattern using remote sensing and GIS: A case study of Kolkata, India. *International Journal of Remote Sensing,* 30, 4733–4746.

Cohen, B. (2006). Urbanization in developing countries: Current trends, future projections, and key challenges for sustainability. *Technology in Society*, 28(1–2), 63–80.

Das, S., and Angadi, D. P. (2021). Assessment of urban sprawl using landscape metrics and Shannon's entropy model approach in town level of Barrackpore sub-divisional region, India. *Modeling Earth Systems and Environment*, 7(2), 1071–1095.

Das, S., and Angadi, D. P. (2022). Land use land cover change detection and monitoring of urban growth using remote sensing and GIS techniques: A micro-level study. *GeoJournal*, 87(3), 2101–2123.

Fichera, C. R., Modica, G., and Pollino, M. (2011). GIS and remote sensing to study urban-rural transformation during a fifty-year period. *In Computational Science and Its Applications-ICCSA 2011: International Conference, Santander, Spain, June 20–23, 2011. Proceedings, Part I 11* (pp. 237–252). Springer Berlin Heidelberg.

Ganaie, T. A., Jamal, S., and Ahmad, W. S. (2021). Changing land use/land cover patterns and growing human population in Wular catchment of Kashmir Valley, India. *GeoJournal*, 86(4), 1589–1606.

Haregeweyn, N., Fikadu, G., Tsunekawa, A., Tsubo, M., and Meshesha, D. T. (2012). The dynamics of urban expansion and its impacts on land use/land cover change and small-scale farmers living near the urban fringe: A case study of Bahir Dar, Ethiopia. *Landscape and urban planning*, 106(2), 149–157.

Hassan, M. M. (2017). Monitoring land use/land cover change, urban growth dynamics and landscape pattern analysis in five fastest urbanized cities in Bangladesh. *Remote Sensing Applications: Society and Environment*, 7, 69–83.

Hassan, M. M. and Nazem, M. N. I. (2016). Examination of land use/land cover changes, urban growth dynamics, and environmental sustainability in Chittagong city, Bangladesh. *Environment, Development and Sustainability*, 18(3), 697–716.

Hassan, Z., Shabbir, R., Ahmad, S. S., Malik, A. H., Aziz, N., Butt, A., and Erum, S. (2016). Dynamics of land use and land cover change (LULCC) using geospatial techniques: a case study of Islamabad Pakistan. *SpringerPlus*, 5(1), 812.

Hassen, E. E., and Assen, M. (2018). Land use/cover dynamics and its drivers in Gelda catchment, Lake Tana watershed, Ethiopia. *Environmental Systems Research*, 6(1), 1–13.

Hind, M., M'hammed, S., Djamal, A., and Zoubida, N. (2022). Assessment of land use–land cover changes using GIS, remote sensing, and CA–Markov model: A case study of Algiers, Algeria. *Applied Geomatics*, 14(4), 811–825.

Islam, M. S. (2021). Assessing the dynamics of land cover and shoreline changes of Nijhum Dwip (Island) of Bangladesh using remote sensing and GIS techniques. *Regional Studies in Marine Science*, 41, 101578.

Kesgin, B., and Nurlu, E. (2009). Land cover changes on the coastal zone of Candarli Bay, Turkey using remotely sensed data. *Environmental Monitoring and Assessment*, 157, 89–96.

Liping, C., Yujun, S., and Saeed, S. (2018). Monitoring and predicting land use and land cover changes using remote sensing and GIS techniques—A case study of a hilly area, Jiangle, China. *PLoS One*, 13(7), e0200493.

Liu, Z., He, C., Zhou, Y. and Wu, J. (2014). How much of the world's land has been urbanized, really? A hierarchical framework for avoiding confusion. *Landscape Ecology*, 29(5), 763–771.

Mallupattu, P. K., and Sreenivasula Reddy, J. R. (2013). Analysis of land use/land cover changes using remote sensing data and GIS at an Urban Area, Tirupati, India. *The Scientific World Journal*, 2013(1), 268623.

Meshesha, T. W., Tripathi, S. K., and Khare, D. (2016). Analyses of land use and land cover change dynamics using GIS and remote sensing during 1984 and 2015 in the Beressa Watershed Northern Central Highland of Ethiopia. *Modeling Earth Systems and Environment*, 2, 1–12.

Mishra, V. N., and Rai, P. K. (2016). A remote sensing aided multi-layer perceptron-Markov chain analysis for land use and land cover change prediction in Patna district (Bihar), India. *Arabian Journal of Geosciences*, 9, 1–18.

Mondal, B., Das, D. N., and Bhatta, B. (2017). Integrating cellular automata and Markov techniques to generate urban development potential surface: A study on Kolkata agglomeration. *Geocarto international*, 32(4), 401–419.

Mugiraneza, T., Ban, Y. and Haas, J. (2019). Urban land cover dynamics and their impact on ecosystem services in Kigali, Rwanda using multi-temporal Landsat data. *Remote Sensing Applications: Society and Environment*, 13, 234–246.

Muhammad, R., Zhang, W., Abbas, Z., Guo, F., and Gwiazdzinski, L. (2022). Spatiotemporal change analysis and prediction of future land use and land cover changes using QGIS MOLUSCE plugin and remote sensing big data: A case study of Linyi, China. *Land*, 11(3), 419.

Mundia, C. N. and Aniya, M. (2005). Analysis of land use/cover changes and urban expansion of Nairobi city using remote sensing and GIS. *International Journal of Remote Sensing*, 26(13), 2831–2849.

Poyil, R. P., and Misra, A. K. (2015). Urban agglomeration impact analysis using remote sensing and GIS techniques in Malegaon city, India. *International Journal of Sustainable Built Environment*, 4(1), 136–144.

Ramachandra, T. V., Aithal, B. H., and Sanna, D. D. (2012). Insights to urban dynamics through landscape spatial pattern analysis. *International Journal of Applied Earth Observation and Geoinformation*, 18, 329–343.

Rawat, J. S., and Kumar, M. (2015). Monitoring land use/cover change using remote sensing and GIS techniques: A case study of Hawalbagh block, district Almora, Uttarakhand, India. *The Egyptian Journal of Remote Sensing and Space Science*, 18(1), 77–84.

Rawat, J. S., Biswas, V., and Kumar, M. (2013). Changes in land use/cover using geospatial techniques: A case study of Ramnagar town area, district Nainital, Uttarakhand, India. *The Egyptian Journal of Remote Sensing and Space Science*, 16(1), 111–117.

Rimal, B., Zhang, L., Keshtkar, H., Haack, B. N., Rijal, S., and Zhang, P. (2018). Land use/land cover dynamics and modeling of urban land expansion by the integration of cellular automata and Markov chain. *ISPRS International Journal of Geo-Information*, 7(4), 154.

Samal, D. R., and Gedam, S. S. (2015). Monitoring land use changes associated with urbanization: An object based image analysis approach. *European Journal of Remote Sensing*, 48(1), 85–99.

Sarif, M. O., and Gupta, R. D. (2022). Spatiotemporal mapping of land use/land cover dynamics using remote sensing and GIS approach: A case study of Prayagraj City, India (1988–2018). *Environment, Development and Sustainability*, 24(1), 888–920.

Shawul, A. A., and Chakma, S. (2019). Spatiotemporal detection of land use/land cover change in the large basin using integrated approaches of remote sensing and GIS in the Upper Awash basin, Ethiopia. *Environmental Earth Sciences*, 78(5), 141.

Singh, S., and Kumar, D. (2021). Investigation of rainfall variability of the southern part of Uttarakhand using entropy theory. *Indian Journal of Soil Conservation*, 49(2), 83–88.

Singh, S., Kumar, A., and Kumara, S. (2023b). Assessment of drought in meteorological data using SPI and SPEI indicators for sustaining agricultural productivity in the Agra division of Uttar Pradesh, India. *In Advancement of GI-Science and Sustainable Agriculture: A Multi-dimensional Approach* (pp. 285–303). Cham: Springer Nature Switzerland.

Singh, S., Kumar, D., Kumar, A., and Kuriqi, A. (2023a). Entropy-based assessment of climate dynamics with varying elevations for hilly areas of Uttarakhand, India. *Sustainable Water Resources Management*, 9(4), 130.

Singh, S., Kumar, D., Vishwakarma, D. K., Kumar, R., and Kushwaha, N. L. (2024). Seasonal rainfall pattern using coupled neural network-wavelet technique of southern Uttarakhand, India. *Theoretical and Applied Climatology*, 155(6), 5185–5201.

Subasinghe, S., Estoque, R. C., and Murayama, Y. (2016). Spatiotemporal analysis of urban growth using GIS and remote sensing: A case study of the Colombo Metropolitan Area, Sri Lanka. *ISPRS International Journal of Geo-Information*, 5(11), 197.

Thapa, R. B. and Murayama, Y. (2010). Drivers of urban growth in the Kathmandu valley, Nepal: Examining the efficacy of the analytic hierarchy process. *Applied Geography*, 30(1), 70–83.

Vinayak, B., Lee, H. S., and Gedem, S. (2021). Prediction of land use and land cover changes in Mumbai City, India, using remote sensing data and a multilayer perceptron neural network-based Markov chain model. *Sustainability*, 13(2), 471.

Vinoth Kumar, J. A., Pathan, S. K., and Bhanderi, R. J. (2007). Spatio-temporal analysis for monitoring urban growth–a case study of Indore city. *Journal of the Indian Society of Remote Sensing*, 35, 11–20.

Wakode, H. B., Baier, K., Jha, R. and Azzam, R. (2014). Analysis of urban growth using Landsat TM/ETM data and GIS-a case study of Hyderabad, India. *Arabian Journal of Geosciences*, 7(1), 109–121.

Weng, Q., and Lu, D. (2009). Landscape as a continuum: An examination of the urban landscape structures and dynamics of Indianapolis City, 1991–2000, by using satellite images. *International Journal of Remote Sensing*, 30(10), 2547–2577.

Wubie, M. A., Assen, M., and Nicolau, M. D. (2016). Patterns, causes and consequences of land use/cover dynamics in the Gumara watershed of lake Tana basin, Northwestern Ethiopia. *Environmental Systems Research*, 5, 1–12.

Xu, H. Q., Wang, X. Q., and Xiao, G. R. (2000). A remote sensing and GIS integrated study on urbanization with its impact on arable lands: Fuqing City, Fujian Province, China. *Land Degradation and Development*, 11(4), 301–314.

Yuan, F. (2010). Urban growth monitoring and projection using remote sensing and geographic information systems: A case study in the twin cities metropolitan area, Minnesota. *Geocarto International*, 25(3), 213–230.

Zhai, H., Lv, C., Liu, W., Yang, C., Fan, D., Wang, Z., and Guan, Q. (2021). Understanding spatio-temporal patterns of land use/land cover change under urbanization in Wuhan, China, 2000–2019. *Remote Sensing*, 13(16), 3331.

Zhang, H., Zhou, L. G., Chen, M. N., and Ma, W. C. (2011). Land use dynamics of the fast-growing Shanghai Metropolis, China (1979–2008) and its implications for land use and urban planning policy. *Sensors*, 11(2), 1794–1809.

3 Application of Remote Sensing and GIS for the Study of Climate Change on Glacier Retreat

Shubhi Khare, Deepak Kumar, and Priyadarshi Maurya

3.1 INTRODUCTION

Any large mass of ice that crystallizes over the surface of Earth due to any form of precipitation is known as a glacier. About 700,000 km^2, that is 270,000 square miles of earth, is estimated to be covered with glaciers, most of which are in the Antarctica (Pink, 2016). Being primarily found on poles, almost all the world's continents except Australia possess glaciers. Based on morphology, thermal characteristics, and behavior, glaciers are classified into two major categories, namely continental and alpine glaciers. Continental glaciers are the large ice sheets present on the surface of Earth. Several parts of North America and numerous National parks were shaped by ice sheets during the last ice age, which was approximately 20,000 years ago, of which only two are present in the modern day and are located in Greenland and Antarctica (Windnagel et al., 2023). Alpine glaciers, also known as Mountain or Valley glaciers, are glaciers formed in mountain regions, surrounded by mountain terrains and carving their way down the mountain slopes by flowing under their own weight (Zemp et al., 2006). Alpine glaciers are also referred to as Cirque glaciers due to their formation in small bowl-like structures in mountain laps (Barr and Spagnolo, 2015). Glaciers that are present today are only the remnants of the glaciers present during the ice age. The thick sheets of ice that used to be present across most continents of the world advanced and retreated to the polar region several times. The continental glaciers remained constant in Antarctica and Greenland, whereas the alpine glaciers retreated to the highest peaks of the world (Sommer et al., 2020).

The most prominent reason for this retreat of the glaciers is the change in the global climate. Rising temperatures across the globe are accelerating the melting rate of ice, which in turn is leading to the glaciers retreating to higher altitudes. Long-term data on glacier retreat were observed for four U.S. glaciers, and it was observed that the glaciers started to lose mass roughly over the last decade, but the shrinkage of the glaciers started back in the 1950s. From 2000 to 2019, the total lost mass of glaciers across the world accounted for 267 billion tonnes of ice per year (Ndehedehe, 2023). The rate of ice lost per year was 227 billion tonnes per year between 2000 and 2004,

DOI: 10.1201/9781003596080-3

which accelerated to a rate of 298 billion tonnes per year between 2015 and 2019 (Hugonnet et al., 2021). The most affected and the fastest melting glaciers due to global climate change are the continental glaciers present in Alaska, the Alps, and Iceland (Kochtitzky and Copland, 2022; Sommer et al., 2020). The alpine glaciers in the Pamir mountains, the Hindu Kush mountains, and the Himalayan range are also severely impacted by the current global climate situation (Goerlich et al., 2020; Naegeli et al., 2022; Romshoo et al., 2020). Research also exhibited that the rate of melting lagged in some regions as a result of the current global climatic situation. The glaciers present on Greenland's east coast and in Scandinavia and Iceland contribute to this case and are expected to result from weather anomalies in the North Atlantic, leading to higher precipitation and lower temperatures between 2010 and 2019 (Blunden and Boyer, 2022).

Glaciers are one of the most critical indicators of climate change. Climate change is detected by ice retreat from the glacier, measured through mass balance approaches. The records of glacier inventory have been compiled since the 1970s and have initially tracked over 100,000 glaciers covering an area of approximately 240,000 km². In addition, satellite records indicate that around 445,000 km² is still covered with ice and may be affected by climate change. Glaciers not only serve as a climate change indicator, but are also the prominent source of freshwater in the downstream areas as well as the primary supporter of freshwater ecosystems (Madrigal-Martínez et al., 2022). The increased rate of glacier retreat is one of the most horrifying threats to the downstream areas as the consequences of the retreat can be devastating. Water, food, energy security, and environmental management will all be affected adversely due to the consequent retreat of glaciers. The availability of water provided by glaciers in the downstream areas will decrease along with the shrinkage of glacier mass. It will subsequently affect the availability and reliability of water resources throughout the year in the short term. This reduction in meltwater flux will be reflected in a disturbed hydrological regime of river basins, causing potential stress in glacier-fed rivers for irrigation, hydropower generation, and municipal water supply purposes. As glaciers melt, sediments, nutrients, and pollutants stored in ice will be released in the water moving downstream to the rivers, leading to the problems of sedimentation and turbidity downstream which can impair water treatment processes and degrade aquatic habitats also due to the alteration in water temperature and flow regimes, thus causing an impact on the distribution and abundance of aquatic species.

3.2 TRADITIONAL MASS BALANCE APPROACH FOR MEASURING GLACIER RETREAT

The assessment of the health of the glacier is carried out by the mass balance method. The mass balance method calculates the net ice gain or loss from the glacier for a particular period. Therefore, the glacier system in a specific period may be considered the product of the net gain of ice in the system and the net loss of ice from the system (Geibel et al., 2022). If the net ice gain value rises to the net ice loss value, the glacier system will advance, or in the contrary case, it will recede. The mass balance is the difference between the gain and loss of ice from the system, and it can be measured seasonally or annually. There are various approaches through which the mass balance of a glacier system can be calculated. These approaches are shown in Figure 3.1 and are explained thoroughly.

The geodetic method estimates the volume change (*dV*) of a glacier and then converts it into mass change with the help of the surface density of the glacier at different parts, that is, the surface density of ice (ρ_i) and surface density of water (ρ_w); and average surface area (*A*) of the glacier. The mathematical equation employed to calculate geodetic mass balance is represented by Equation 3.1 (Mukherjee et al., 2023). This method was not used much due to its various limitations, but it was applied complementarily to the glaciological method.

$$B = \frac{dV}{A} \cdot \frac{\rho_i}{\rho_w} \tag{3.1}$$

The glaciological method employs the measurement of change in surface levels of a glacier between two different time intervals. This change can be measured through several ways, including stakes drilled manually into the glaciers. The difference between the two levels (gain or loss) when multiplied by the near-surface density of the glacier indicates the mass balance at that point. Due to its high reliability, the density is usually measured in a snow pit dug to a reference surface. By calculating the mass balance at various reference sites, the total mass balance of the glacier can be obtained by either through a plot of mass balance as a function of elevation and the plot of the area of the glacier with elevation or using the contour map approach in which total mass balance (*B*) is obtained using Equation 3.2 (Angchuk et al., 2021):

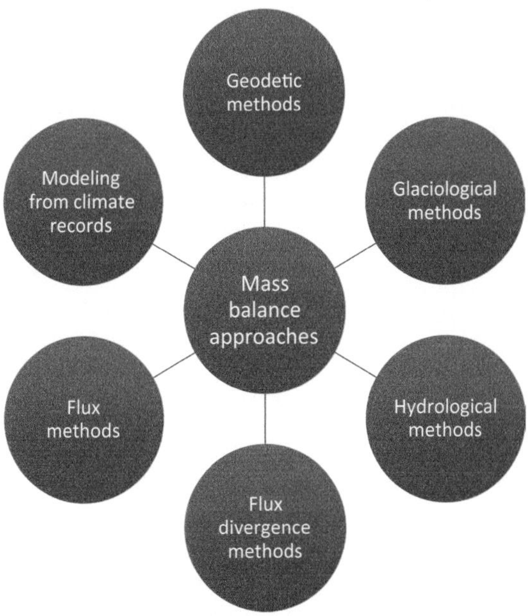

FIGURE 3.1 Mass balance approaches used for system modeling.

$$B = \sum b_n \Delta S_n \tag{3.2}$$

where, ΔS_n is the area of the glacier over which the net mass balances b_n applies (Kumar et al., 2021).

The hydrological mass balance approach is used to analyze the deflection in the volume of glaciers. The approach involves the contribution of various hydrological phenomena such as runoff, evaporation, and precipitation. The net mass balance of a glacier using this approach is addressed using Equation 3.3 which involves the parameters net balance of the glacier B_n, the mean specific accumulation over the glacier S_a, the total runoff from the glacier R, and the amount of water, ice and snow lost by evaporation E (Negi et al., 2022):

$$B_n = S_a - R - E \tag{3.3}$$

There is always an uncertainty to every empirical approach and the same is the case with this approach as well as the uncertainty in hydrological mass balance (U_{MB}) approach can be obtained using Equation 3.4 (Schmidt et al., 2023):

$$U_{MB} = \sqrt{\left(\frac{\partial MB}{\partial T} \times U_p\right)^2 + \left(\frac{\partial MB}{\partial T} \times U_E\right)^2 + \left(\frac{\partial MB}{\partial T} \times U_R\right)^2} \tag{3.4}$$

where U_P is the uncertainty of station precipitation, U_E is the uncertainty of evaporation, U_R is the uncertainty of runoff, and MB is the mass balance.

The vertical movement of ice in an upward or downward direction with respect to the surface change and rate of temporal change of the glacier is generally modeled using the ice flux divergence approach. It is also sometimes known as emergence or submergence velocity and is generally defined as the difference between the supplied ice and the lost ice from a glacier with respect to a particular position. When ice gets lost from the glacial surface, it is referred to as submergence velocity, and when ice gets supplied to the glacier, it is considered negative and is called emergence velocity. Van Tricht et al. (2021) considered ice thickness and surface velocity as a function of ice flux divergence, which can be calculated with the help of Equation 3.5 (Gao et al., 2022):

$$\nabla.\vec{q} = F\left(u_{s,x}\frac{\partial H}{\partial x} + u_{s,y}\frac{\partial H}{\partial y} + H\frac{u_{s,x}}{\partial x} + H\frac{u_{s,y}}{\partial y}\right) \tag{3.5}$$

where F is the depth-averaged glacier velocity ratio (here, it is 0.9); u_s is the glacier surface velocity; H is the glacier thickness; $\frac{\partial H}{\partial x}, \frac{\partial H}{\partial y}, \frac{u_{s,x}}{\partial x}$ and $\frac{u_{s,y}}{\partial y}$ are the gradients of glacier thickness and surface velocity in the x and y directions.

The movement of ice within a glacier system under the influence of gravity and other internal stresses acting within the glacier is referred to as glacier flux.

The movement thus generated is a balance between inputs that are the accumulation of snow and outputs that are the ablation or melting of ice. The ice flux gate approach is a method that can be employed to analyze the relationship between the ice floor and the altitude of the glacier to model the mass balance (Kochtitzky et al., 2022). Glacier flux is an important aspect in understanding the lacey dynamics and behavior of glaciers and glaciology. Understanding of glacier flux involves understanding of various processes, namely the internal deformation that is the creeping of deformed ice due to its weight; Basal sliding, which means deformation of the glacier into subglacial formations and then gliding over the glacier bed surface; and deformation of the bed of the glacier into subglacial formations and then contributing to flow. The flux gate approach works on identifying the cross section of a glacier, measuring ice flow velocity at different cross sections, and inferring the average surface mass balance between the flux gates. The sum of ice flux on all segments can be calculated with the formula given in Equation 3.6 (Xu et al., 2022):

$$Q = \sum_{i=1}^{N} \rho_i H_i l_i \overline{U_i} \tag{3.6}$$

where N is the total number of segments and i is the segment number, $\rho_{i,} H_{i,} l_{i\ and} \overline{U_i}$ are the components of ice density, ice thickness, segment width, and depth-averaged velocity perpendicular to the ice gate in segment i, respectively. The depth-averaged velocities $\overline{U(z)}$ at different gate positions can be calculated by the formula mentioned in Equation 3.7:

$$\overline{U(z)} = U_s - \left(\frac{2A}{n+1} . \tau_b^n . H + \frac{2A}{n+2} . \tau_b^n . H \right) \tag{3.7}$$

where A is the creep coefficient with value $A = 9 \times 10^{-25}$ s^{-1} Pa^{-3}, n is the flow law exponent that is, $n = 3$, H is the ice thickness and $\tau_b = f' \tau_d$, in which τ_d is the shear stress, f' denotes a scale factor (usually of order one).

3.3 ROLE OF REMOTE SENSING IN ASSESSMENT OF GLACIER RETREAT

Traditional methods for glacier mass balance measurement were primarily dependent on the accessibility of the glacier and were significantly difficult to carry out on large areas with complex topography and environment. Consequently, a large number of glaciers remained unassessed until the 1990s (Zhang et al., 2024). With the advancement of remote sensing technology in the 1990s, there has been a significant development in the global cryosphere change assessment due to the comprehensive, dynamic, and rapid advantages provided by the remote sensing technology (Veettil and Kamp, 2017). Supported by the remote sensing technology, there are mainly three methods for glacier mass balance, namely satellite gravimetry, geometric method, and the Input–Output Method (IOM). As the name suggests, the satellite gravity method utilizes the changes in the Earth's gravity field to infer the mass changes of glaciers and ice sheets under the influence of global climate change. For the same purpose, the GRACE satellite uses a K-band ranging system to accurately measure the distance between two satellites, enabling the measurement of Earth's time-varying gravity field (Velicogna and Wahr, 2013). The geometric methods for glacier mass balance equip the use of satellite data to measure the change in the elevation of glaciers and the its extent to get an idea of mass change over time. These methods include techniques like satellite altimetry, stereo imagery, and bistatic SAR interferometry. The IOM of glacier mass balance approach, with the use of satellite data GPS and InSAR observations, compares the input, that is, the accumulation of snow and ice, and outputs, which include ablation or ice flux, with the use of satellite data, GPS and inside observations (Shepherd et al., 2012).

Several satellites are used to assist the glacier mass balance throughout the globe. The cardinal ones are: altimetry satellite, gravity satellite, optical and infrared satellites. The altimetry satellites include ICESat-1 and ICESat-2 used for precisely detecting height of the earth surface along with the glaciers employing laser technique (Wang et al., 2011; Brown et al., 2023); CryoSat-2 uses radar technology to assess changes in surface elevations and snow depth (Morris et al., 2022); and ERS-1, ENVISat is used to measure change in glacier surface elevation using radar altimetry (Wu et al., 2024). The gravity satellites used for assessing mass balance are GRACE and GRACE-FO, used for measuring variation in the Earth's gravity field along with the changes in mass of glaciers and ice sheets; and CHAMP and GOCE. MODIS, ASTER, and Landsat are the optical and infrared satellites used for the assessment of snow cover, albedo, surface temperature, and glacier extent data (Deoli et al., 2021; Racoviteanu et al., 2021; Herreid, 2021; Zhang et al., 2021). The details of the satellites used for assessing glacial mass balance are shown in Table 3.1.

The process of satellite data processing has been shown in Figure 3.2. As depicted in the figure, the Digital Elevation Model (DEM) profile is usually created with the

TABLE 3.1
Satellite and Its Sensors for Glacier Mass Balance Observations

S. No.	Satellite	Sensors	Spatial Resolution	Bands
1	Landsat	Visible, NIR and SAR	30 m	Red, green, blue, SWIR
2	Sentinel	SAR, MSI	20 m	SAR C-band, VNIR, SWIR
3	IRS	LISS IV	5.8 m	TIR, visible and SWIR band
4	Terra	MODIS	250–500 m	Visible, NIR, SWIR
5	Terra (EOS AM-1)	ASTER	15–30 m	TIR, 3N, 3B, VNIR
6	ICESat	ATLAS	11–20 m	-
7	CryoSat	SIRAL	300 m	Ku band
8	ERS-1	InSAR	30 m	C-band
9	ENVISat	ASAR	30–150 m	C-band
10	GRACE	K-band ranging system	Several hundred kilometers	–

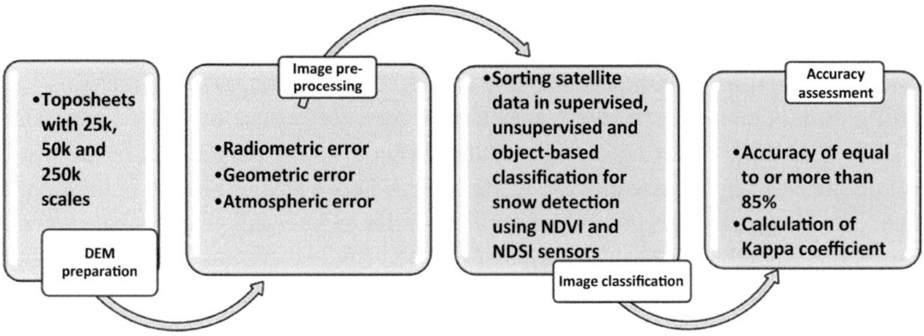

FIGURE 3.2 Processes involved in the processing of satellite data.

help of toposheets prepared by the Survey of India (SOI), mainly at three different scales, that is, 25,000, 50,000, and 250,000. The next step is image processing, in which the errors in the data are minimized, for which processes like orthorectification, image correction, and enhancement are adopted, requiring the input data to have accurate sensor geometry and a DEM model with a value of root mean square error as small as possible. The raw satellite data could possess mainly three types of errors, namely radiometry and atmospheric errors (Zhang et al., 2022). The error that is generated due to the failure of grids of sensors that store digital numbers against the reflected brightness from the source is termed the radiometric error. The errors that get generated in the satellite imagery with respect to the relative position of location between satellite sensor and the earth surface, irregular topographic features, unstable sensor platform or viewing path of sensors are known as geometric errors; and when the true reflectance value of the object of interest gets hampered by the amount of noises (the layers of atmosphere causing refraction and scattering like phenomena and carrying small particles, air pollutants and aerosols), there generate an error in the satellite imagery known as atmospheric error. The next step is image classification, which constitutes one of the most important steps in the processing of data (Zhao et al., 2022). The input satellite images are classified into supervised, unsupervised, and object-based classification, after which indices like Normalized Difference Snow Index (NDSI) and Normalized Difference Vegetation Index (NDVI) are used to effectively extract the snow and vegetation features from the imagery. The last step is the assessment of accuracy, which can be carried out with the help of the Kappa coefficient, which signifies how much error has been reduced in classifying images. The overall accuracy of remote sensing imagery should be more than or equal to 85% for it to be accepted (Florath et al., 2022; Turpo Cayo et al., 2022).

3.4 RECENT STUDIES OF GLACIER RETREAT

A study was carried out by Tiwari et al. in 2021 in the Chipa glacier of the Indian Himalayan region. The mass balance of the glacier was calculated by preparing the DEM using topographic sheets and a triangulated irregular network of the glacier region. It was found in the study that the annual retreat of the Chipa glacier is approximately 6 m per year, and the mass of the glacier is declining at the rate of 5.27×10^9

kg per year. This brought the study to the conclusion that the Indian Himalayan region is highly sensitive to changes in global climatic conditions, and the study pressed the issue that the Himalayan region is being hit by climatic changes at an alarming rate. A differential diagnosis strategy was opted for by a study done by Forte et al. (2021) to study the actual meaning of high scattering faces that emerged due to the use of ground penetrating radar within the Eastern Gran Zebrù glacier located in the Central Alps. The study used the ground-penetrating radar to interpret geophysical information of the glacier. The study concluded that the high scattering zones in the study glacier were most likely to be a mixture of ice and debris, which could have possibly formed due to the shrinkage of the glacier. The study also concluded that the GPR technique has significant importance in studying the various internal parameters of the glacier.

A study of Namcha Barwa-Gyala Peri (NBGP) was carried out by Guo et al. (2021) using a multiple hierarchical object- and rule-based classification (MHORC) scheme combined with the Landsat time series images, which was proposed for enhanced assessment of the complex debris-covered glacier. The method was adopted due to its capability of handling multi-source data simultaneously so as to obtain the distinction between objects with similar spectra through the development of multiple hierarchical rule-based classifications. Landsat images were collected from 1987 to 2019 and the cloud condition (less than 10%) was filtered out. The glacier extent was assessed with the help of images collected in the snow-free region, and the fluctuations in the glacier were monitored using the images collected in the snowfall season. The results of the study confirmed the reduction in the glacier area with area change velocity to be –0.63 km^2 per year between 1999 and 2003 and –2.95 km^2 per year between 2003 and 2015. The study also showed that there was a total of 65 glaciers covering the total area of 426.7km^2, among which 21.5% in number, that is, 76% of the total area included valley glaciers. The study concluded that MHORC is effective in delineating debris-covered glaciers, and the Landsat time series has significant potential to assess global glacial environment evolution.

In a study by Thakur et al. (2023), the transformation of the Gangotri glacier terminus in the Central Himalayas was assessed employing a remote sensing technique and using multi-sensor data from 2017 to 2021. The analysis revealed that the snout of the Gangotri glacier lost an area of approximately 0.08 km^2 from 2017 to 2021 in the central part of the glacier. In addition, according to the Sentinel 2 data, the surface velocity near the snout of the glacier was approximately 8 m per year, and the recession in the snout of the glacier was around 270 m between 2017 and 2021. The analysis concluded that the glacier below the meeting point of the Chaturangi glacier melted at a slower rate than the ice above the meeting point. These findings can play a vital role in understanding the response of the Gangotri glacier to global climate change, which in turn can result in the formation of various ephemeral lakes, which could generate a negative feedback loop.

A study of cryosphere research in different regions of the Greenland Ice Sheet using the Segment Anything Model (SAM) released by meta-AI has been conducted by Shankar et al. in (2024). SAM, a self-supervised learning model trained on unlabeled training data, is used to generate semantic segmentation masks, which are heavily needed to precisely understand the change in various parameters of glaciology, such as glacial termini, supraglacial lakes, iceberg crevasses, etc. Sentinel 1 and Sentinel 2 imagery was taken from Google Earth Engine with a spatial resolution of 20 × 22 m in interferometric wide swath; the planet imagery was obtained from PlanetScope data portal at a spatial resolution of 3 m; time-lapse imagery was fetched from Stardot Technology CAM-SEC5-B, and Landsat-4, -5, and -8 images obtained from United States Geological Survey (USGS) Earth Explorer were used in the current study. The study concluded that SAM can provide a significant approach in implementing deep learning in glaciology by achieving accuracy and efficiency at the same time. With the help of SAM, high-resolution data sets can be created by using optical and SAR remote sensing imagery simultaneously. For areas with large coverage, the model can be used by dividing the image into subsets by creating grids to precisely study the various glacial logical features, like lakes and icebergs.

A study was carried out by Xue et al. (2021) to verify the feasibility of the Central Tibetan Plateau glacier and to detect the short-term changes in Xia Dongkemadi (XDKMD) glacier, a south-west facing glacier with an area of 1705 km^2, and Ganglongjiama (GLJM) glacier, which is located in the northeast corner of the Ganglongjiama with the Yangtze River as a natural source, by conducting four high-resolution Unmanned Aerial Vehicle (UAV) surveys over the glaciers in 2017 and 2019. To depict the short-term changes in the two glaciers, Landsat images over the last 30 years were assessed, and a 10-year interval between the two selected images was chosen due to the coarse spatial resolution of the images. The changes in the length, area, and elevation of the glaciers at different times were obtained by creating the Ortho mosaics and digital elevation models from the images provided by UAVs. The result of the study highlighted the fact that the XDKMD glacier retreated with an average of 3.9 m along with a terminal decrease of 2.3% and area reduction of 0.0032 km^2. The Western site of XDKMD glacier retreated more than the eastern side with the thinning of the ablation zone from approximately 2 to 3 m, and the minimum ice loss from the glacier was 3 m minimum and 8.5 m maximum. The GLJM glacier was assessed with various deep freebases due to the continuation of erosion by meltwater from the past several years. The area of the glacier lost was found to be 0.0529 km^2 during the last 2 years, with a rate of 3.6% per year and an average retreat length of 49.7 m from 2017 to 2019. The elevation of the glacier changed by more than 16 m due to the glacier's shrinkage.

A study by Li et al. (2024) of Jiongpu Co, one of the largest glacier lakes in the Tibetan plateau, along with its bathymetry and area evolution assessment, was carried out to determine the ice loss in the glacier from 1976 to 2021 by constructing multisource DEM datasets. The result of the study showed that the Jiongpu Co glacier expanded from 1.19 ± 0.09 km^2 to 5.34 ± 0.07 km^2 with a subsequent increase in volume from 0.09 ± 0.004 GT to 0.66 ± 0.03 GT from 1976 to 2021. The subaqueous equivalent ice loss from the Jiongpu Co glacier was 0.32 ± 0.01 GT water, which resulted in an underestimated ice loss of 0.06 GT, that is, 19% as compared with the previous evaluations. From 2000 to 2020, the total ice loss of the Jiongpu Co glacier was 1.52 ± 0.37 GT, and more than one-third of the ice loss was related to lake expansion. For further study on glacier retreat, the following studies can be considered: Raup et al. (2015), Solomina et al. (2015), Kaab et al. (2016), Mal et al. (2016), Nie et al. (2017), Rai et al. (2017), Baumhoer et al. (2018), Mir and Majeed (2018), Patel et al. (2019), Li et al. (2019), Rashid et al. (2020), Rashid et al. (2021), Wood et al. (2021), Kaushik et al. (2022), Rajat et al. (2022), Bosson et al. (2023), He et al. (2023), Banerjee et al. (2024), Izagirre et al. (2024), Li et al. (2025).

3.6 CONCLUSION

Glaciers are one of the most important sources of freshwater on Earth. The constant and accelerated retreat of most of the glacier across the globe due to an increase in the global temperature has raised a very important concern with respect to the level of water present in the ocean, as well as the formation of glacial lakes, which may be a cause of flooding in the mountains as a result of glacial lake outburst. The present study tries to convey the facts with respect to the global situation of the glaciers and their response to the global rise in temperature, and presents the methods that we use traditionally to assess the change in the glaciers. With the introduction of remote sensing, it has become quite convenient to assess the change in the glaciers located in those locations, where human access is almost unfeasible. The present study also proclaims the role of remote sensing in assessing change in the glaciers and enlists various satellite and sensors used for the purpose. The study also tries to present numerous studies that have been done in this area with the help of remote sensing technology.

REFERENCES

Angchuk, T., Ramanathan, A., Bahuguna, I. M., Mandal, A., Soheb, M., Singh, V. B., Mishra, S., and Vatsal, S. (2021). Annual and seasonal glaciological mass balance of Patsio Glacier, western Himalaya (India) from 2010 to 2017. *Journal of Glaciology*, 67(266), 1137–1146.

Banerjee, A., Kang, S., Guo, W., Meadows, M. E., Wang, W., Sengupta, D., and Zhang, T. (2024). Glacier retreat and lake outburst floods in the central Himalayan region from 2000 to 2022. *Natural Hazards*, 120(6), 5485–5508.

Barr, I. D., and Spagnolo, M. (2015). Glacial cirques as palaeoenvironmental indicators: Their potential and limitations. *Earth-Science Reviews*, 151, 48–78.

Baumhoer, C. A., Dietz, A. J., Dech, S., and Kuenzer, C. (2018). Remote sensing of Antarctic glacier and ice-shelf front dynamics—A review. *Remote Sensing*, 10(9), 1445.

Blunden, J., and Boyer, T. (2022). State of the climate in 2021. *Bulletin of the American Meteorological Society*, 103(8), S1–S465.

Bosson, J. B., Huss, M., Cauvy-Fraunié, S., Clément, J. C., Costes, G., Fischer, M., Poulenard, J., and Arthaud, F. (2023). Future emergence of new ecosystems caused by glacial retreat. *Nature*, 620(7974), 562–569.

Brown, M. E., Arias, S. D., and Chesnes, M. (2023). Review of ICESat and ICESat-2 literature to enhance applications discovery. *Remote Sensing Applications: Society and Environment*, 29, 100874.

Deoli, V., Kumar, D., Kumar, M., Kuriqi, A., and Elbeltagi, A. (2021). Water spread mapping of multiple lakes using remote sensing and satellite data. *Arabian Journal of Geosciences*, 14, 1–15.

Florath, J., Keller, S., Abarca-del-Rio, R., Hinz, S., Staub, G., and Weinmann, M. (2022). Glacier monitoring based on multi-spectral and multi-temporal satellite data: A case study for classification with respect to different snow and ice types. *Remote Sensing*, 14(4), 845.

Forte, E., Santin, I., Ponti, S., Colucci, R. R., Gutgesell, P., and Guglielmin, M. (2021). New insights in glaciers characterization by differential diagnosis integrating GPR and remote sensing techniques: A case study for the Eastern Gran Zebrù glacier (Central Alps). *Remote Sensing of Environment*, 267, 112715.

Gao, L., Yang, X., Qi, J., and Chen, W. (2022). Glacier Mass Balance Pattern and Its Variation Mechanism in the West Kunlun Mountains in Tibetan Plateau. *Remote Sensing*, 14(11), 2634.

Geibel, L., Huss, M., Kurzböck, C., Hodel, E., Bauder, A., and Farinotti, D. (2022). Rescue and homogenization of 140 years of glacier mass balance data in Switzerland. *Earth System Science Data*, 14(7), 3293–3312.

Goerlich, F., Bolch, T., and Paul, F. (2020). More dynamic than expected: an updated survey of surging glaciers in the Pamir. *Earth System Science Data and Discussions*, 2020, 1–26.

Guo, S., Du, P., Xia, J., Tang, P., Wang, X., Meng, Y., and Wang, H. (2021). Spatiotemporal changes of glacier and seasonal snow fluctuations over the Namcha Barwa–Gyala Peri massif using object-based classification from Landsat time series. *ISPRS Journal of Photogrammetry and Remote Sensing*, 177, 21–37.

He, Q., Kuang, X., Chen, J., Hao, Y., Feng, Y., Wu, P., and Zheng, C. (2023). Glacier retreat and its impact on groundwater system evolution in the Yarlung Zangbo source region, Tibetan Plateau. *Journal of Hydrology: Regional Studies*, 47, 101368.

Herreid, S. (2021). What can thermal imagery tell us about glacier melt below rock debris?. *Frontiers in Earth Science*, 9, 681059.

Hugonnet, R., McNabb, R., Berthier, E., Menounos, B., Nuth, C., Girod, L., Farinotti, D., Huss, M., Dussaillant, I., Brun, F., and Kääb, A. (2021). Accelerated global glacier mass loss in the early twenty-first century. *Nature*, 592(7856), 726—731.

Izagirre, E., Revuelto, J., Vidaller, I., Deschamps-Berger, C., Rojas-Heredia, F., Rico, I., Alonso-González, E., Gascoin, S., Serrano, E., and López-Moreno, J. I. (2024). Pyrenean glaciers are disappearing fast: state of the glaciers after the extreme mass losses in 2022 and 2023. *Regional Environmental Change*, 24(4), 172.

Kääb, A., Winsvold, S. H., Altena, B., Nuth, C., Nagler, T., and Wuite, J. (2016). Glacier remote sensing using Sentinel-2. Part I: Radiometric and geometric performance, and application to ice velocity. *Remote sensing*, 8(7), 598.

Kaushik, S., Singh, T., Joshi, P. K., and Dietz, A. J. (2022). Automated mapping of glacial lakes using multisource remote sensing data and deep convolutional neural network. *International Journal of Applied Earth Observation and Geoinformation*, 115, 103085.

Kochtitzky, W., and Copland, L. (2022). Retreat of Northern Hemisphere marine-terminating glaciers, 2000–2020. *Geophysical Research Letters*, 49(3), e2021GL096501.

Kochtitzky, W., Copland, L., Van Wychen, W., Hugonnet, R., Hock, R., Dowdeswell, J. A., ... and Navarro, F. (2022). The unquantified mass loss of Northern Hemisphere marine-terminating glaciers from 2000–2020. *Nature Communications*, 13(1), 5835.

Kumar, R., Singh, S., Singh, A., Kumar, R., Singh, S., and Randhawa, S. S. (2021). Surface mass balance analysis at Naradu Glacier, Western Himalaya, India. *Scientific Reports*, 11(1), 12710.

Li, Y. J., Ding, Y. J., Shangguan, D. H., and Wang, R. J. (2019). Regional differences in global glacier retreat from 1980 to 2015. *Advances in Climate Change Research*, 10(4), 203–213.

Li, D., Shangguan, D., Han, T., Butt, A. Q., Pan, B., Cao, B., Wang, M., Wang, R., and Li, Y. (2024). An ice loss evaluation of lake-terminating glaciers based on lake bathymetry—A case study of the Jiongpu Glacier. *Remote Sensing*, 16(16), 3027.

Li, T., Hofer, S., Moholdt, G., Igneczi, A., Heidler, K., Zhu, X. X., and Bamber, J. (2025). Pervasive glacier retreats across Svalbard from 1985 to 2023. *Nature Communications*, 16(1), 705.

Madrigal-Martínez, S., Puga-Calderón, R. J., Bustínza Urviola, V., and Vilca Gómez, Ó. (2022). Spatiotemporal changes in land use and ecosystem service values under the influence of glacier retreat in a high-Andean environment. *Frontiers in Environmental Science*, 10, 941887.

Mal, S., Singh, R.B., and Schickhoff, U. (2016). Estimating recent glacier changes in Central Himalaya, India, using remote sensing data. In: Singh, R., Schickhoff, U., Mal, S. (eds) *Climate Change, Glacier Response, and Vegetation Dynamics in the Himalaya*. Springer, Cham. https://doi.org/10.1007/978-3-319-28977-9_11

Mir, R. A., and Majeed, Z. (2018). Frontal recession of Parkachik Glacier between 1971–2015, Zanskar Himalaya using remote sensing and field data. *Geocarto International*, 33(2), 163–177.

Morris, A., Moholdt, G., Gray, L., Schuler, T. V., and Eiken, T. (2022). CryoSat-2 interferometric mode calibration and validation: A case study from the Austfonna ice cap, Svalbard. *Remote Sensing of Environment*, 269, 112805.

Mukherjee, K., Menounos, B., Shea, J., Mortezapour, M., Ednie, M., and Demuth, M. N. (2023). Evaluation of surface mass-balance records using geodetic data and physically-based modelling, Place and Peyto glaciers, western Canada. *Journal of Glaciology*, 69(276), 665–682.

Naegeli, K., Franke, J., Neuhaus, C., Rietze, N., Stengel, M., Wu, X., and Wunderle, S. (2022). Revealing four decades of snow cover dynamics in the Hindu Kush Himalaya. *Scientific Reports*, 12(1), 13443.

Ndehedehe, C. (2023). Understanding the anthropocene. In *Hydro-Climatic Extremes in the Anthropocene* (pp. 3–22). Cham: Springer International Publishing. https://doi.org/10.1007/978-3-031-37727-3_1.

Negi, H. S., Gusain, H. S., Dhahiya, T., and Kumar, A. (2022). Hydrological mass balance study of Siachen glacier, East Karakoram. *Journal of Earth System Science*, 131(1), 33.

Nie, Y., Sheng, Y., Liu, Q., Liu, L., Liu, S., Zhang, Y., and Song, C. (2017). A regional-scale assessment of Himalayan glacial lake changes using satellite observations from 1990 to 2015. *Remote Sensing of Environment*, 189, 1–13.

Patel, A., Prajapati, R., Dharpure, J. K., Mani, S., and Chauhan, D. (2019). Mapping and monitoring of glacier areal changes using multispectral and elevation data: A case study over Chhota-Shigri glacier. *Earth Science Informatics*, 12(4), 489–499.

Pink, R. M. (2016). People's republic of China: Confronting catastrophic drought and pollution. In *Water Rights in Southeast Asia and India* (pp. 139–169). New York: Palgrave Macmillan US. https://doi.org/10.1057/9781137504234_7.

Racoviteanu, A. E., Nicholson, L., and Glasser, N. F. (2021). Surface composition of debris-covered glaciers across the Himalaya using linear spectral unmixing of Landsat 8 OLI imagery. *The Cryosphere*, 15(9), 4557–4588.

Rai, P. K., Mishra, V. N., Singh, S., Prasad, R., and Nathawat, M. S. (2017). Remote sensing-based study for evaluating the changes in glacial area: A case study from Himachal Pradesh, India. *Earth Systems and Environment*, 1, 1–13.

Rajat, S., Singh, B. R., Prakash, C., and Anita, S. (2022). Glacier retreat in Himachal from 1994 to 2021 using deep learning. *Remote Sensing Applications: Society and Environment*, 28, 100870.

Rashid, I., Majeed, U., Jan, A., and Glasser, N. F. (2020). The January 2018 to September 2019 surge of Shisper Glacier, Pakistan, detected from remote sensing observations. *Geomorphology*, 351, 106957.

Rashid, I., Majeed, U., Najar, N. A., and Bhat, I. A. (2021). Retreat of Machoi Glacier, Kashmir Himalaya between 1972 and 2019 using remote sensing methods and field observations. *Science of the Total Environment*, 785, 147376.

Raup, B. H., Andreassen, L. M., Bolch, T., and Bevan, S. (2015). Remote sensing of glaciers. *Remote Sensing of the Cryosphere*, 123–156. England: John Wiley & Sons, Ltd.

Romshoo, S. A., Fayaz, M., Meraj, G., and Bahuguna, I. M. (2020). Satellite-observed glacier recession in the Kashmir Himalaya, India, from 1980 to 2018. *Environmental Monitoring and Assessment*, 192, 1–17.

Schmidt, L. S., Schuler, T. V., Thomas, E. E., and Westermann, S. (2023). Meltwater runoff and glacier mass balance in the high Arctic: 1991–2022 simulations for Svalbard. *The Cryosphere*, 17(7), 2941–2963.

Shankar, S., Stearns, L. A., and van der Veen, C. J. (2024). Semantic segmentation of glaciological features across multiple remote sensing platforms with the Segment Anything Model (SAM). *Journal of Glaciology*, 70, e4.

Shepherd, A., Ivins, E. R., A, G., Barletta, V. R., Bentley, M. J., Bettadpur, S., ... and Zwally, H. J. (2012). A reconciled estimate of ice-sheet mass balance. *Science*, 338(6111), 1183–1189.

Solomina, O. N., Bradley, R. S., Hodgson, D. A., Ivy-Ochs, S., Jomelli, V., Mackintosh, A. N., Nesje, A., Owen, L.A., Wanner, H., Wiles, G. C., and Young, N. E. (2015). Holocene glacier fluctuations. *Quaternary Science Reviews*, 111, 9–34.

Sommer, C., Malz, P., Seehaus, T. C., Lippl, S., Zemp, M., and Braun, M. H. (2020). Rapid glacier retreat and downwasting throughout the European Alps in the early 21st century. *Nature Communications*, 11(1), 3209.

Thakur, P. K., Pandey, P., and Chauhan, P. (2023). Assessment of dynamics of frontal part of Gangotri glacier, India, from 2017 to 2021 using remote sensing data. *Journal of the Indian Society of Remote Sensing*, 51(4), 691–698.

Turpo Cayo, E. Y., Borja, M. O., Espinoza-Villar, R., Moreno, N., Camargo, R., Almeida, C., Hopfgartner, K., Yarleque, C., and Souza Jr, C. M. (2022). Mapping three decades of changes in the tropical Andean glaciers using landsat data processed in the Earth engine. *Remote Sensing*, 14(9), 1974.

Van Tricht, L., Huybrechts, P., Van Breedam, J., Vanhulle, A., Van Oost, K., and Zekollari, H. (2021). Estimating surface mass balance patterns from unoccupied aerial vehicle measurements in the ablation area of the Morteratsch–Pers glacier complex (Switzerland). *The Cryosphere*, 15(9), 4445–4464.

Veettil, B. K., and Kamp, U. (2017). Remote sensing of glaciers in the tropical Andes: A review. *International Journal of Remote Sensing*, 38(23), 7101–7137.

Velicogna, I., and Wahr, J. (2013). Time-variable gravity observations of ice sheet mass balance: Precision and limitations of the GRACE satellite data. *Geophysical Research Letters*, 40(12), 3055–3063.

Wang, X., Cheng, X., Gong, P., Huang, H., Li, Z., and Li, X. (2011). Earth science applications of ICESat/GLAS: A review. *International Journal of Remote Sensing*, 32(23), 8837–8864.

Windnagel, A., Hock, R., Maussion, F., Paul, F., Rastner, P., Raup, B., and Zemp, M. (2023). Which glaciers are the largest in the world?. *Journal of Glaciology*, 69(274), 301–310.

Wood, M., Rignot, E., Fenty, I., An, L., Bjørk, A., van den Broeke, M., Cai, C., Kane, E. ...and Zhang, H. (2021). Ocean forcing drives glacier retreat in Greenland. *Science Advances*, 7(1), eaba7282.

Wu, F., Gourmelon, N., Seehaus, T., Zhang, J., Braun, M., Maier, A., and Christlein, V. (2024). Contextual hookformer for glacier calving front segmentation. *IEEE Transactions on Geoscience and Remote Sensing,* 62, 1–15.

Xu, D., Tang, X., Yang, S., Zhang, Y., Wang, L., Li, L., and Sun, B. (2022). Revisiting ice flux and mass balance of the Lambert Glacier–Amery Ice Shelf System using multi-remote-sensing datasets, East Antarctica. *Remote Sensing*, 14(2), 391.

Xue, Y., Jing, Z., Kang, S., He, X., and Li, C. (2021). Combining UAV and Landsat data to assess glacier changes on the central Tibetan Plateau. *Journal of Glaciology*, 67(265), 862–874.

Zemp, M., Haeberli, W., Hoelzle, M., and Paul, F. (2006). Alpine glaciers to disappear within decades?. *Geophysical Research Letters*, 33(13), L13504.

Zhang, J., Jia, L., Menenti, M., Zhou, J., and Ren, S. (2021). Glacier area and snow cover changes in the range system surrounding tarim from 2000 to 2020 using google earth engine. *Remote Sensing*, 13(24), 5117.

Zhang, B., Wu, Y., Zhao, B., Chanussot, J., Hong, D., Yao, J., and Gao, L. (2022). Progress and challenges in intelligent remote sensing satellite systems. *IEEE Journal of Selected Topics in Applied Earth Observations and Remote Sensing*, 15, 1814–1822.

Zhang, G., Carrivick, J. L., Emmer, A., Shugar, D. H., Veh, G., Wang, X., Labedz, C., Mergili, M., Mölg, N., Huss, M., Allen, S., Sugiyama, S., and Lützow, N. (2024). Characteristics and changes of glacial lakes and outburst floods. *Nature Reviews Earth & Environment*, 5(6), 447–462.

Zhao, Q., Yu, L., Du, Z., Peng, D., Hao, P., Zhang, Y., and Gong, P. (2022). An overview of the applications of earth observation satellite data: Impacts and future trends. *Remote Sensing*, 14(8), 1863.

4 Land Use Pattern Changes and Their Impact on Sustainable Agriculture Using Remote Sensing and GIS

Shekhar Singh, Krishnakant Nivesh, and Ravi Kant Bhardwaj

4.1 INTRODUCTION

Land-use patterns have evolved significantly over time, influenced by population growth, urbanization, economic activities, climate change, and policy decisions. These shifts have substantial implications for sustainable agriculture, which depends on the availability and quality of land resources. Sustainable agricultural practices aim to fulfil current food demands while preserving the ability of future generations to cultivate food, making strategic land use planning a crucial aspect of sustainable development (Foley et al., 2005). Remote Sensing (RS) and Geographic Information Systems (GIS) have become indispensable for assessing and monitoring land use transformations. These technologies provide detailed spatial and temporal data, enabling researchers and policymakers to evaluate the impact of land conversion on agriculture, identify degradation trends, and develop informed land use strategies (Lambin and Meyfroidt, 2011). As challenges such as deforestation, soil erosion, and diminishing arable land continue to rise, RS and GIS applications offer critical insights into optimizing land use for sustainable agricultural practices.

The conversion of farmland into urban and industrial zones poses a significant threat to food security and environmental stability. Expanding cities and infrastructure often encroach upon fertile lands, disrupting agricultural operations and diminishing production capacity (Seto et al., 2002). Additionally, shifts from traditional to commercial farming, monoculture practices, and overgrazing contribute to soil degradation and loss of biodiversity (Matson et al., 1997). A comprehensive understanding of these changes is essential for managing land resources effectively to balance economic development with environmental sustainability. As rainfall has a direct impact on agricultural productivity, with regular patterns supporting healthy crop yields and irregular patterns leading to droughts and floods, the combined effect of land use and land cover (LULC) changes and rainfall variability poses a serious

DOI: 10.1201/9781003596080-4

threat to food security and rural livelihoods (Singh and Kumar, 2021; Singh et al., 2023a; Singh et al., 2023b; Singh et al., 2024). Land use alterations also impact water availability, soil health, and climate conditions critical for sustainable agriculture. For example, deforestation for agricultural expansion can lead to soil erosion and reduced water retention, while excessive irrigation and chemical use can degrade soil quality (Pimentel et al., 1995). Addressing these concerns requires advanced spatial analysis and strategic planning, facilitated by RS- and GIS-based research.

RS-based technology involves capturing satellite imagery and aerial photographs to monitor land use and environmental conditions. Conversely, GIS is a spatial analysis tool that integrates geographic data for effective decision-making and helps in the management of water demand (Gururani et al., 2023). The combined application of RS and GIS facilitates land cover mapping, trend identification, and predictive modelling for future land-use scenarios (Zhu et al., 2022). One of the key advantages of using RS and GIS in land-use planning is their ability to analyze large-scale changes over extended periods. Historical satellite imagery aids in detecting transformation patterns such as deforestation, urban sprawl, and shifts in agricultural practices. By employing machine learning algorithms and classification techniques, researchers can categorize land cover types and assess their impact on agricultural productivity (Rane et al., 2023). These insights empower policymakers to provide different policies toward a sustainable future (Singh and Goyal, 2025), implement zoning regulations, conservation strategies, and sustainable farming methods to mitigate adverse effects. The mechanism of data acquisition by a satellite has been shown in Figure 4.1.

The importance of sustainable agricultural practices has gained global attention due to growing concerns over food security, climate change, and environmental

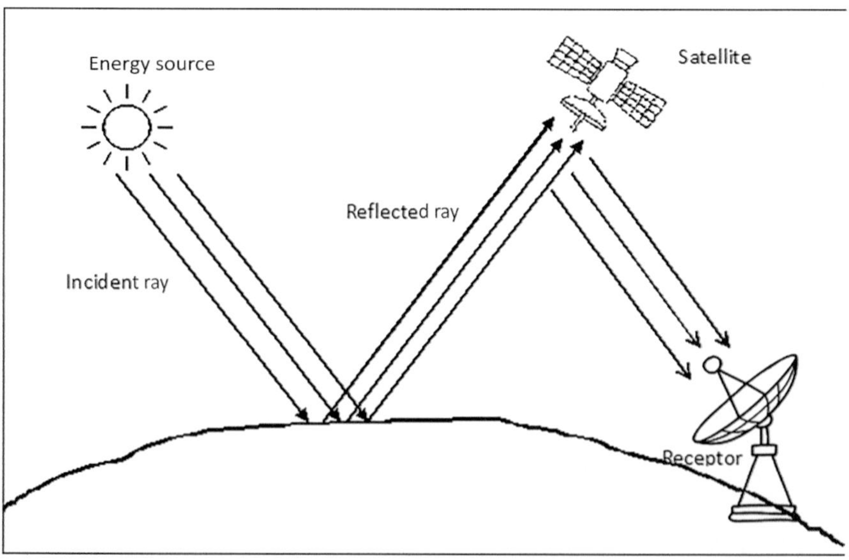

FIGURE 4.1 Process of data acquisition by remote sensing.

degradation. Sustainable agriculture integrates ecological principles with farming techniques to maximize resource efficiency while minimizing environmental harm (Mulla, 2013; Tilman et al., 2002). Understanding land use changes and their implications for sustainable agriculture is essential for effective land management. RS and GIS technologies provide valuable tools for monitoring these changes, supporting data-driven decision-making, and ensuring responsible land resource management. This study aims to analyze land use changes and their effects on agriculture, offering insights to policymakers and stakeholders to promote sustainable development.

4.2 HISTORICAL PERSPECTIVE OF AGRICULTURAL LAND USE CHANGE DYNAMICS

The analysis of land use changes and their impact on agriculture has been extensively studied, focusing on RS and GIS applications. This section reviews relevant literature chronologically to provide a comprehensive understanding of the subject.

Fazal (2000) briefly explained the decrease in agricultural land due to urban growth in Saharanpur City between 1988 and 1998, employing RS and GIS. To quantify this change, the study used aerial photographs and satellite images from the IRS-1C satellite with 5.8, combined with ground surveys. This methodology enabled detailed mapping of land-use transformations, specifying the locations and characteristics of lost agricultural areas. The research emphasizes how these techniques facilitate the observation of land-use changes. Cities in India are growing quickly. Back in 1901, only a small part of the population (about 11%) lived in cities. However, by 2011, that number had jumped to almost a third (around 31%). One big thing that happened when cities grew is that farmland turned into other things, like buildings or roads. In this study by Seto et al. (2002), they examined how the land was being used by examining different types of land cover. Landsat Thematic Mapper (TM) images were used to see how the land's brightness, greenness, and wetness changed over time. They used these changes to create maps showing different types of land and how they changed. They found that a lot of farmland was being turned into cities. The amount of agricultural land decreased by 11% between 1988 and 1996. They checked their maps by going to the actual places and found that our maps were accurate (about 93.5%).

Preserving farmland is one of China's primary goals. Keeping track of land use is essential for managing and using land wisely. Zhao et al.(2004) used satellite images from digital technology to help develop a system to detect changes in land in the Yellow River area. Four different methods were tested to find land changes. Overall, finding just one best way was hard because the land had mixed signals caused by soil changes and different land uses. Two types of methods, visual change detection and classification result overlay, were better than other methods. They worked well when RS data was available during the right season, and different land types could be distinguished. The results showed that this area's farmland went down by 5,321.8 hectares from 1987 to 1998. This means about 483.8 hectares of farmland were lost yearly, mostly in the central paddy fields and dry land areas in the northeast. The remotely sensed data used in this study were TM data with a 30 m pixel size. Abd El-Kawy et al. (2011) used LULC, which helps measure changes over time. They analyzed four satellite images from 1994, 1995, 2005, and 2009 to see how the land changed, especially in the western Nile delta. The results showed that the classification was improved by combining images and interpreting them visually. This helped identify land changes like irrigation, development of private land, and drained or dried areas. The main change was a decrease in barren land and an increase in agricultural land. Between 1984 and 1999, the barren land decreased by 28%, farmland increased by 14%, and built-up areas grew by 5%. The study also found evidence of land degradation, such as salt-affected soils and land drying out. These changes are mainly due to human activities, like farming and construction. The findings suggest that land policies should be improved to better manage and protect land resources.

Bauchi is a city in northeastern Nigeria that has grown greatly because many people have moved there quickly. For the past 10 years, Bauchi has remained a small city with few people. But now, Bauchi is growing faster than most cities in northern Nigeria. So, it is essential to see how much the city has changed. To study this change, Kafi et al. (2014) utilized remotely sensed data from Landsat 7 Enhanced Thematic Mapper Plus (ETM+) and Landsat 8 to see how the land in Bauchi city had changed. Landsat 7 pictures from 2003 and Landsat 8 images from 2013 were obtained for the study. This study used three ways to find the areas that changed. These ways were: supervised, unsupervised, and post-classification comparison. While the unsupervised classification was less effective, the supervised classification produced good results with about 93.5% and 89.7% accuracy for the 2003 and 2013 images, respectively. Kumar and Agrawal (2019) wanted to discover how farmland changed. They used computer mapping (GIS) and looked at pictures taken from space (satellites). They also went to the area and collected information. They studied how farmland changed in different parts of the Allahabad area over 18 years, from 2000 to 2018. Landsat images of the study period, that is, from 2000 to 2018, were downloaded and then pre-processed. Specifically, the study used images from Landsat Enhanced Thematic Mapper Plus (ETM+), TM, and Operational Land Imager (OLI) sensors for the years 2000, 2010, and 2018, respectively.

Most of Egypt's good farmland is lost because cities are growing quickly. This is a normal change in many countries, but in Egypt, it is a big problem because farmland is limited. The study's main goal by Abd EL-kawy et al. (2019) was to see how farming land and other types of land were changing because of urban expansion, and to predict what will happen in the future in cities like Alexandria and El-Beheira. The study used satellite images from 1987, 2001, 2015, and 2019 to identify land types such as farms, gardens, roads, and water. They used a method called maximum likelihood classification. The study also used land cover data to make the results more accurate and reduce mistakes. The researchers made yearly maps to compare how the land changed from 1987 to 2019. They found that land use changed significantly, with about 95% accuracy. They also predicted land use for 2030, 2040, and 2080. The results showed that farmland has decreased, while urban areas have grown. Specifically, farmland shrank by about 11%, and cities expanded by about 12%. Farm farmland could decrease by 11.5% in 2050 and by 24.5% in 2100. Urban areas might increase by 12% in 2050 and 25% in 2100. Yan et al. (2020) showed that forests and farmland are the primary resources on Earth, and they showed how healthy the environment was in a region. In this study, Landsat 5 TM and Landsat 8 OLI images were used. Images from 1990 to 2017 were used to extract information on forests in Malaysia and analyze the spatial-temporal changes of forests and agricultural lands. It was found that Malaysia's natural forests shrank by 441 million hectares, a 21% decrease. Many forests were turned into plantations in Peninsular Malaysia and or secondary forests in East Malaysia. During the same time, farmland increased by 55.7%, with rice fields growing slightly and plantations expanding greatly. In Peninsular Malaysia, rice fields are mainly along the north-central coast and the Kelantan Delta. In East Malaysia, plantations are mainly along the coast. Large areas in Kelantan and Kedah could become rice farms in the future, and there is also potential for expansion in Sarawak and Sabah. However, as farming expands, the conflict between agricultural development and forest protection is growing.

The study by Karakus et al. (2020) was done in Kilis city and its nearby areas. The region has a lot of farming potential. The goal was to examine how land use has changed over time, based on temperature and rainfall trends from 1985 to 2019. The study also aimed to suggest ways to keep farming areas sustainable. The research found that between 1985 and 2019, settlements (built-up areas) increased, while farmland decreased. In addition, the analysis showed that temperatures went up and rainfall decreased. Because of this, it is essential to use land wisely to reduce the effects of global warming. In this situation, decisions based on ecology and land use are needed to protect the environment. Yonaba et al. (2021) looked at patterns of land use change and predicted future changes in the landscape at the watershed level in semi-arid northern Burkina Faso. It focused on three types of land: natural vegetation, degraded soils, and farmed areas. The study used data from satellite images taken in 1986, 1999, 2009, and 2017. It also checked seasonal changes by monitoring radio signals from 2015 to 2017. The best time to map land use was from October to November. To predict future land use, a special type of computer model called a Multilayer Perceptron (MLP) was used, which was tested with data from 1999 to 2009 and checked that against a 2017 map. The model then predicted land use for 2030 and 2050, considering population growth. The results showed that soil degradation was a primary reason for increase in farmed areas between 1986 and 2017. The maps predicted that cultivated areas would grow by about 10.4% by 2030 and 22.7% by 2050.

Gondwe et al. (2021) studied how LULC changed in Blantyre City over 20 years (from 1999 to 2019) using Landsat 7 ETM+ images from 1999 and 2010 and Landsat 8 OLI images from 2019. They used a computer method called Artificial Neural Network (ANN) to identify different types of land use in these pictures. They also checked to make sure their land use maps were accurate. By comparing the land use maps from 1999 and 2019, they found that both built-up and agricultural land increased. Specifically, built-up areas grew by about 28.5 km^2 yearly, and farmland increased by about 35.8 km^2 yearly. On the other hand, the amount of bare land, forest, grassland, and waterbody areas all decreased. These land use changes happened because the city grew, more people moved there, the economy changed, and even the climate played a role. Kumar et al. (2021) studied satellite images and computer programs (RS and GIS) to monitor land use changes in the Jhansi area of Uttar Pradesh, India, between

2000 and 2020. Satellite images from Landsat-5 TM for 2001, Landsat-7 ETM+ for 2010, and Landsat 8 (OLI) for 2020 were used to extract LULC classes. The researchers categorized the land into six types: farmland, empty land, built-up areas, water, and forests. Their analysis revealed significant shifts over the two decades. Farmland, built-up areas, and water bodies increased, while empty land and forests decreased. Notably, a large portion of the empty land was converted into farmland. These findings provided valuable information for researchers, policymakers, and land managers to make informed decisions about land use planning in the Jhansi district.

Ahmed et al. (2022) looked at how LULC changed in the Mathura district between 2014 and 2018 using satellite images and computer mapping. They looked at different types of land, such as open areas with few plants, farmland, empty land, built-up areas, and water bodies, and found significant changes in the amount of farmland and open areas with few plants. The study showed that cities and towns in the area grew from about 234 km^2 to 332 km^2 by 2018, while other developed land decreased. This growth mainly happened toward the west and south of the area, and also moving up in the south and east. Mathura is a special religious place, so we must balance building up the area and protecting its natural environment and social importance. The commercial data (Landsat ETM+) with a spatial resolution of 30 m was downloaded from the Earth Resources Observation Systems data center (United States Geological Survey (USGS)) to detect the change in the study region. Alam and Ahamed (2022) looked at how land use changed in Bangladesh from 1995 to 2021 over 26 years within 25 km of the main river systems, and also predicted which areas might be vulnerable for farming land, mainly due to flooding. The study covered an area of about 64,556 km^2 and used river water data and land maps to understand these changes. This study used satellite images and computer programs to classify land into five types: water bodies, bare land, vegetation, farming land, and urban areas. They found that farming land and urban areas grew in most areas, while forests and natural vegetation decreased. Specifically, farming land increased by about 3.65%, 2.18%, and 2.55% in three regions, but overall, farmland decreased by 11.68%. Some areas, like around Khulna, parts of Rajshahi, Mymensingh, Sylhet, Barishal, and Chittagong, were found to be highly vulnerable to flooding, mainly because they have a lot of farmland and are prone to flooding. About 44% of the Rangpur area was not vulnerable.

Gupta et al. (2023) presented how LULC changed over 30 years (1990–2020) in the Mathura district of Uttar Pradesh, India, by using satellite images from Landsat-5, Landsat-7, and Landsat-8 and analyzed the data with GIS data. They sorted the images into different LULC categories and then studied how these categories changed over time. They also checked how accurate their mapping was by dividing the land into six main types: urban areas, water bodies, vegetation, wheat fields, mustard fields, and other crops. The study showed significant changes in LULC over the past 30 years. The accuracy of their mapping was good, ranging from about 70% to 84% in different years. Usually, it was hard to track these kinds of changes over time using traditional methods, but by using satellite images taken at different times, they could get a clear picture of how things have changed. The study found that the primary land use in the area was vegetation. There was less farming activity between 2000 and 2010, which led to more land being covered by plants. The study by Seyam et al. (2023) focused on LULC patterns in Mymensingh (Bangladesh), a rapidly growing industrial region.

This study aimed to analyze these changes to support sustainable land management in the future. Researchers used satellite images from Landsat 7 and 8 satellites and processed them with ArcGIS 10.8 to study five main types of land use over the past 20 years. The analysis showed an overall accuracy of 87.2% for 2002 and 89.6% for 2022. A study shows that the total agricultural land has decreased by 25.28 km^2, primarily driven by rapid urbanization and industrial development. This rapid change highlights the need for careful planning and sustainable development strategies to manage its impact effectively. Landsat 7 ETM+ images from 17 February 2002 at a resolution of 30 m and Landsat 8 OLI data from 27 May 2017 were used to classify land use/cover. The Pearl River Delta area in China is growing very fast economically. The government made some rules in the late 1970s and early 1980s that caused a lot of land to be changed.

Urbanization makes more people work in non-farming jobs. It also changes land cover, switching from farmland to non-farm areas, gradually over time. The study by Abebe et al. (2023) examined how land use changed in Dukem town using satellite images and GIS technology from 2003 to 2019. Data was collected by satellite images taken from far away. The study found that the area covered by built-up land (like buildings and roads) increased a lot, from about 698.06 hectares in 2003 to about 3,091.67 hectares in 2019, an increase of about 16.45%. At the same time, farmland decreased a lot, from about 2,764.37 hectares in 2003 to about 999.05 hectares in 2019, a decrease of 65.12%. This shows that urban growth has greatly reduced agricultural land. Moisa et al. (2023) looked at how land use changed over time in a specific area called the Upper Anger watershed in Western Ethiopia. To do this, the study used satellite pictures from 1990, 2003, and 2020 to see the different types of land and how green the area was. They found that the amount of farmland grew by a lot (707.4 km^2), while the amount of forest shrank (179.9 km^2) between 1990 and 2020. This means that farmland and bare land got bigger each year, and grassland and forest got smaller. The area also became less green overall during that time because there was more farmland, bare land, and places where people lived. About 530 km^2 of grassland and 241.5 km^2 of forest were turned into other types of land.

Roy et al. (2024) looked at LULC changes in Dhaka City, Bangladesh, from 2010 to 2021 and predicted what might happen in 2030 and 2040. Landsat satellite images were used to analyze land changes, and results showed that built-up areas increased from 31% to 42% between 2010 and 2021. Meanwhile, agricultural land decreased from 19 % to 15%. Between 2010 and 2021, these areas are expected to continue changing, with built-up areas growing and the natural regions shrinking. The study predicted that the value of ES could drop from about \$172.66 to \$160.23 million over 30 years. The increase in water bodies, vegetation, and farmland by 2040 suggests that efforts are needed to recover and protect natural resources through sustainable practices. Blissag et al. (2024) studied how land use had changed in the Hodna basin (an area of 26,000 km^2) from 2000 to 2020. The researchers used Landsat images to do this, and they also predicted how land use might change by 2050. They used a method called Maximum Likelihood Supervised Classification (MLSC) on the Landsat data every 5 years between 2000 and 2020. This method was quite accurate, with an overall accuracy of 85%. We used a computer method called ANN, and machine learning to predict future changes. They also used a technique called cellular automata (CA) to predict what the land would look like in 2030, 2040, and 2050. This prediction method was also reasonably accurate, with an overall kappa of 68.5%. The results showed that between 2000 and 2020, the amount of agricultural land decreased by 2.16%, while the amount of sandy land increased by 9.45%, which means the desert area is growing. Between 2000 and 2050, the amount of bare land decreased by 10% in 2015, then increased by 15% in 2020, and is expected to increase by 8% by 2050, reaching 73%. The mixed and dense forest grew slightly from 0.7% in 2000 to 1.66% in 2050. The area covered by buildings increased by 150%, from 0.86% in 2000 to 2.5% in 2020, while the amount of open forest stayed about the same, covering 4–6% of the area. Farmland changed differently from bare land because of new farming policies.

Rangelands, which make up 40% of Earth's land and surface, are very important for the environment and people's livelihoods worldwide. In Punjab, Pakistan, these lands face enormous pressures from farming and climate change. A study by Feng et al.(2025) examined how rangelands in Punjab's Khushab and Bhakkar districts changed from 2000 to 2020 using satellite images and other data. The researchers used different methods to classify land cover, analyze land surface temperature, and study vegetation. They aimed to understand the condition of these lands and achieved more than 85% accuracy in their results. The findings show that the area of rangelands decreased a lot, from 1,338.1 km² in 2000 to 798.7 km² in 2020, mainly because the land was converted into farmland. The analysis also found that vegetation increased in some areas, but the land was still affected by stress periods, which impacted plant growth. Vegetation conditions varied over time, with improvements during some periods and declines during others. These changes highlight how climate change and land use affect rangelands and make them more vulnerable. Bundelkhand is a semi-dry area in Central India. Droughts happen because of irregular rainfall, uneven land shape, and poor water conservation. A study of the past 20 years (2000–2019) found 10 years with severe drought, 5 with normal rainfall, and 5 with too much rain. Different water-saving methods were used and studied to deal with these issues. Using land cover maps and RS and GIS-based methods, researchers examined land use patterns to help plan for sustainable land management in the future. For this study, the Garkhunder-Dabar (GKDD) watershed in Madhya Pradesh, part of central India, was chosen by Singh et al. (2025). To study water flow, the watershed was divided into two parts: treated water and one with untreated water. Satellite images from 1999 and 2019 were downloaded and processed using GIS tools. The land-use classification was done using a method called ERDAS Imagine, which gave an accuracy of about 89–90%. The land changed significantly, with about 90% of the forest and farmland turning into cultivated land and crops.

4.3 LULC ANALYSIS OF THE MATHURA REGION OF INDIA

Being a prominent region in Uttar Pradesh, Mathura spans an area of 3,340 km^2 and is home to a population of 2,547,184 according to the most recent count. With 763 people residing in each square kilometer, this district showcases a vibrant mix of culture and demographics across its 730 villages. It is roughly 130 kilometers from

India's capital, and experiences a wide range of monthly temperatures. Summers typically see averages climb from 36°C to a scorching 47°C, while winters bring a cooler span of 25–32°C. Spanning approximately 3,340 km^2, the district's location can be pinpointed using GPS coordinates between 27°14′ and 27°77′ north latitude and 77°17′ and 78°12′ east longitude. Mapping of the region is covered by Survey of India topographical sheets numbered 54E and 54I and is shown in Figure 4.2. Physiographically, Mathura's terrain is primarily composed of the Ganga's alluvial plains, which are further divided into older, more established plains, and newer floodplains. The Yamuna River serves as the principal water source in this district, acting as a sub-basin of the sacred Ganga River. A segment of the Yamuna sub-basin, influenced by the district's geographical layout and generally sloping from north toward the south-southeast in the neighboring Agra district, defines a distinct physiographic zone.

To examine LULC, a structured approach utilizing satellite imagery within QGIS is employed. Initially, the research zone is demarcated through existing administrative boundary files or by manually outlining the desired area in QGIS. Following this, appropriate satellite images are obtained from the USGS. Selecting cloud-free imagery from relevant periods is crucial for distinguishing various land cover categories, followed by pre-processing. Each land cover class is assigned a unique color, and map elements such as legends, scale bars, and north arrows are incorporated using the QGIS Print Layout function. Spatial statistics, like the area of each land cover type, are computed, and the outcomes are analyzed to understand land-use changes, such as urban development, forest loss, or agricultural growth. These procedures provide the groundwork for interpreting trends and informing land management or policy decisions. The data was categorized into three categories: vegetation, water bodies, and built-up area. The vegetation is designated by green color, the water bodies are defined by blue color, and the built-up areas are designated by red color (Table 4.1).

Spanning the decade from 2015 to 2024, a study was conducted to determine the reduction of vegetation land within the Mathura district of Uttar Pradesh. The land was categorized into three types: built-up, water bodies, and vegetation. The results revealed that vegetated land was the primary land use throughout these 10 years. Water bodies expanded from 1.07% in 2015 to 1.53% in 2024, and vegetation land decreased from 81.24% to 67.59%. Conversely, built-up area increased from 17.68% to 30.89% during the same period, as shown in Figures 4.3 and 4.4, respectively. These results have indicated a connection between the growth of built-up areas and water bodies and the reduction of vegetated land. The findings suggested that the decrease in farmland and the vegetated regions has increased built-up area (Figure 4.4). As a result, the gradual loss of agricultural land is strongly linked to endangering food security and health impacts, and the ecosystem's health will be significantly affected. The rise in population also contributed to urban expansion. The analysis covering the LULC change from 2015 to 2024 has been shown in Figure 4.3. Satellite imagery has been categorized into an optimal number of classes to forecast changes within the study area. A substantial shift has occurred in the vegetation of the study area.

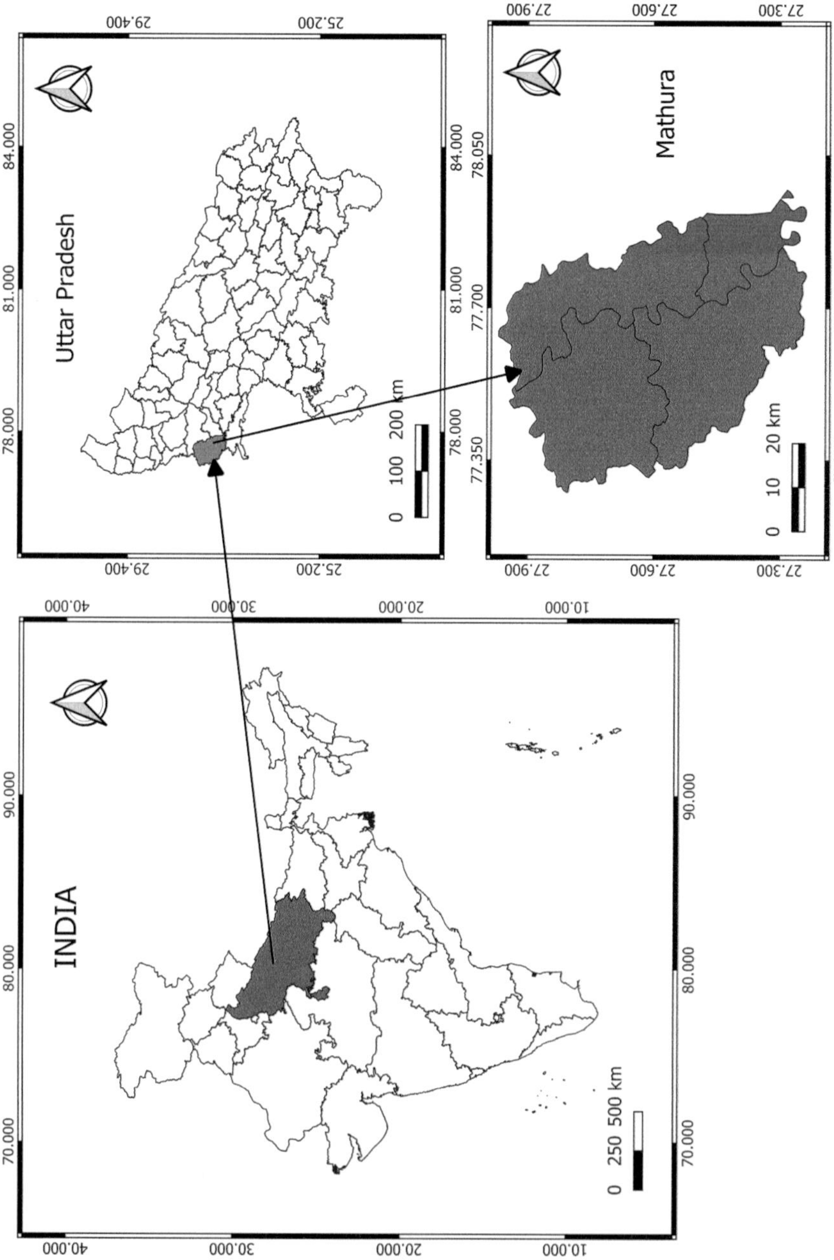

FIGURE 4.2 The map shows the study area (Mathura district), Uttar Pradesh, India.

TABLE 4.1
Details of Satellite Data Used in This Study

S. No.	Year	2015	2024
1	Image	Landsat 8	Landsat 8
2	Path	146	146
3	Row	41	41
4	Senser	L8 OLI/TIRS C2 L2	L8 OLI/TIRS C2 L2
5	Resolution	30 × 30	30 × 30
6	Bands	11	11
7	Acquisition Date	27 January 2015	12 January 2024
8	Source	USGS Earth Explorer	USGS Earth Explorer

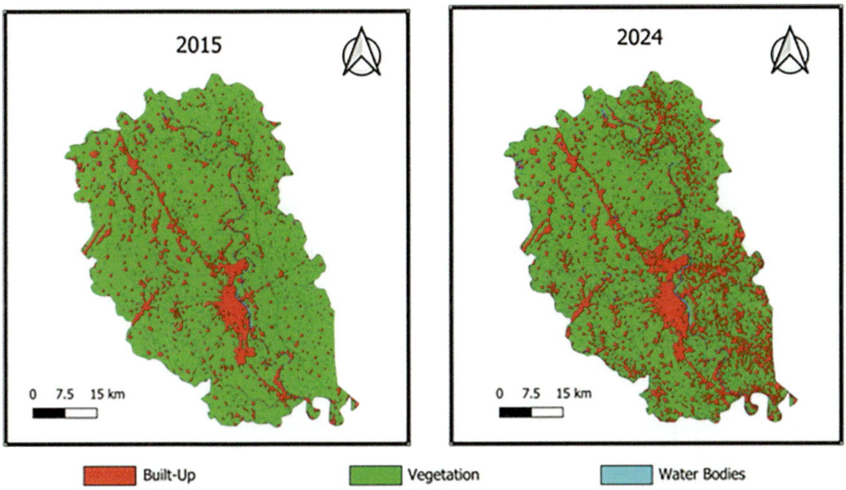

FIGURE 4.3 Map showing land use and land cover in the year 2015 and 2024.

4.4 CONCLUSION

LULC maps provide a precise idea of the utility of the landscape. The spatial and temporal distribution of the various land-use utilities is marked by the LULC maps. The continuously shifting trend of agricultural area to built-up area significantly stresses the lack of natural resource management. The study presented various previously done research wherein it was depicted that agricultural land use in multiple landscapes globally is decreasing drastically. The study also indicated a study on the Mathura district of Uttar Pradesh, which reduced the area under agriculture from 2015 to 2020. The study concludes that there is an urgent need to shift the focus to the decreasing agricultural area to develop sustainable natural resource management.

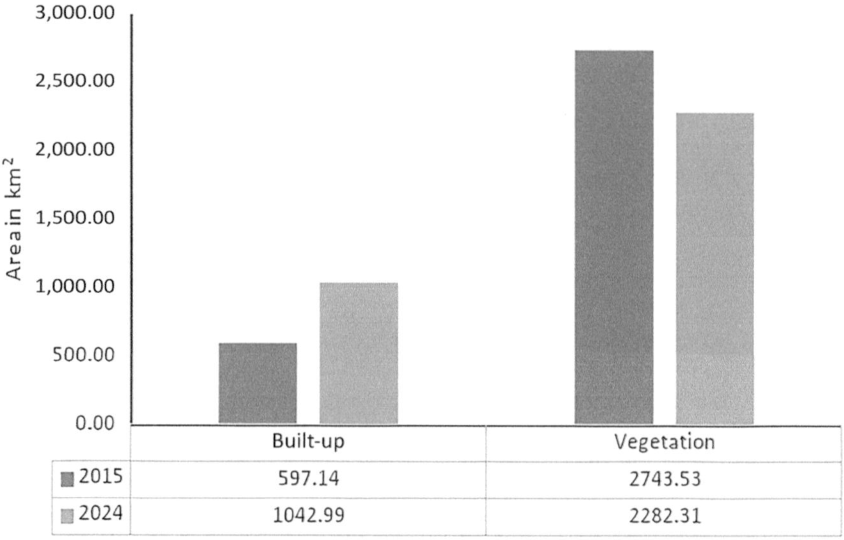

FIGURE 4.4 Land use dynamics of Mathura district.

REFERENCES

Abd El-kawy, O. R., Ismail, H. A., Yehia, H. M., and Allam, M. A. (2019). Temporal detection and prediction of agricultural land consumption by urbanization using remote sensing. *The Egyptian Journal of Remote Sensing and Space Science*, 22(3), 237–246.

Abd El-Kawy, O. R., Rød, J. K., Ismail, H. A., and Suliman, A. S. (2011). Land use and land cover change detection in the western Nile delta of Egypt using remote sensing data. *Applied geography*, 31(2), 483–494.

Abebe, B. G., Gemeda, B. S., and Sisay, G. (2023). Analysing land use land cover (LULC) dynamics by using remote sensing and GIS techniques: The case of Dukem town, Oromia special zone. *Geocarto International*, 38(1), 2210523.

Ahmed, S., Khurshid, S., Ahmad, R., Sultan, W., Khan, M. M. A., Mansor, H. E., James, E., and Shamsuddin, M. K. N. (2022). Land use/land cover change analysis using geospatial technology in Mathura district, Uttar Pradesh, India. In *AIP Conference Proceedings, AIP Publishing* (Vol. 2454, No. 1). Kelantan, Malaysia

Alam, K. F., and Ahamed, T. (2022). Assessment of land use land cover changes for predicting vulnerable agricultural lands in river basins of Bangladesh using remote sensing and a fuzzy expert system. *Remote Sensing*, 14(21), 5582.

Blissag, B., Yebdri, D., and Kessar, C. (2024). Spatiotemporal change analysis of LULC using remote sensing and CA-ANN approach in the Hodna basin, NE of Algeria. *Physics and Chemistry of the Earth*, Parts A/B/C, 133, 103535.

Fazal, S. (2000). Urban expansion and loss of agricultural land-a GIS based study of Saharanpur City, India. *Environment and Urbanization*, 12(2), 133–149.

Feng, L., Naz, I., Quddoos, A., Zafar, Z., Gan, M., Aslam, M., Hussain, Z. K., Soufan, W., and Almutairi, K. F. (2025). Exploring rangeland dynamics in Punjab, Pakistan: Integrating LULC, LST, and remote sensing for ecosystem analysis (2000–2020). *Rangeland Ecology & Management*, 98, 377–388.

Foley, J. A., et al. (2005). Global consequences of land use. *Science*, 309(5734), 570–574.

Gondwe, J. F., Lin, S., and Munthali, R. M. (2021). Analysis of land use and land cover changes in urban areas using remote sensing: Case of Blantyre City. *Discrete Dynamics in Nature and Society*, 2021(1), 8011565.

Gupta, P., Singh, S. K., Gupta, P., Kanga, S., and Mishra, V. N. (2023). Application of remote sensing and GIS techniques for identification of changes in land use and land Cover (LULC): A case study. *Indian Journal of Science & Technology*, 16(46), 4456–4468.

Gururani, D. M., Singh, S., Joshi, H., Kumar, Y., Kumar, A., Bohra, M. S., and Mehta, P. (2023). Assessment of groundwater prospects zones using RS, GIS, and MIF methods. In *Surface and Groundwater Resources Development and Management in Semi-Arid Region: Strategies and Solutions for Sustainable Water Management* (pp. 317–335). Cham*: Springer International Publishing*.

Kafi, K. M., Shafri, H. Z. M., and Shariff, A. B. M. (2014, June). An analysis of LULC change detection using remotely sensed data; A Case study of Bauchi City. In *IOP conference series: Earth and environmental science* (Vol. 20, No. 1, p. 012056). IOP Publishing.

Karakus, C. B., Demiroglu, D., and Karadag, A. A. (2020). Evaluation of land use/land cover change (lulcc) and meteorological data in terms of sustainable agriculture in kilis, Turkey. *Journal of Environmental Protection and Ecology*, 21(4), 1186–1193.

Kumar, P., Dobriyal, M., Kale, A., and Pandey, A. K. (2021). Temporal dynamics change of land use/land cover in Jhansi district of Uttar Pradesh over past 20 years using LANDSAT TM, ETM+ and OLI sensors. Remote Sensing Applications: *Society and Environment*, 23, 100579.

Kumar, V., and Agrawal, S. (2019). Agricultural land use change analysis using remote sensing and GIS: A case study of Allahabad, India. *The International Archives of the Photogrammetry, Remote Sensing and Spatial Information Sciences*, 42, 397–402.

Lambin, E. F., and Meyfroidt, P. (2011). Global land use change, economic globalization, and the looming land scarcity. *Proceedings of the National Academy of Sciences*, 108(9), 3465–3472.

Matson, P. A., et al. (1997). Agricultural intensification and ecosystem properties. *Science*, 277(5325), 504–509.

Moisa, M. B., Busha Hinkosa, L., Negasa, G. F., Olika, G., Ijigu, T. E., Wedajo, Y. N., Gurmessa, M. M., Deribew, K. T., Gemeda, D. O. (2023). GIS and remote sensing based analysis of land use and land cover change in the upper anger watershed, western Ethiopia. *Geology, Ecology, and Landscapes*, 1–10.

Mulla, D. J. (2013). Twenty-five years of remote sensing in precision agriculture: Key advances and remaining knowledge gaps. *Biosystems Engineering*, 114(4), 358–371.

Pimentel, D., et al. (1995). Environmental and economic costs of soil erosion and conservation benefits. *Science*, 267(5201), 1117–1123.

Rane, N. L., Choudhary, S. P., Giduturi, M., and Pande, C. B. (2023). Remote sensing (RS) and geographical information system (GIS) as a powerful tool for agriculture applications: Efficiency and capability in agricultural crop management. *International Journal of Innovative Science and Research Technology*, 8, 264–274.

Roy, S. K., Alam, M. T., Mojumder, P., Mondal, I., Kafy, A. A., Dutta, M., Ferdous, M.N., Mamun, M. A. A., Mahtab, S. B. (2024). Dynamic assessment and prediction of land use alterations influence on ecosystem service value: A pathway to environmental sustainability. *Environmental and Sustainability Indicators*, 21, 100319.

Seto, K. C., Woodcock, C. E., Song, C., Huang, X., Lu, J., and Kaufmann, R. K. (2002). Monitoring land-use change in the Pearl River Delta using Landsat TM. *International Journal of Remote Sensing*, 23(10), 1985–2004.

Seyam, M. M. H., Haque, M. R., and Rahman, M. M. (2023). Identifying the land use land cover (LULC) changes using remote sensing and GIS approach: A case study at Bhaluka in

Mymensingh, Bangladesh. *Case Studies in Chemical and Environmental Engineering*, 7, 100293.

Singh, K. N., Singh, R., Rajput, P., Bhatt, R., de Oliveira, M. W., and de Freitas Santos, D. (2025). Analysing land use land cover (LULC) changes using remote sensing (RS) and GIS approach in Bundelkhand Region of Central India. *observatório de la economía latinoamericana*, 23(3), e9397–e9397.

Singh, S., and Goyal, M. K. (2025). A review of India's water policy and implementation toward a sustainable future. *Journal of Water and Climate Change*, 16(2), 493–510.

Singh, S., and Kumar, D. (2021). Investigation of rainfall variability of the southern part of Uttarakhand using entropy theory. *Indian Journal of Soil Conservation*, 49(2), 83–88.

Singh, S., Kumar, A., and Kumara, S. (2023b). Assessment of drought in meteorological data Using SPI and SPEI indicators for sustaining agricultural productivity in the Agra division of Uttar Pradesh, India. In *Advancement of GI-Science and Sustainable Agriculture: A Multi-dimensional Approach* (pp. 285–303). Cham: Springer Nature Switzerland.

Singh, S., Kumar, D., Kumar, A., and Kuriqi, A. (2023a). Entropy-based assessment of climate dynamics with varying elevations for hilly areas of Uttarakhand, India. *Sustainable Water Resources Management*, 9(4), 130.

Singh, S., Kumar, D., Vishwakarma, D. K., Kumar, R., and Kushwaha, N. L. (2024). Seasonal rainfall pattern using coupled neural network-wavelet technique of southern Uttarakhand, India. *Theoretical and Applied Climatology*, 155(6), 5185–5201.

Tilman, D., et al. (2002). Agricultural sustainability and intensive production practices. *Nature*, 418(6898), 671–677.

Yan, J., Gao, S., Xu, M., and Su, F. (2020). Spatial-temporal changes of forests and agricultural lands in Malaysia from 1990 to 2017. *Environmental Monitoring and Assessment*, 192, 1–16.

Yonaba, R., Koïta, M., Mounirou, L. A., Tazen, F., Queloz, P., Biaou, A. C., Niang, D., Zoure, C., Karambiri, H., Yacouba, H. (2021). Spatial and transient modelling of land use/land cover (LULC) dynamics in a Sahelian landscape under semi-arid climate in northern Burkina Faso. *Land Use Policy*, 103, 105305.

Zhao, G. X., Lin, G., and Warner, T. (2004). Using Thematic Mapper data for change detection and sustainable use of cultivated land: A case study in the Yellow River delta, China. *International Journal of Remote Sensing*, 25(13), 2509–2522.

Zhu, Z., Qiu, S., and Ye, S. (2022). Remote sensing of land change: A multifaceted perspective. *Remote Sensing of Environment*, 282, 113266.

5 Assessment of Diara Land under Koshi River in Bihar Using Remote Sensing and GIS Techniques

Akash Raj, Rohit Kumar Gupta, S. K. Nirala, and Manish Kumar

5.1 INTRODUCTION

The term "Diara" has been derived from the word "Diya," meaning earthen lamp. The Diaras of eastern India, namely the Brahmaputra Diaras, the Ganga Diaras, the Saryu Diaras and the Mahanadi Diaras, have the same commonalities among them. Rivers originating from the Himalayas, flowing into northern Bihar, carry huge sediment loads and are notorious for frequent changes to their course (Sinha, 2008).

Diara lands are extensively found in Uttar Pradesh, Bihar, Assam, and Orissa. Accordingly, Uttar Pradesh has flood-prone Diara lands of 1.5 million hectares in various river systems, namely Ganga, Yamuna, Ghaghra, Saryu, Sharda, Ramganga, Kalindi and Gomti. In the state of Bihar, the area under Diara lands has been reported to be 0.9 million hectares. Diara lands are, in general, associated with two natural calamities, viz., flood and drought. During the rainy season, flood waters inundate the Diara lands for varying periods in different locations depending on their climatic conditions and physiography. Drainage congestion in Diara lands aggravates the problems in increasing the duration of inundation and thereby reducing the length of the growing periods of crops. The transportation network in Diara lands has usually been found to be poor. After the monsoon season, the water from the riverbeds retreats to its channel, leaving large areas dry. These areas of land are generally left unused. It may be selected as an entry point to promote inclusive economic growth for the benefit of the landless people. The subterranean moisture seeped from adjacent rivers and streams, making the upper layers of land more suitable for growing early vegetable crops. Diara farming is a pro-poor focused program for the rural community to increase household income and to improve the food security of landless and land-poor households in India.

Koshi River Diara is a flood plain located in the northeastern region of Bihar, India. It is formed by the Koshi River, which is also known as the Sorrow of Bihar,

DOI: 10.1201/9781003596080-5

due to its tendency to flood and cause devastation to the surrounding areas. The Koshi River Diara is a vast area of low-lying land that stretches for over 200 kilometers along the banks of Koshi River. It is a fertile region that is ideal for agriculture, with farmers growing crops such as rice, wheat, and maize. Despite its fertility, the Koshi River Diara is prone to flooding, particularly during the monsoon season when heavy rains cause the river to swell and burst its banks. The floods can cause widespread damage to crops, homes, and infrastructure, and have been responsible for the loss of many lives over the years. In recent times, efforts have been made to mitigate the impact of flooding in the Koshi River Diara, with the construction of embankments and other flood measures, but these efforts have not been entirely successful, and the region remains vulnerable to floods.

Diara land is often considered an ecosystem and is seen to have great potential from an agricultural point of view. Soils of Diara land occurring in alluvial and flood plains are young and stratified with AC profiles (A horizon directly over C horizon). They are characterized by light color, sand to sandy loam texture, single grain to massive structure, slightly alkaline reaction, low electrical conductivity (EC) and high base saturation. The soils were also found to be highly siliceous and low in iron and aluminum oxides. The variation in CaO and K_2O contents is attributed to the nature of parent materials (Diwakar and Singh, 1994). Flood-affected areas in the Koa catchment have been mapped using remote sensing satellite data (IRS LISS III, 1999 and Landsat Thematic Mapper (TM), 1995), according to Jain et al. (2005). The ecosystem decline was noticed in the Niger basin, with emphasis on the Niger and Mali sides of the river, by using remote sensing and Geographical Information System (GIS) techniques. They developed a new framework for analyzing the declining health of riverine ecosystems and also designed a geo-spatial tool for riverine ecosystem management and impact analysis (Twumasi and Merem, 2007). Delineation of the Diara land from a merged image of IRS-ID LISS-III and Panchromatic (PAN) was done using a visual image interpretation technique and ground information. For change detection at 2-year intervals, two sets of data, viz. IRS-1D, LISS III data of November 8, 1999, and November 27, 2001 and IRS-1D, LISS III data from February 6, 2000 and March 7, 2002 were used. There was a considerable change in land use/land cover classes over Diara land within the short time frame (2-year interval), and remote sensing was found to be a suitable technique for land use land cover change detection (Kumar et al., 2008). Diara land was assessed in the Bhagalpur district of Bihar using remote sensing and GIS tools. Out of 19 blocks in Bhagalpur district, only six were identified as old Diara land (203.26 km^2) and 13 blocks were identified as new Diara land (869.78 km^2). The increment in Diara land area was from 7% to 33% of the geographical area of Bhagalpur and the maximum new area comes under Gopalpur block, that is, 107.22 km^2 (Kumar et al., 2013). Spatial and temporal details of the river Ganga in Fatehpur district was studied in term of change in features area available in the study area over a 15-year (2004–2019) period using RS and GIS tools and techniques, around 4.90 km^2 of area deep water body has degraded which will be a major concern for the upcoming future to all who directly or indirectly depends upon the Ganga (Yadav et al., 2019). Vulnerabilities in Diaras struggling with floods in the Gandak River Basin, West Champaran district, between February and July 2016 in four villages (Udas et al., 2021).

Remote sensing and GIS techniques were used for the development of Tal and Diara Land for sustainable agriculture in Central Bihar (Choudhary et al., 2019). Evolution of river course and morphometric features of the River Ganga (Sarif et al., 2021). A case study of Tal and Diara lands of Bhagalpur district (Bihar) was made by using multi-spectral satellite data, viz. IRS LISS III (Pre Monsoon), MODIS V6 and SRTM DEM, for visual interpretation and mapping of flood-affected Tal and Diara lands (Vimal et al., 2021). GIS-based Flood Risk Mapping of Kosi River Basin, Bihar, India, was carried out by ArcGIS 10.3 software for classifying 10 districts and 76 blocks of the basin region into risk areas (Kumar et al., 2022).

5.2 STUDY AREA

The Diara land of the Koshi River of the Bihar region is selected for study purposes (Figure 5.1). The Koshi is a transboundary river which flows through China, Nepal and India. It drains the northern slopes of the Himalayas in Tibet and the southern slopes in Nepal. From a major confluence of tributaries north of the Chatra Gorge onward, the Koshi River is also known as Saptakoshi (Nepali, saptakoshī) for its seven upper tributaries. These include the Tamur River, originating from the Kanchenjunga area in the east, and the Arun River and Sun Kosi from Tibet. The Koshi River passes through several districts of Bihar. These districts are Supaul, Saharsa, Madhubani, Madhepura, Khagaria, Bhagalpur, Purnea, and Katihar. The catchment area of the Koshi River in Bihar is 11,410 km^2 out of the total area of 74,030 km^2. The length of the Koshi River in Bihar is 233 km.

5.3 DEMARCATION OF DIARA LAND IN THE KOSHI RIVER (BIHAR)

Diara land continues to remain largely unexplored. The present study attempts to fill this gap by demarcating the Koshi Diara land of Bihar (Figure 5.2). An attempt was made to delineate the Diara land of the Koshi River using satellite imagery using Google Earth Pro for 3 years (2005, 2014, and 2022). Different steps are involved in delineation:

Step I: First, the Diara lands were identified in Google Earth Pro.
Step II: The identified Diara lands in satellite imagery were digitized using the add polygon tool in Google Earth Pro.
Step III: Different KMZ files are made after digitization.
Step IV: Conversion of the KMZ file to a layer file in the ArcGIS 10.8.2 environment
Step V: The union of these files is done using ArcGIS 10.8.2
Step VI: Area Estimation of Diara land, which involved the following:

- Area calculation was done by using the Attribute table in ArcGIS 10.8.2.
- The geographical coordinate system was set as UTM 1984 45 N.
- In the attribute table, a new field was added name area.

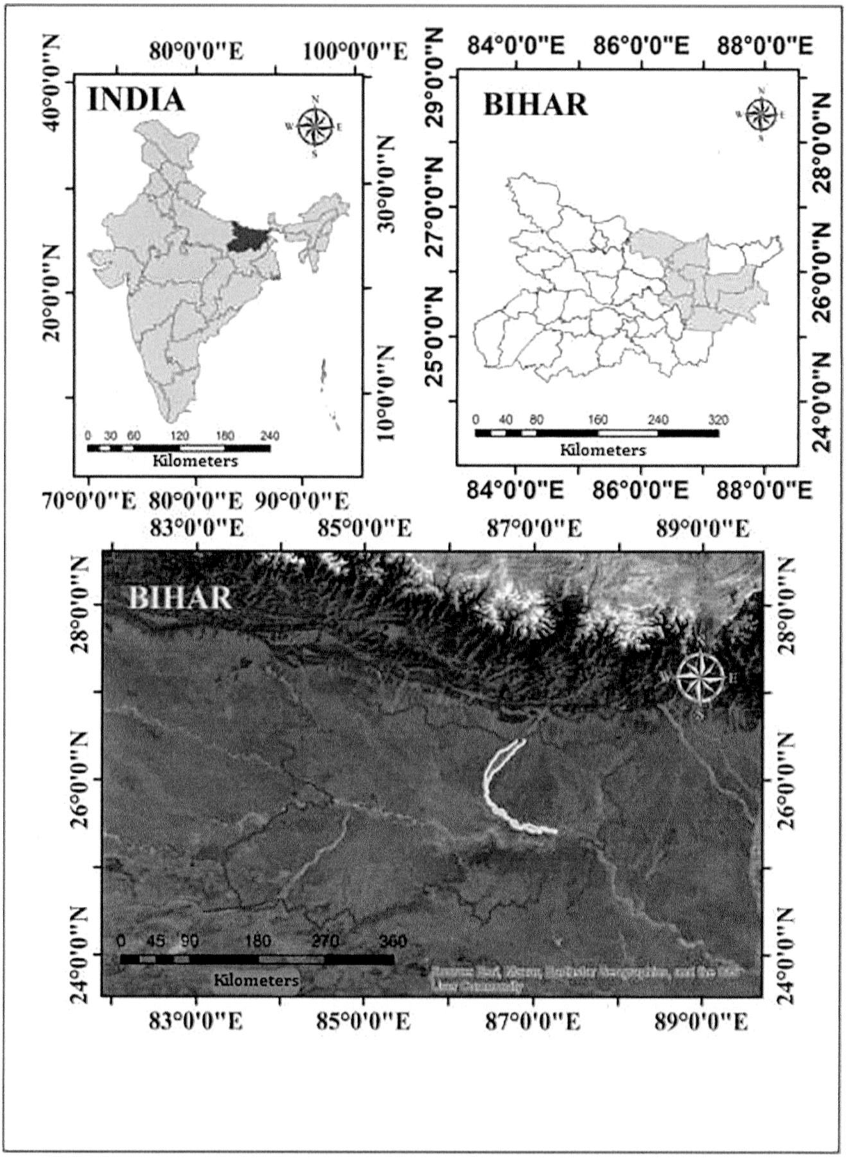

FIGURE 5.1 Location map of the study area.

- Estimation of the area was done by right-clicking on the area field and selecting calculate geometry (Figure 5.3).

Step VII: After the estimation of the Diara land, the Map layout was done and exported to a suitable format.

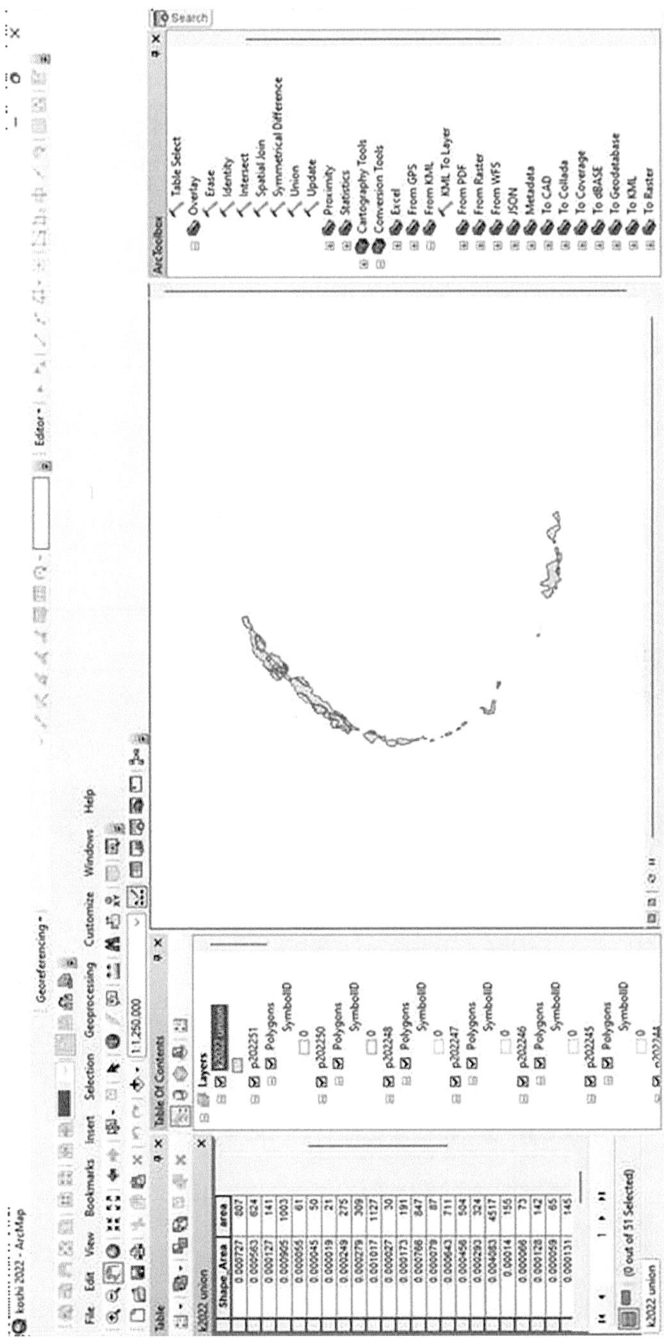

FIGURE 5.2 Union of Diara land for the Koshi River.

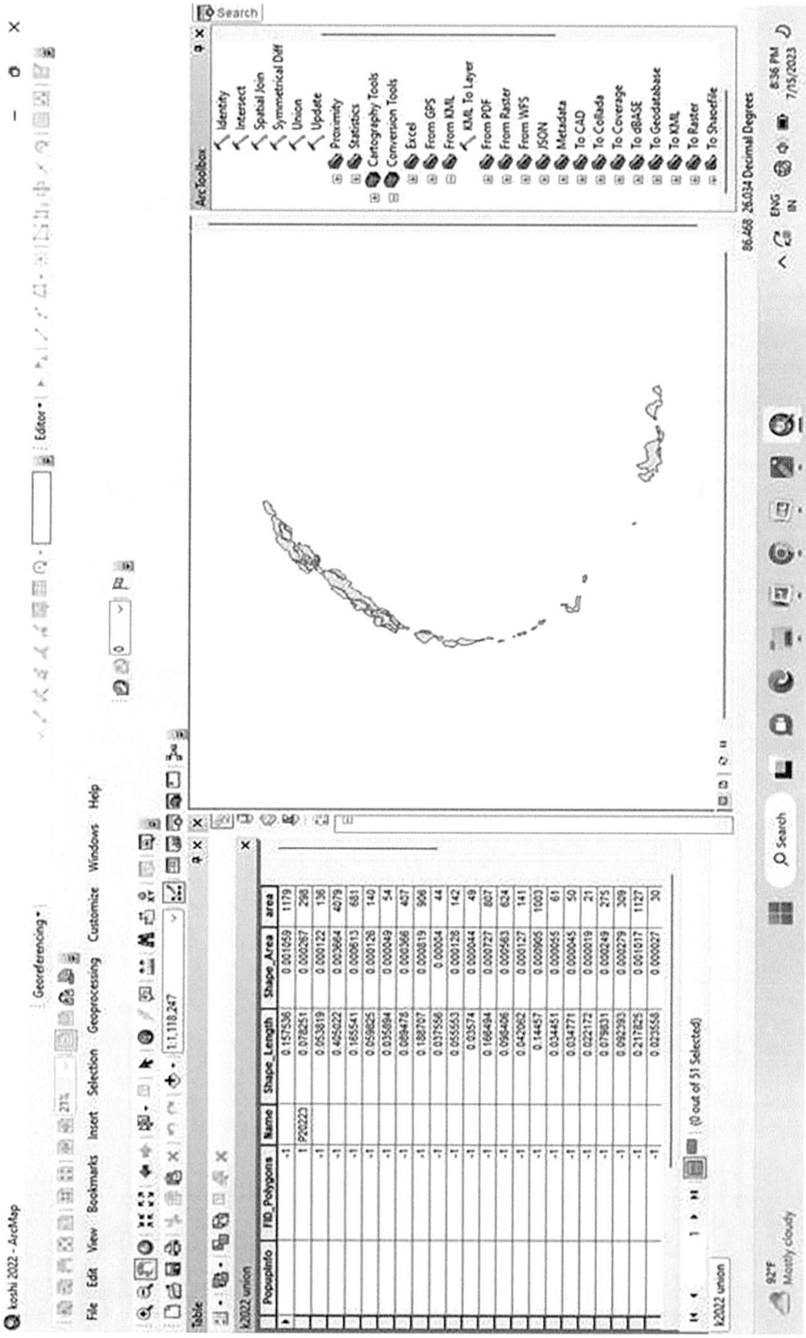

FIGURE 5.3 Area estimation of Diara land for the Koshi River.

5.4 RESULTS AND DISCUSSION

5.4.1 Estimation of The Area of Diara Land in the Koshi River for the Year 2005

The Diara land map of the Koshi River for the year 2005 was prepared in the Google Earth Pro and ArcGIS environments. District-wise Diara land of the Koshi River during 2005, as shown in Figure 5.4, represents the Diara land throughout the Koshi River bed. Based on the map, the area of Diara land for different districts was also evaluated, as shown in Table 5.1. The results showed that the Diara land for Supaul was the largest area (14,890 ha) in 2005, followed by Saharsa (3,397 ha), Bhagalpur (1,991 ha), Khagaria (832 ha), Purnea (619 ha), Madhubani (587 ha), and Katihar (459 ha). The map also suggested that the upstream (upper half) of the river course constituted more Diara land than the lower half of the river length. It was found that Supaul constituted 65.37% of the total area of Diara land found during 2005, followed by Saharsa 14.91%, Bhagalpur 8.74%, Khagaria 3.65%, Purnea 2.71%, Madhubani 2.57%, and Katihar 2.01%.

5.4.2 Estimation of the Area of Diara Land in the Koshi River During 2014

In the Google Earth Pro and ArcGIS environments, the Diara land map of the Koshi River for the year 2014 was developed, as shown in Figure 5.5. The area of Diara land for the various districts was also assessed using the map, and the results are shown in Table 5.1. The findings showed that the maximum area of Diara land was found in Supaul (9,507 ha), and the minimum area of Diara land was found in Khagaria (175 ha) and Katihar (175 ha). The Diara land area for Bhagalpur, Saharsa, Purnea, and Madhepura was 4,617 ha, 834 ha, 617 ha, and 601 ha, respectively. It was found that Supaul constituted 57.94% of the total area of Diara land found during 2014, followed by Bhagalpur (28.14%), Saharsa (5.08%), Purnea (3.76%), Madhepura (3.66%), Khagaria and Katihar (1.06%).

5.4.3 Estimation of the Area of Diara Land in the Koshi River During 2022

In the Google Earth Pro and ArcGIS environments, the Diara land map of the Koshi River for the year 2022 was created, as shown in Figure 5.6. The area of Diara land for the various districts was also assessed using the map, and the results are shown in Table 5.1. The findings showed that the maximum area of Diara land was found in Supaul (17,057 ha), and the minimum area of Diara land was found in Katihar (44 ha). The Diara land area of Bhagalpur (5,967 ha), Saharsa (3,876 ha), Madhubani (885 ha), Purnea (634 ha), Khagaria (161ha), Madhepura (86 ha). It was found that Supaul constituted 59.41% of the total area of Diara land found during 2022, followed by Bhagalpur (20.78%), Saharsa (13.50%), Madhubani (3.08%), Purnea (2.20%), Khagaria (0.56%), Madhepura (0.29%), and Katihar (0.15%).

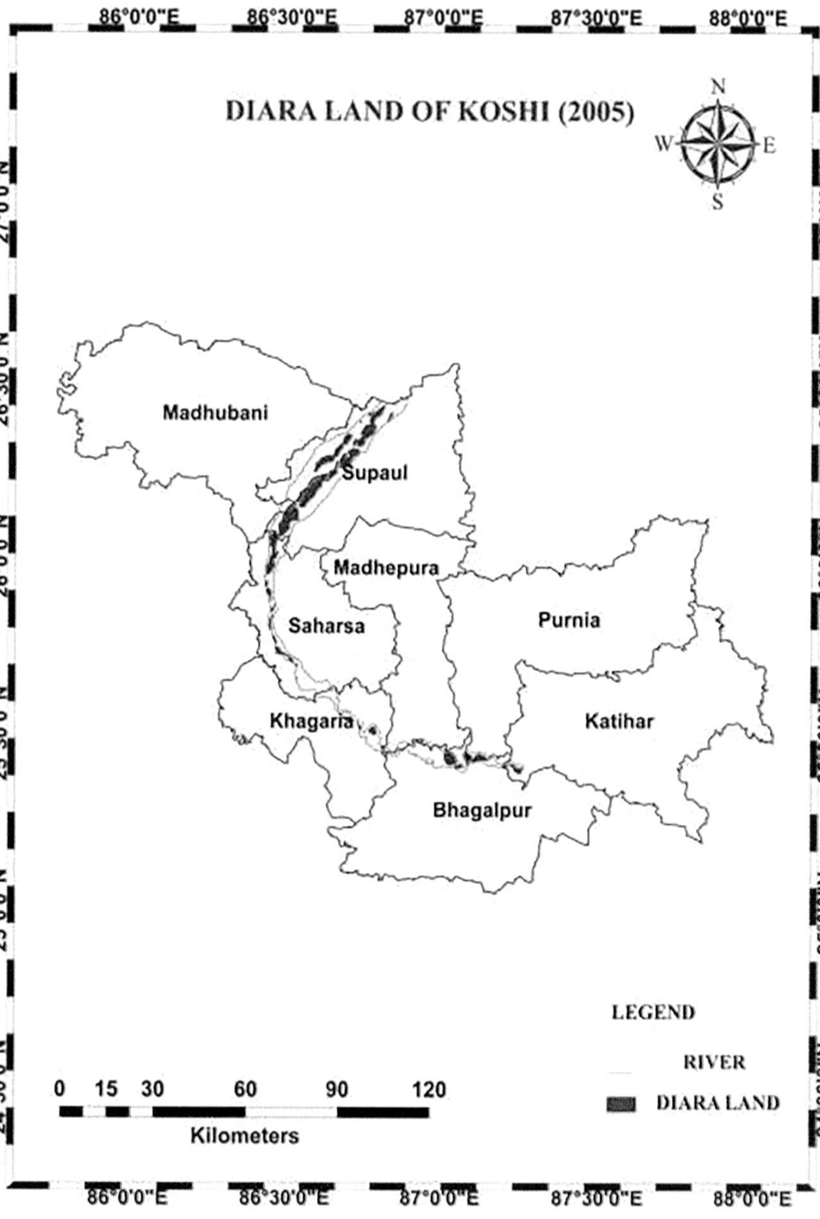

FIGURE 5.4 Diara land of the Koshi River during 2005 in Bihar.

5.4.4 Comparison of Diara Land Area from 2005 to 2014

It was found that the total area of Diara land in 2005 was 22,775 ha, and 16,406 ha in 2014. Thus, from the estimated area of Diara land, it was found that the decrease in Diara land area from 2005 to 2014 was about 6,369 ha. Supaul had the biggest area of

TABLE 5.1
Changing the Area of Diara Land (ha) in Different Districts in Different Years

	Year 2005		Year 2014		Year 2022	
Districts	**Diara Area (ha)**	**% Area**	**Diara Area (ha)**	**% Area**	**Diara Area (ha)**	**% Area**
Supaul	14,890	65.37	9,507	57.53	17,057	59.41
Saharsa	3,397	14.91	834	5.05	3,876	13.50
Madhubani	587	2.57	0.0	0.0	885	3.08
Khagaria	832	3.65	175	1.06	161	0.56
Bhagalpur	1,991	8.74	4,617	27.94	5,967	20.78
Purnea	619	2.72	617	3.73	634	2.20
Madhepura	0.0	0.0	601	3.64	86	0.29
Katihar	459	2.04	175	1.06	44	0.15
Total	**22,775**		**16,526**		**28,710**	

Diara land in 2005, accounting for 14,890 ha (65.37%), followed by Saharsa (3,397 ha, 14.91%), Bhagalpur (1,991 ha, 8.74%), Khagaria (832 ha, 3.65%), Purnea (619 ha, 2.71%), Madhubani (587 ha, 2.57%), and Katihar (459 ha, 2.01%). In Madhepura, no Diara land was found.

Diara land has a 16,406 ha total area as of 2014. Supaul had the biggest area of Diara land in 2014, accounting for 9,507 ha (57.94%) followed by Bhagalpur (4,617 ha, 28.14%), Saharsa (834 ha, 5.08%), Purnea (617 ha, 3.76%), Madhepura (601 ha, 3.66%), and Katihar Khagaria (175 ha, 1.06%). In Madhubani, no Diara land was found.

During 2005–2014, it was found that there was a decrease in the total area of the Diara land. In Supaul, the area decreased by about 9,353 ha; in Saharsa, the area decreased by about 2,563 ha; in Bhagalpur, the area increased by about 2,626 ha; in Purnea, the area decreased by about 2 ha; in Madhepura, the area increased by about 14 ha; in Khagaria, the area decreased by about 657 ha; and in Katihar, a decrease of about 284 ha was found.

5.4.5 Comparison of Diara Land Area from 2014 to 2022

It was found that the total area of Diara land in 2014 was 16,406 ha, and 28,710 ha in 2022. Thus, from the estimated area of Diara land, it was found that the increase in Diara land area from 2014 to 2022 was about 12,304 ha. Supaul had the biggest area of Diara land in 2014, accounting for 9,507 ha (57.94%), followed by Bhagalpur (4,617 ha, 28.14%), Saharsa (834 ha, 5.08%), Purnea (617 ha, 3.76%), Madhepura (601 ha, 3.66%), and Katihar Khagaria (175 ha, 1.06%). In Madhubani, no Diara land was found.

Diara land has a 28,710 ha total area as of 2022. Supaul has the biggest area of Diara land in 2022, accounting for 17,057 ha (59.41%), followed by Bhagalpur (5,967 ha, 20.78%), Saharsa (3,876 ha, 13.50%), Madhubani (885 ha, 3.08%), Purnea

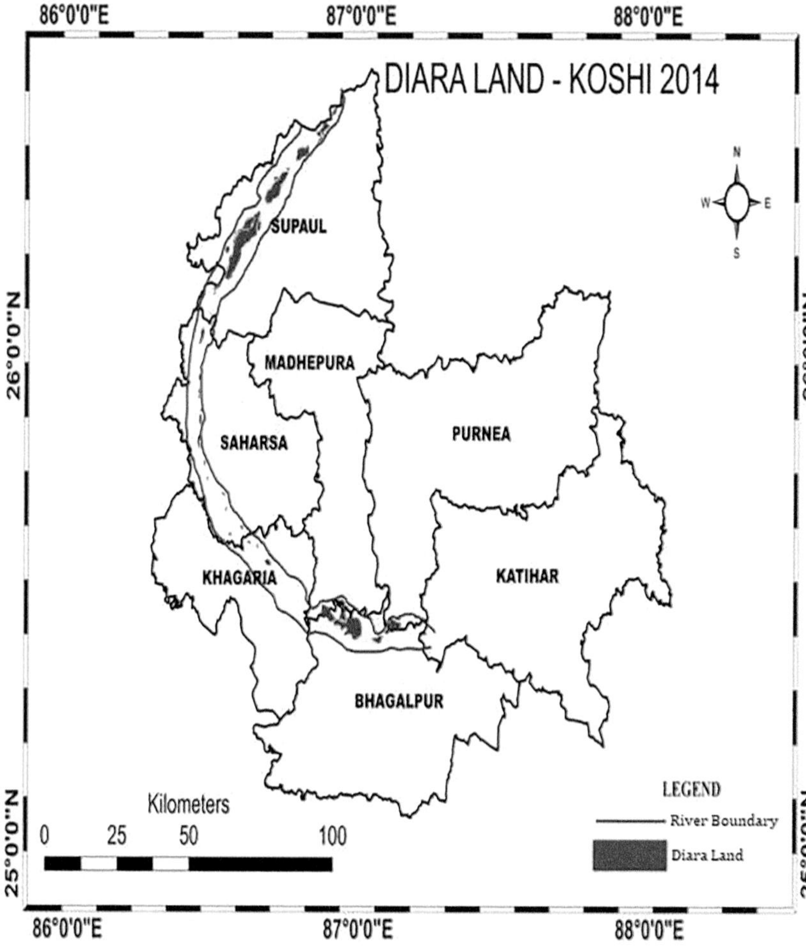

FIGURE 5.5 Diara land of the Koshi River during 2014 in Bihar.

(634 ha, 2.20%), Madhepura (86 ha, 0.29%), Khagaria (161 ha, 0.56%), and Katihar (44 ha, 0.15%).

From 2014 to 2022, an increase in the total area of Diara land was found. The increase of Diara Land for Supaul, Saharsa, Bhagalpur, Purnea, and Madhubani districts was 7,550 ha, 3,042 ha, 1,350 ha, 17 ha, and 885 ha, respectively. The decrease in Diara land was observed for Madhepura (515 ha), Khagaria (6 ha), and Katihar (131 ha).

5.4.6 Comparison of Diara Land Area from 2005 to 2022

It was found that the total area of Diara land in 2005 was 22,775 ha, and 28,710 ha in 2022. Thus, from the estimated area of Diara land, it was found that the increase in

FIGURE 5.6 Diara land of Koshi River in 2022 in Bihar.

Diara land area from 2014 to 2022 was about 5,935 ha. Supaul had the biggest area of Diara land in 2005, accounting for 14,890 ha (65.37%), followed by Saharsa (3,397 ha, 14.91%), Bhagalpur (1,991 ha, 8.74%), Khagaria (832 ha, 3.65%), Purnea (619 ha, 2.71%), Madhubani (587 ha, 2.57%), and Katihar (459 ha, 2.01%). In Madhepura, no Diara land was found.

Diara land has a 28,710 ha total area as of 2022. Supaul has the biggest area of Diara land in 2022, accounting for 17,057 ha (59.41%), followed by Bhagalpur (5,967 ha, 20.78%), Saharsa (3,876 ha, 13.50%), Madhubani (885 ha, 3.08%), Purnea (634 ha, 2.20%), Madhepura (86 ha, 0.29%), Khagaria (161 ha, 0.56%), and Katihar (44 ha, 0.15%).

During the period from 2005 to 2014, it was found that the total area of the Diara land increased in different districts. In Supaul, the area increased by about 2,167 ha; in Saharsa, the area increased by about 479 ha. In Bhagalpur, the area increased by about 3,976 ha. In Purnea, the area increased by about 15 ha; in Madhepura, the area increased by about 86 ha; in Madhubani, it increased by about 298 ha; in Khagaria , the Diara land area decreased by about 671 ha, and in Katihar, the area decreasd by about 415 ha. The comparison of the change in Diara land area during the years 2005, 2014, and 2022 is shown graphically in Figure 5.7.

Based on the study, the Google Earth Pro and ArcGIS 10.8.2 environment were effective in the estimation of the Diara land for the Koshi River in Bihar. The results

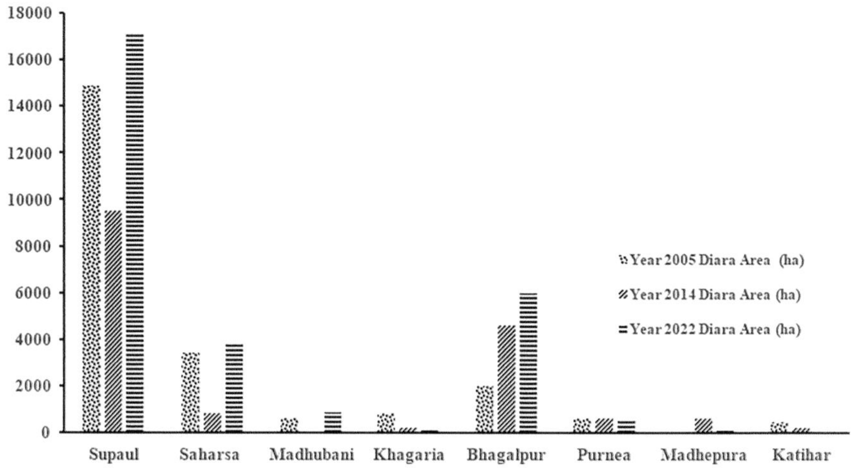

FIGURE 5.7 Graphical representation of changes in the area of Diara land in different years in different districts.

depicted the temporal and spatial changes of the Diara land. In 2005, the total area of Diara land was 22,775 ha, whereas in 2014, the total Diara land was 16,406 ha, and in 2022, the total Diara land was 28,710 ha. Supaul district was found to contribute maximum to Diara land during all 3 years, which is 14,890 ha (65.37%) in 2005, 9,507 ha (57.94%) in 2014, and 17,057 ha (59.41%) in 2022.

5.5 CONCLUSIONS

The present study estimated the Diara land for the Koshi River using Google Earth Engine and ArcGIS 10.8.2. Based on the analysis, Supaul has the largest area of Diara land among other districts in all 3 years (2005, 2014, and 2022). In 2022, Diara land area was the maximum among the 3 years (2005, 2014, and 2022). The area of Diara land in 2014 declined because of the decline in the area of Diara land in every district of Bihar except Bhagalpur. In Bhagalpur district, there is a substantial increase observed in the Diara land area during all 3 studied years (2005, 2014, and 2022). The area of Diara land in Supaul, Saharsa, was seen as a similar alteration. In the Purnea district, there was no significant change in area. The study reveals that there is a substantial increase or decrease in the area of Diara land every year because these lands are vulnerable to floods and Droughts. The area of Diara land in 2005 was 22,775 ha, which decreased to 16,406 ha in 2014, and then increased to 28,710 ha in 2022. Most of these Diara lands are considered wasteland and barren. However, the huge Diara land area has great potential from an agricultural point of view and it can be harnessed for the benefit of the people inhabiting these areas. As crops and cattle require land to grow and provide food, land is a crucial resource for agriculture. Hence, there is an urgent requirement for diligent planning to effectively and sustainably utilize the available resources for their development, and such areas can be utilized in a better way for the development of agro-forestry orchards and recreation zones.

REFERENCES

Choudhary, S.K., Kumar, R., Gupta, S.K., Kumar, A. and Vimal, B.K. (2019). Development of tall and Diara land for sustainable agriculture in Central Bihar, India. *Current Journal of Applied Science and Technology*, 35(5), pp.1–13.

Diwakar, D.P.S. and Singh, R.N. (1994) Characterization of soils of Diarti land occurring in Gangetic plains of Bihar. *Journal of the Indian Society of Soil Science*, 42(3), pp.420–425.

Jain, S.K., Singh, R.D., Jain, M.K. and Lohani, A.K (2005). Delineation of flood- prone areas using remote sensing techniques. *Water Resources Management*, 19, pp.333–347.

Kumar, V., Kumar, R., Vimal, B.K. and Kumar, M. (2013). Assessment of Diara land under Bhagalpur district using remote sensing and GIS tools. *Journal of Applied and Natural Science*, 5(1), 213.

Kumar, V., Kumar, S., Yadav, S., Kumar, S. and Kumar, A (2022). Diara cultivation of cucurbits. *Just Agriculture Multidisciplinary e-Newsletter*, e- ISSN: 2582-8223 Vol.2 No 7, MAR 2022.

Kumar, G., Sahoo, R.N., Tomar, R.K., Bhavanarayana, M., Gupta, V.K., Rao, C.S., Sehgal, V.K. and Chakraborty, D. (2008) Land use land cover change in active flood plain using satellite remote sensing. *Journal of Agricultural Physics*, 8, pp.22–28.

Sarif, N., Siddiqui, L., Islam, S., Parveen, N. and Saha, M. (2021). Evolution of river course and morph metric features of the River Ganga: A case study of up and down Faraka Barrage. *International Soil and Water Conservation Research*, 9, p.578–590.

Sinha, R. (2008) The Great avulsion of Koshi on 18 August 2008, Indian institute of technology Kanpur. *Current Sciece*, 97(3), p.429–433.

Twumasi, Y.A. and Merem, E.C. (2007). Using remote sensing and GIS in the analysis of ecosystem decline along the River Niger Basin: The case of Mali and Niger. *International Journal of Environmental Research and Public Health*, 4(2), pp.173–184.

Udas, P.B., Prakash, A. and Goodrich, C.G. (2021). 3 Gendered vulnerabilities in Diaras. *Engendering Climate Change*, p.38–57.

Vimal, B.K., Srinivasaraghvan A., Kumar, A., Kumari, R., Ghanshyam, Kumari, M. and Priya, J. (2021). RS-GIS based mapping of an impact of flood inundation on cultivated lands: A case study of Tal and Diara lands of Bhagalpur district (Bihar), India. *The Pharma Innovation Journal*: SP-10(11), p.270–274.

Yadav, M., Mehta, D. and Vaishya, R.C. (2019). Estimating vacant land parcels on the Ganga riverbed near Fatehpur district, Uttar Pradesh using GIS and Remote Sensing. *39th INCA International Congress*, Dehradun (India), 18th–20th December, 2019.

6 Google Earth Engine-Based Hypsometric Evaluation and Statistical Moments Analysis of Large River Basins

Srija Roy, Shubham Dixit, Abhilash Gogineni, and Shekhar Singh

6.1 INTRODUCTION

The hypsometric curve (HC) is the graphical representation of relative area versus relative elevation of a region, which is used to categorize landforms into stages based on their erosional characteristics. This curve can take three forms, that is, convex, S-shaped, and concave, and these correspond to the youthful, mature, and old stages, respectively, of geomorphic development. The hypsometric integral (HI), which is the area under the HC, is a key metric used to infer these stages. A high HI value near 1 (100%) indicates a youthful stage with a convex HC, while an HI value around 0.50 (50%) suggests a mature stage. However, variations in the HC shape can occur even with similar HI values, influenced by factors such as erosion and tectonic forces. Hypsometric analysis is crucial for understanding the geomorphology, hydrogeology, and sediment transport in a river basin. The HI and HC help assess the erosional stages of the basin, which in turn guides soil and water conservation efforts. The curve's shape and the HI value are influenced by factors like basin size, topography, climate, drainage patterns, tectonics, and lithology. Recent advancements in remote sensing (RS) and Geographic Information Systems (GIS) have improved the accuracy and efficiency of hypsometric analysis by using high-resolution digital elevation models (DEMs). Tools like CalHypso, an ArcGIS and QGIS extension, facilitate the automatic extraction and analysis of HCs and their statistical moments, though challenges remain in terms of scalability and computation time for large river basins. Cloud-based geo-processing platforms like Google Earth Engine (GEE) offer solutions to these challenges by providing access to extensive geospatial datasets and advanced computational capabilities. The integration of GEE with tools like CalHypso allows for efficient and scalable hypsometric analysis, making it possible to characterize river basins of any size with greater ease and accuracy.

DOI: 10.1201/9781003596080-6

6.1.1 Statistical Characteristics of the HC and Its Density Distribution

The HC can be interpreted as a cumulative distribution function that is monotonic with few inflections (Harlin, 1978; Roy and Chintalacheruvu, 2024a). It can be represented by a continuous polynomial function (Equation 6.1) of the form:

$$f(x) = a_0 x + a_1 x + a_2 x^2 + a_3 x^3 + ... + a_n x^n \quad (6.1)$$

where $f(x)$ value is the relative altitude (h/H) and the x value is the relative area (a/A) (Figure 6.1) of the polynomial with $n = 2$ or 3 order sufficiently describing the catchment or a drainage basin (Ahmad et al. 2018). The HC can be fitted to these lower-order polynomials, which are significant because they share characteristics with the cumulative distribution of a random variable equal to zero to the left of $x = 0$, equal to 1 to the right of $x = 1$, and are monotonically increasing. By calculating the statistical moments and the center of gravity of the curve, changes in the morphology of the drainage basin can be described, particularly indicating relative erosional changes across different zones. The coefficients a_0, a_1, a_2, … are calculated through regression of the curve.

In Figure 6.1, the HC takes the value of $f(x)$ within a unit square defined by the area and altitude axes (Pérez-Peña et al. 2009; Roy and Chintalacheruvu, 2024b). The (HI is defined as the area under the HC (the region "R" under the curve) and is computed by integrating $f(x)$ within the bounds of this unit square. This integration can be expressed as a summation formula, as outlined by Harlin (1978). HC parameters are especially significant when comparing river basins with similar HI values but different HC shapes. To quantify these differences, statistical moments of hypsometric skewness (HSK), hypsometric kurtosis (HKU), density skewness (DSK), and density kurtosis (DKU) are used. These moments describe the shape of the HC's probability distribution in comparison to a normal distribution. The first moment (μ_1) represents the x-mean or centroid $f(x)$ within the unit square. The second moment μ_2, or variance, is centered around the x-mean, with the square root of the variance being the standard deviation ($\sigma = \sqrt{\mu_2}$) The third moment (μ_3) and fourth moment (μ_4) describe the skew and kurtosis, respectively. Because skew and kurtosis are dimensioned, their dimensionless coefficients are used instead—(μ_3/σ_3) (i.e., HSK) and HKU (μ_4/σ_4) (i.e., HKU). The equations for all these four statistical moments can be referred from the literature by Harlin (1978); Pérez-Peña et al. (2009); Ahmad et al. (2018); Ahmed and Rao, 2016. Similarly, a new function of $g(x)$ which is the derivative of $f(x)$ can be defined as the density function of the HC. Applying the same methodology as used for function $f(x)$, the moments of $g(x)$ were defined along with the derivation of the coefficients of skewness and kurtosis to be termed DSK and DKU (Equation 6.2):

$$g(x) = f'(x) = a_1 + 2a_2 x + 3a_3 x^2 \quad (6.2)$$

The values of HSK and HKU are indicative of erosion conditions within a river basin. Specifically, HSK reflects headward erosion in the upper reaches, while HSK provides insight into erosional conditions across both the upper and lower reaches of

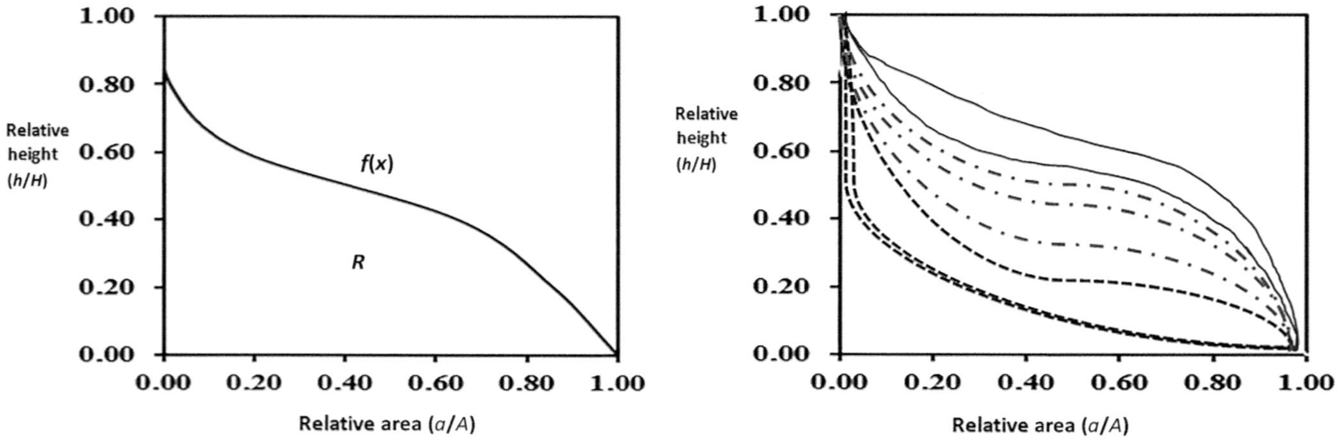

FIGURE 6.1 HC characteristics.

the basin. Additionally, DSK and DKU are used to analyze slope changes within the basin, with DSK focusing on the overall slope variation and DKU on the mid-basin slopes. These statistical moments allow for a detailed characterization of the HC and the morphological features of a river basin. By analyzing changes in these moments, it is possible to distinguish between basins that have similar HI values but different HC shapes. This makes HSK, HKU, DSK, and DKU particularly valuable in hypsometric analysis for understanding the geomorphic processes at play and the stages of erosion in different river basins.

6.1.2 Hot-Spot and Cold-Spot Analysis of Hypsometric Parameters

The spatial distribution of hypsometric parameters can be further examined using the Getis and Ord (1992) statistic in ArcMap 10.3. This hot-spot analysis tool calculates spatial statistics to assess the presence of spatial autocorrelation among features. It uses standardized *z*-scores and associated *p*-values to determine spatial patterns, helping to accept or reject the null hypothesis of spatial autocorrelation. *Z*-scores indicate the presence or absence of spatial clustering based on standard deviation. *P*-values are used to determine the statistical significance of these patterns, with a threshold of $p \leq 0.05$ being considered significant in this analysis. The *z*-scores and *p*-values identify spatial clusters of high or low values (hot spots or cold spots) relative to their surrounding values. It is important to note that an individual's high or low value is only significant if it is part of a cluster of similar values. This spatial clustering is a key aspect of exploratory spatial data analysis (ESDA) and provides more detailed, visually interpretable results. Typically, the hot-spot analysis of hypsometric parameters reveals patterns of spatial clustering within a river basin, which can be crucial for understanding geomorphic processes and landscape evolution.

6.2 MATERIALS AND METHODS

The methodology outlined in this study is depicted in Figure 6.2, which shows the flow diagram of materials and methods used. The process involves the use of pre-processed and pre-mosaicked Shuttle Radar Topography Mission (SRTM v3) DEM with a resolution of 1-arc second and standardized pre-delineated sub-basin boundary polygon layers as inputs in the CalHypso tool within QGIS 2.8.9. CalHypso, a Python script (Pérez-Peña et al., 2009; Roy and Chintalacheruvu, 2024b), is employed in the GIS interface to generate HCs and calculate their statistical moments. The spatial datasets used in the study, detailed in Table 6.1, include the NASA SRTM "Plus" DEM, which is accessed through GEE using JavaScript API. The DEM undergoes void-filling and pre-processing using open-source datasets like ASTER, GDEM2, GMTED2010, and NED. An alternative method for watershed delineation is proposed, where HydroSHEDS v1 BASINS data is overlaid on the SRTM "Plus" DEM to prepare and download sub-basin maps at various scales using GEE. HydroBASINS provides vectorized polygons of sub-basin boundaries globally, allowing for efficient and scalable watershed analysis.

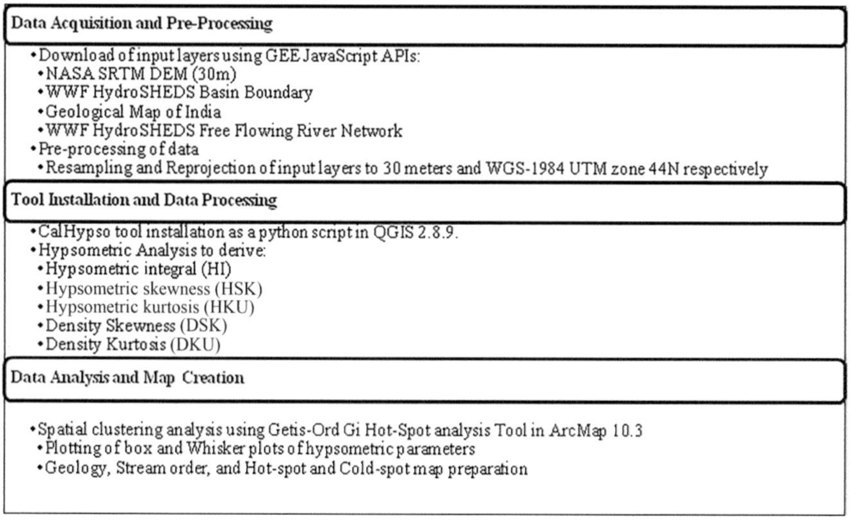

FIGURE 6.2 Flow diagram of methods and materials used in the present study.

TABLE 6.1
Spatial Datasets and Their Sources Accessed Using JavaScript API of GEE

Sl. No.	Spatial Layer	Dataset Provider	Reference
1	World Wildlife Fund (WWF) HydroSHEDS Basins	WWF	Lehner et. al, 2008.
2	WWF HydroSHEDS Free Flowing Rivers Network v1 15 arc-second	WWF	Lehner et.al, 2008. .
3	NASA SRTM Plus (1 arc second approximately).	NASA/JPL SRTM	Farr, 2011
4	Geological Map of India	U.S. Geological Survey, Central Energy Resources Team, Craig J. Wandrey.	-

This methodology simplifies and expedites the traditional time-consuming processes of DEM processing and sub-basin delineation by utilizing GEE-based resources. The CalHypso tool further streamlines hypsometric analysis by automating the generation of HCs and the calculation of statistical moments directly within QGIS, offering advantages over MATLAB-based tools like TecDEM by integrating directly into the GIS environment. This integration allows for the use of pre-existing spatial analysis algorithms and enhanced data visualization, making the process efficient and scale-independent without sacrificing accuracy. Additionally, the study explores the spatial distribution of hypsometric parameters through hot-spot and cold-spot analysis using the Getis and Ord (1992) statistic in ArcMap 10.3, revealing patterns of spatial clustering within the river basin.

6.2.1 Study Area

The derived methodology was applied and verified on the GRB which is the largest Indian river basin. The GRB is prone to active erosional and deformational processes, making it susceptible to flash floods, landslides, and seismic activities. Therefore, evaluating the geomorphic stages of the GRB's sub-basins is crucial for identifying zones with higher erosion and deformation activity (Anand and Pradhan, 2019; Roy and Chintalacheruvu, 2024a). In this study, the ensemble of GEE and CalHypso was used for hypsometric evaluation of the GRB. This approach was tested on the GRB to derive HCs and calculate statistical moments. The location map of the Ganga River Basin (GRB) is shown in Figure 6.3. The GRB spans from latitudes 22.45°N to 31.47°N and longitudes 73.37°E to 89.31°E, covering a total geographical area of 838,082 km². The Ganga River originates from the Gangotri Glacier near Gombak in Uttarkashi, Uttarakhand, as the Bhagirathi River, and is named the Ganga after its confluence with the Alaknanda River at DevPrayag, Uttarakhand. The river rises from an elevation of 7,010 m above mean sea level and flows 2,525 km before emptying into the Bay of Bengal (Figure 6.3).

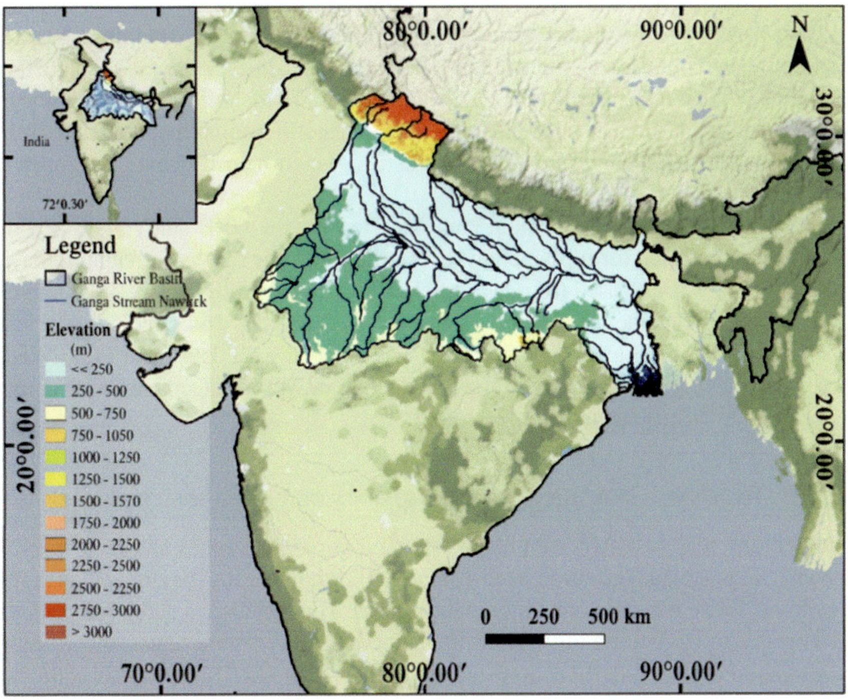

FIGURE 6.3 Location map of GRB.

The Ganga River is a key component of the Ganga–Brahmaputra–Meghna (GBM) river system, one of Asia's major river systems, and it serves as the lifeline for approximately 500 million people, making it the most populous river basin in the world (Chowdhury and Ward 2004; Maswood and Hossain 2016). The river drains several Indian states, including Uttarakhand, Himachal Pradesh, Haryana, Delhi, Uttar Pradesh, Madhya Pradesh, Rajasthan, Chhattisgarh, West Bengal, and Bihar. Its major tributaries include the Yamuna, Ramganga, Ghaghara, Gandak, Chambal, and Son. The delta region of the GRB, known as the Sundarbans, is one of the largest deltas globally and a critical biodiversity hot spot. It is also one of the most densely populated areas in India. The delta regions are increasingly at risk due to population growth, changing land cover patterns, and inherent terrain characteristics. Thus, to address these challenges, the suggested methodology was applied to perform hypsometric analysis on 15 sub-basins of the GRB: Upper Ganga, Upper Yamuna, Ramganga, Middle Yamuna, Chambal, Lower Yamuna, Tons, Sone, Ganga upstream of Gomti Confluence, Gomti, Ghaghara, Ganga Gomti to Ghaghara Confluence, Gandak, Damodar, and Bhagirathi. This sub-basin-wise evaluation of erosion conditions in the GRB is crucial for effective resource management and conservation efforts in the region.

The preparation of input files for the study was conducted using GEE, a cloud-based geospatial analysis platform. GEE provides access to extensive catalogs of geospatial and satellite data, which can be processed using JavaScript APIs. Through this platform, the boundary layer of the GRB and other necessary datasets of DEM and sub-basin boundary layers were extracted. The process involved reprojecting and resampling the input layers, followed by deriving sub-basin-HC, HI, and the statistical moments of the curves: HSK, HKU, DSK, and DKU. To identify spatially significant areas within the basin, hot-spot and cold-spot zones of hypsometric parameters were evaluated using Getis and Ord (1992) statistics, which helped demarcate zones of high and low-value clustering.

SRTM Plus' DEM was overlaid and extracted for the GRB study area, and the data was downloaded for further analysis. Additionally, the sub-basin boundary layer for the Ganga was accessed from HydroSHEDS v1 BASINS/hybas_x using the GEE JavaScript API. The delineated sub-basin map of the GRB, which includes 15 sub-basins, is shown in Figure 6.4. This map was verified with data from India-WRIS and by overlaying it with the geo-referenced Ganga River management board annual report map to ensure accuracy.

6.2.2 Tectonic, Geologic, and Geomorphological Information

The geology map of GRB, with clearly marked geological zones and fault lines, is shown in Figure 6.5a. The northern part of the GRB is bounded by the Himalayan range, a tectonically active and young mountain system. This geological activity makes the GRB prone to slope instability, landslides, floods, and seismic events. According to Anand and Pradhan (2019), neotectonic activities in the GRB significantly influence ongoing weathering and deformation processes. For analytical purposes, the GRB has been divided into 15 sub-basins. The GRB is recognized as the largest alluvial plain

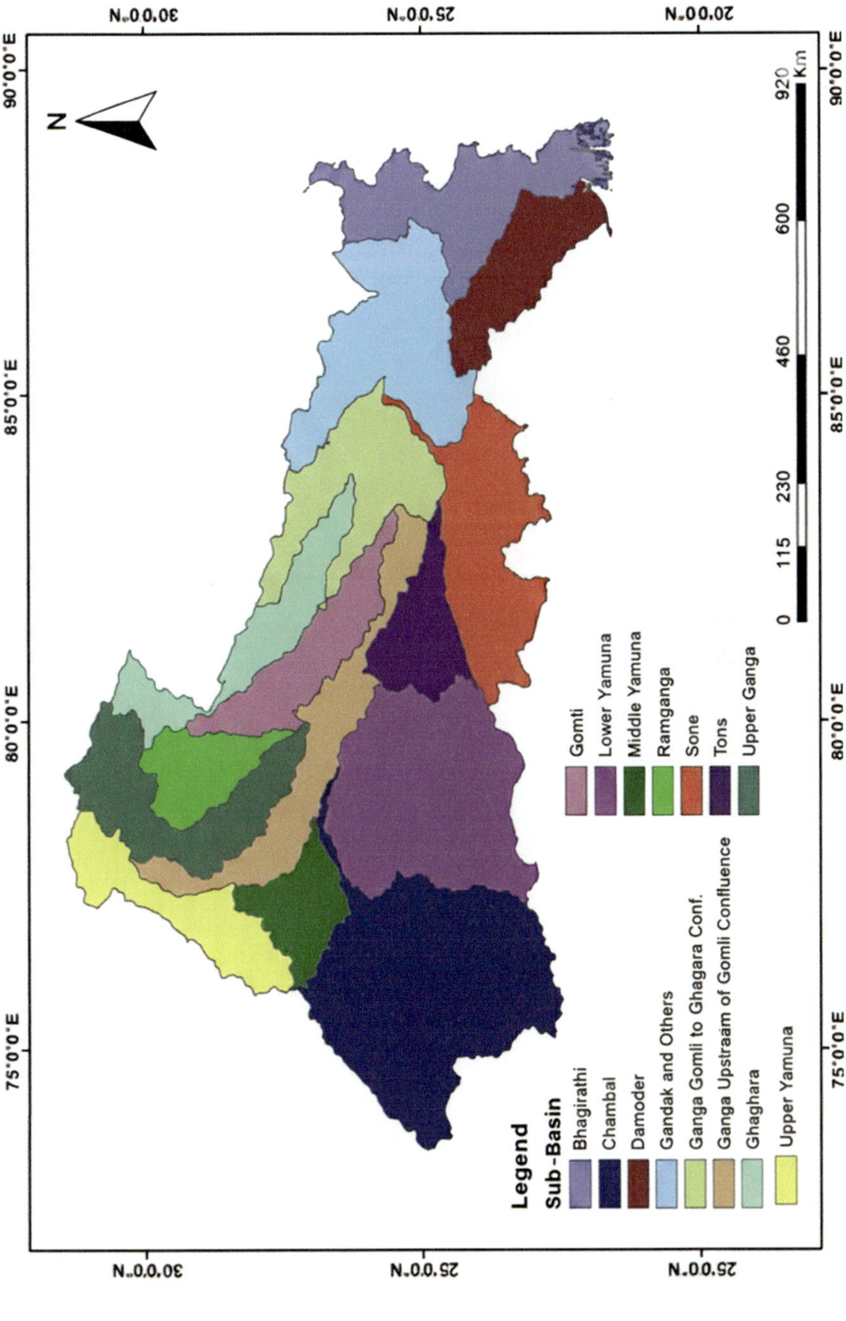

FIGURE 6.4 Delineated sub-basin map of GRB.

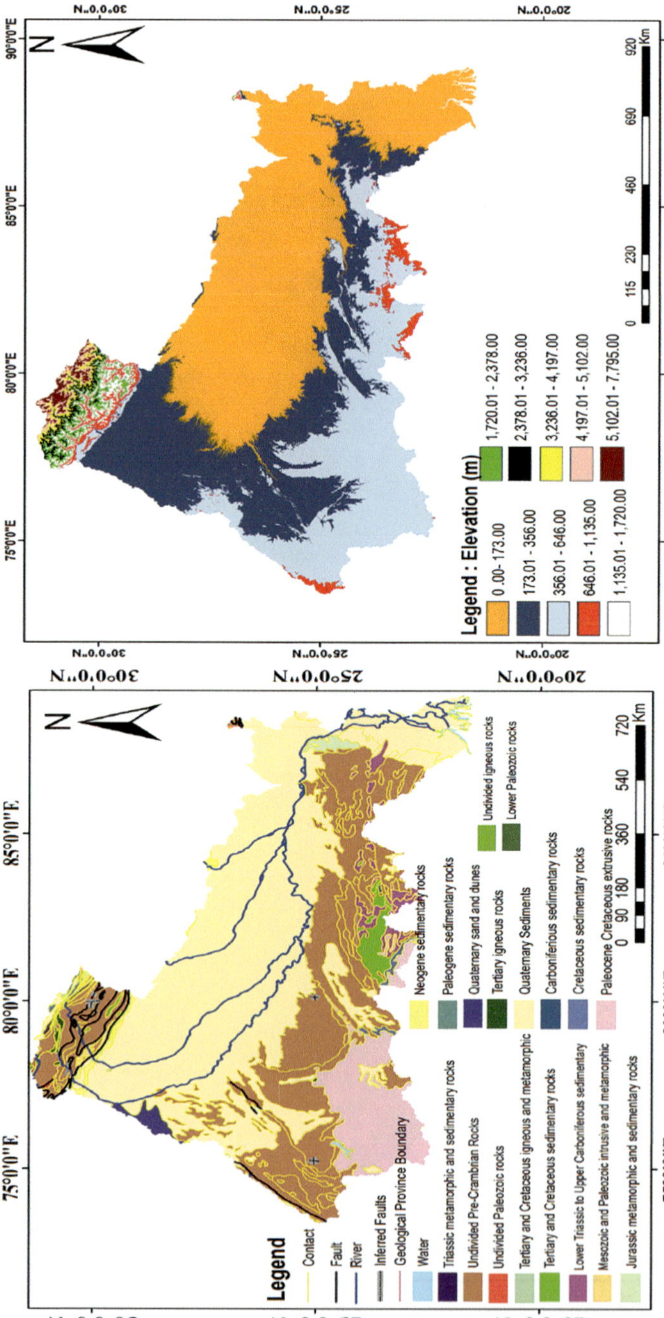

FIGURE 6.5 (a) Geology map of GRB with demarcated geological zones and fault lines. (b) Elevation distribution of GRB.

in the world, primarily composed of alluvial sediments deposited by the Himalayan rivers. This region also contains remnants of weathered sediments from numerous rivers fed by the Himalayas and the North Indian craton, dating back to the late Pleistocene–Holocene epoch (Prakash et al. 2017). The GRB's northern boundary, as shown in Figure 6.5a, consists of the Himalayan and Himalayan Foreland geological regions, with the Damodar and Ganga–Brahmaputra delta regions located to the east. The geological composition of the GRB includes Quaternary sediments in the Himalayan Foreland region, while the Himalayan region and areas south of the Himalayan Foreland are predominantly composed of undivided Precambrian rocks. Figure 6.5b illustrates the elevation distribution within the GRB in India, which ranges from 0 to 7,795 m, with a mean slope of 4.10 degrees. Figure 6.6a provides further details on the slope distribution across the GRB, highlighting the varied topographical features that contribute to the basin's complex geomorphology.

6.3 RESULTS AND DISCUSSION

6.3.1 Hypsometric Analysis

The CalHypso tool was integrated into the QGIS environment to derive sub-basin-wise HCs and calculate curve statistics for the GRB. The input data included NASA's SRTM "Plus" DEM and HydroSHEDS sub-basin boundary layers. The HCs generated for the 15 sub-basins of the GRB are shown in Figure 6.6b, while the classification of HI values as detailed in Table 6.2, is based on literature, particularly the work of Anand and Pradhan (2019). According to the adopted classification, the sub-basins were categorized into three geological stages: young ($HI \geq 0.40$), mature ($0.39 \geq HI \geq 0.30$), and old ($HI < 0.30$). Figure 6.7a illustrates the HI values for each sub-basin, and Table 6.3 provides a comprehensive analysis of HI values, curve moments, and geological stages. The shape of the HC, whether concave or convex, indicates the erosional stage of the river basin (Strahler 1952; Pérez-Peña et al. 2009). A more convex curve and higher HI values suggest a younger river basin, while a more concave curve indicates an older stage of geomorphic development. Thirteen of the fifteen sub-basins of the GRB were classified as being in the old stage, with the exception of the Sone and Lower Yamuna sub-basins, which were categorized as mature (Figure 6.7b).

6.3.2 Statistical Moments of the HC of GRB

The HC moments for the GRB were extracted using the CalHypso tool in QGIS. The sub-basin-wise HI values, along with statistical moments such as HSK, HKU, DSK, and DKU are detailed in Table 6.3. As HI alone does not fully capture the complexity of basin development, the statistical moments provide a deeper understanding of the geomorphic processes at play, as shown in Figure 6.8a–c.

HSK values in GRB vary from 1.85 to 11.71, indicating varying levels of asymmetry in the distribution of mass within the sub-basins. Higher HSK values, such as those observed in the sub-basins of Ramganga, Upper Ganga, Upper Yamuna, and Ghaghara (7.23, 7.80, 8.74, and 11.71, respectively), suggest significant headward development of river streams. HSK values approaching zero would indicate a

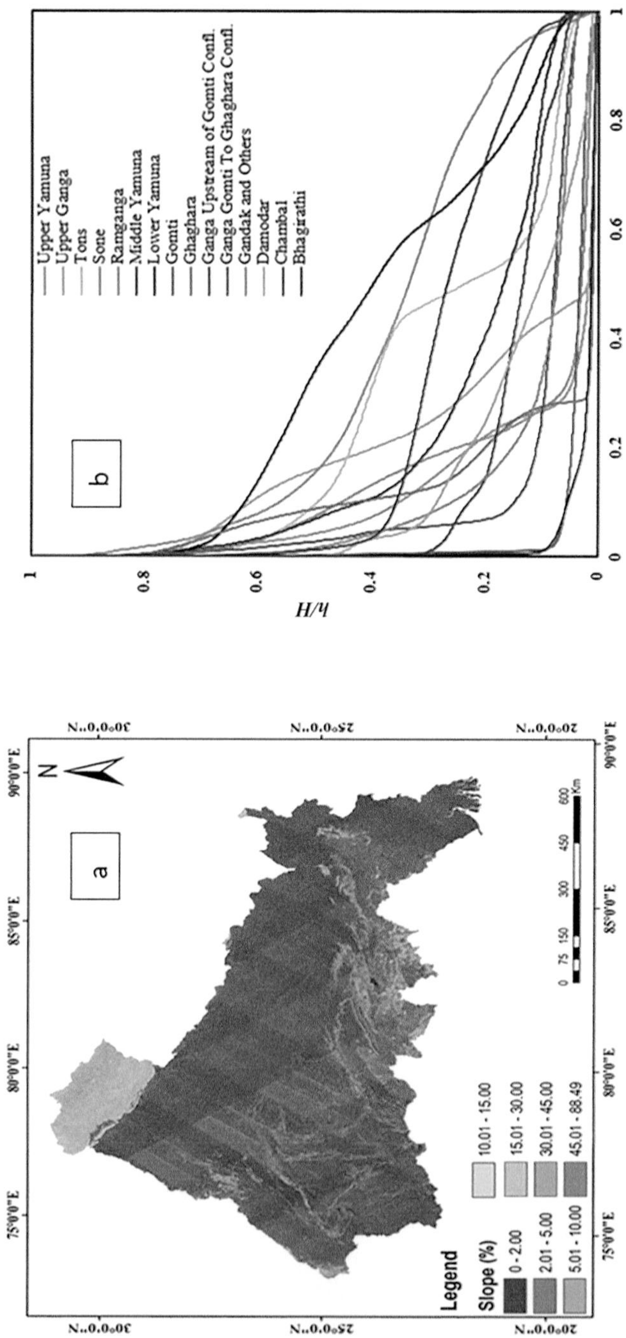

FIGURE 6.6 (a) Slope distribution of GRB. (b) Sub-basin-wise HCs of GRB.

TABLE 6.2
HI-Based Classification of Geological Stages as per Widely Used Literature

Index	Formula	Source	Classes	Dehbozorgi et al. (2010)	Sharma and Mohanty (2018)	Mahmood and Gloaguen (2012)	Anand and Pradhan (2019)
HI	$(E_{mean} - E_{min})/(E_{max} - E_{min})$	Strahler (1952)	Young	>0.50	>0.47	0.51–0.78	≥0.40
			Mature	0.40–0.50	0.46–0.47	0.37–0.50	0.39–0.30
			Old	<0.40	<0.46	<0.37	<0.30

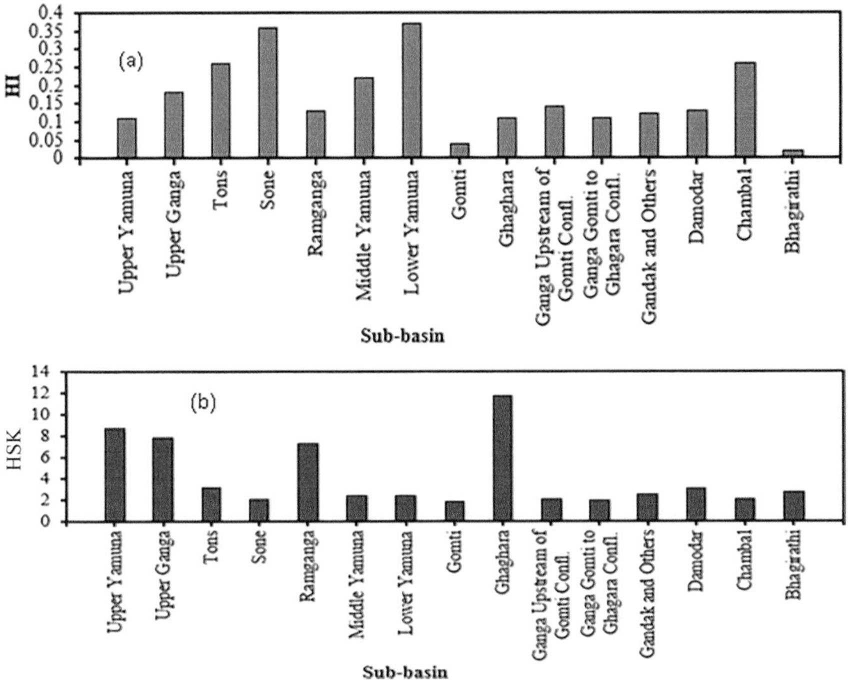

FIGURE 6.7 (a) Sub-basin wise HI values of Ganga River Basin. (b) Sub-basin wise HSK values of Ganga River Basin.

symmetrical mass distribution. HKU values range from 0.33 to 2.94, reflecting the "peakedness" of the distribution. For instance, the Ghaghara sub-basin exhibits the highest HKU value of 2.94, indicating pronounced peaks, while the Chambal sub-basin shows flatter peaks with an HKU value of 0.33. Both HSK and HKU (Table 6.3) values are relatively high across the GRB, signifying advanced basin development and more pronounced highlands.

TABLE 6.3
The HI, Curve Moments, and the Geological Stages for all 15 Sub-Basins of Ganga River Basin

Sub-Basin Name	Area (km²)	HI	Geological Stage	HKU	DKU	Hypsometric Skewness (HSK)	DSK[=
Upper Yamuna	37,651.60	0.11	Old	2.33	2.50	8.74	9.14
Upper Ganga	49,846.20	0.18	Old	1.61	-5.49	7.80	–47.48
Tons	29,654.00	0.26	Old	0.86	0.06	3.10	1.70
Sone	68,295.50	0.36	Mature	0.42	0.37	2.01	1.48
Ramganga	224,929.50	0.13	Old	2.10	2.45	7.23	8.92
Middle Yamuna	25,354.40	0.22	Old	0.71	1.24	2.32	3.19
Lower Yamuna	104,646.00	0.37	Mature	0.55	-0.07	2.40	1.91
Gomti	37,915.20	0.04	Old	0.51	1.04	1.85	2.24
Ghaghara	37,374.00	0.11	Old	2.94	2.60	11.71	9.03
Ganga Upstream of Gomti Confluence	49,985.20	0.14	Old	0.43	0.47	2.02	1.64
Ganga Gomti to Ghaghara Confluence	52,432.70	0.11	Old	0.66	1.42	1.92	3.15
Gandak and Others	73,085.40	0.12	Old	0.88	1.72	2.45	4.49
Damodar	39,596.90	0.13	Old	0.92	0.88	3.04	2.73
Chambal	136,745.00	0.26	Old	0.33	-0.04	1.99	1.41
Bhagirathi	62,715.70	0.02	Old	1.06	1.64	2.65	3.95

The density function approach, which examines the rates of change along the HC, provides a more precise analysis than the cumulative distribution function (CDF). DSK and DKU measure changes in slope and mid-basin slope, respectively. DSK values range from –47.48 in the Upper Ganga to 9.14 in the Upper Yamuna sub-basin. A negative DSK, like that seen in Upper Ganga, typically indicates sapping landforms where erosion begins at the lower reaches and moves upward, highlighting less erosion in the upper reaches of the Himalayan highlands. Conversely, positive DSK values, such as in the Upper Yamuna, indicate significant slope changes in the basin. DKU values tend to increase in the lower reaches of the river, with the Ghaghara and Upper Yamuna sub-basins experiencing the most significant mid-basin slope changes. These variations are influenced by tectonic activity, lithology, and climatic conditions, which actively shape the erosional processes within the GRB.

6.3.3 Hot-Spot and Cold-Spot Analysis of Hypsometric Parameters of GRB

Figure 6.9 displays the results of the hot-spot and cold-spot analysis for HI, statistical moments (HSK, HK, DSK, DKU) across the GRB. The hot spots of HI

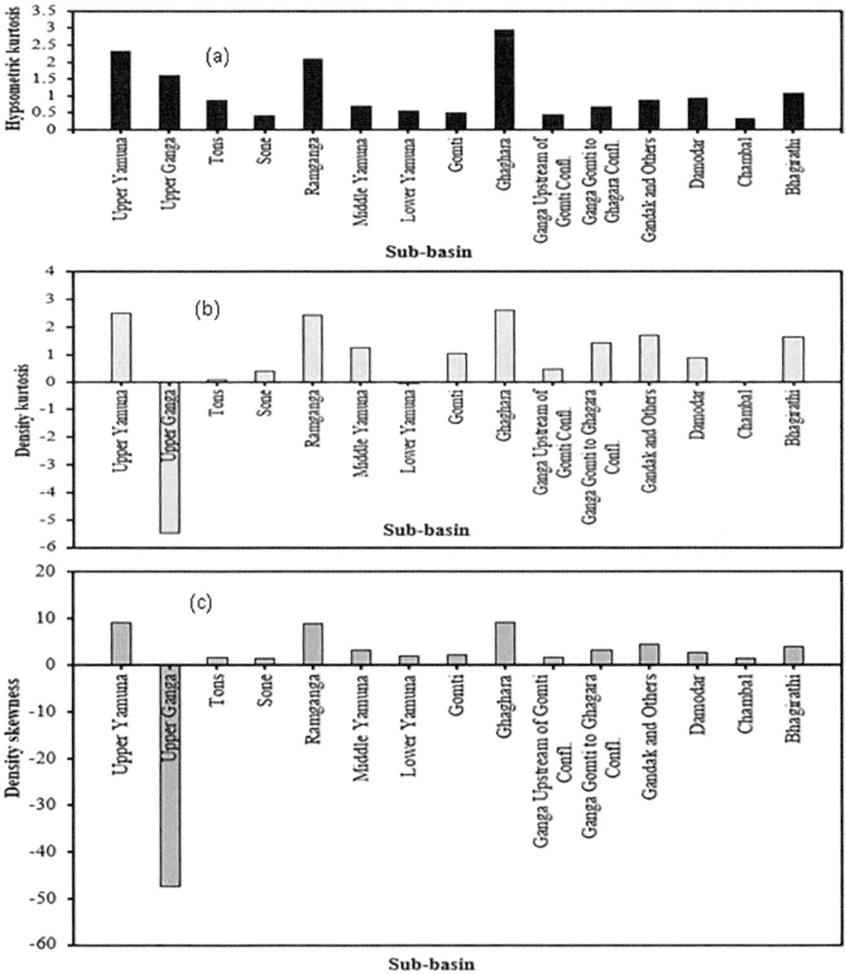

FIGURE 6.8 (a) Sub-basin wise HKU values of GRB. (b) Sub-basin wise DKU values of GRB. (c) Sub-basin wise DSK values of GRB.

values, which are indicative of younger sub-basins, are clustered in the southwestern GRB (Chambal, Lower Yamuna, Tons, Sone, Middle Yamuna), whereas the cold-spot values representing older sub-basins are found in the eastern and northeastern regions. Moreover, for the parameters of HSK and HK values, the northern sub-basins (Upper Ganga, Upper Yamuna, Ramganga, Gomti, Ghaghara), show higher skewness and peakedness, reflecting significant erosion. Lower reaches, however, show flatter profiles with ongoing erosion. Finally, for the parameter of DSK and DKU, significant clustering occurs in the eastern (higher slope change and mid-basin slope) and northern sub-basins (mixed mountainous and flat terrain), indicating varied geomorphic activity. Thus, this analysis underscores the spatial variation in erosion and geomorphic processes within the GRB, aiding in better understanding and management of the basin.

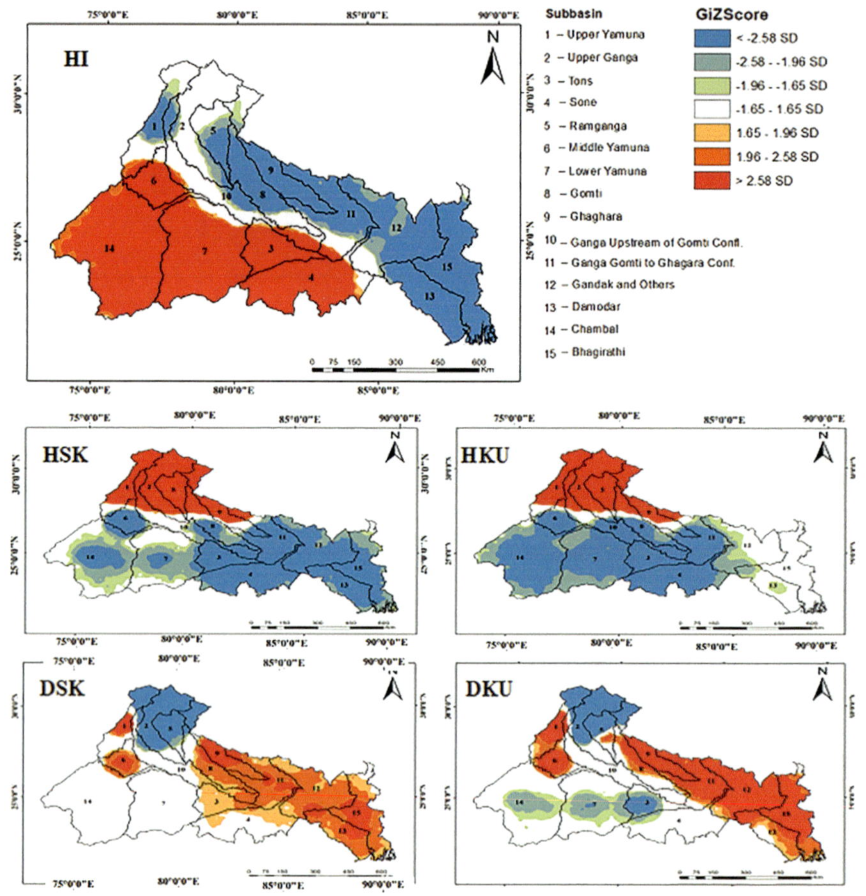

FIGURE 6.9 HI, HSK, HKU, DSK, and DKU values in the hot-spot and cold-spot analysis for GRB.

6.4 CONCLUSIONS

This study presents a scalable methodology for hypsometric analysis of large river basins by integrating GEE with the CalHypso tool. The proposed approach was applied to the GRB, comprising 15 sub-basins, to evaluate the geomorphic development of the basin using the HI and statistical moments of hypsometric and density functions. The results revealed that 13 out of 15 sub-basins fall into the old geomorphic stage (HI < 0.30), while only Sone (HI = 0.36) and Lower Yamuna (HI = 0.37) exhibit characteristics of mature landscapes. This is indicative of relatively higher ongoing erosional activity in the Sone River Basin. The highest HSK value of 11.71 was observed in the Ghaghara sub-basin, which is suggestive of intense headward erosion, followed by Upper Yamuna (HSK = 8.74) and Ramganga (HSK = 7.23). Similarly, Ghaghara also showed the highest HKU (HKU = 2.94), pointing to peaked terrain morphology. In contrast, sub-basins like Chambal (HKU = 0.33) showed flatter terrain profiles, aligning with their more stable geomorphic stage. The DSK ranged from –47.48 (Upper Ganga), suggesting lower reach sapping erosion, to 9.14 (Upper Yamuna), indicating steep upper slope changes. The hot-spot analysis further validated these findings, identifying high-erosion clusters in the northern and eastern sub-basins (e.g., Ghaghara, Ramganga) and more stable, older landscapes in the southwestern zones (e.g., Chambal, Tons). These findings highlight significant spatial heterogeneity in erosional dynamics across the GRB. The proposed methodology drastically reduces computational effort while offering granular, quantitative insights that are critical for informed river basin management, erosion control, and climate resilience planning. This framework is readily adaptable to other large basins globally, making it a valuable tool for both researchers and decision-makers.

REFERENCES

Ahmad, I., Ahmad, Z., Munir, S., Rehman, O. U., Ali, S. R., and Shabbir, Y. (2018). "Characterization of Upper Indus Basin based on DEM hypsometric analysis." *The Nucleus*, 55(1), 8–16.

Ahmed, F., and Rao, K. S. (2016). "Hypsometric analysis of the Tuirini drainage basin: A geographic information system approach." *International Journal of Geomatics and Geosciences*, 6(3), 1685–1695.

Anand, A. K., and Pradhan, S. P. (2019). "Assessment of active tectonics from geomorphic indices and morphometric parameters in part of Ganga basin." *Journal of Mountain Science*, 16(8), 1943–1961.

Chowdhury, M. R., and Ward, N. (2004). "Hydro-meteorological variability in the greater Ganges–Brahmaputra–Meghna basins." *International Journal of Climatology*, 24(12), 1495–1508.

Dehbozorgi, M., Pourkermani, M., Arian, M., Matkan, A. A., Motamedi, H., and Hosseiniasl, A. (2010). Quantitative analysis of relative tectonic activity in the Sarvestan area, central Zagros, Iran. *Geomorphology*, 121(3–4), 329–341.

Farr, T. G. (2011). Remote sensing in geomorphology. In *The SAGE Handbook of Geomorphology* (pp. 210–226). SAGE Publications Ltd., UK.

Getis, A., and Ord, J. K. (1992). "The analysis of spatial association by use of distance statistics." *Geographic Analysis*, 24(3), 189–206.

Harlin, J. M. (1978). "Statistical moments of the hypsometric curve and its density function." *Mathematical Geology*, 10(1), 59–72.

Lehner, B., Verdin, K., and Jarvis, A. (2008). "New global hydrography derived from spaceborne elevation data." *Eos, Transactions American Geophysical Union*, 89(10), 93–94.

Mahmood, S. A., and Gloaguen, R. (2012). Appraisal of active tectonics in Hindu Kush: Insights from DEM derived geomorphic indices and drainage analysis. *Geoscience Frontiers*, 3(4), 407–428.

Maswood, M., and Hossain, F. (2016). "Advancing river modelling in ungauged basins using satellite remote sensing: The case of the Ganges–Brahmaputra–Meghna basin." *International Journal of River Basin Management*, 14(1), 103–117.

Pérez-Peña, J. V., Azañón, J. M., and Azor, A. (2009). "CalHypso: An ArcGIS extension to calculate hypsometric curves and their statistical moments. Applications to drainage basin analysis in SE Spain." *Computers & Geosciences*, 35(6), 1214–1223.

Prakash, K., Singh, S., Mohanty, T., Chaubey, K., and Sing, C. K. (2017). "Morphometric assessment of Gomati River basin, middle Ganga plain, Uttar Pradesh, North India." *Spatial Information Research*, 25(3), 449–458.

Roy, S., and Chintalacheruvu, M. R. (2024a). "Google Earth Engine-based morphometric parameter evaluation and comparative analysis of soil erosion susceptibility using statistical and machine learning algorithms in large river basins." *Earth Science Informatics*, 17(1), 75–97.

Roy, S., and Chintalacheruvu, M. R. (2024b). "Delineating hydro-geologically constrained groundwater zones in the Himalayan River basins of India through an innovative ensemble of hypsometric analysis and machine learning algorithms." *Earth Science Informatics*, 17(1), 501–526.

Roy, S., and Chintalacheruvu, M. R. (2024c). "Enhanced morphometric analysis for soil erosion susceptibility mapping in the Godavari River basin, India: Leveraging Google Earth Engine and principal component analysis." *Indian Society for Hydraulics Journal of Hydraulic Engineering*, 30(2), 228–244.

Sharma, G., and Mohanty, S. (2018). Morphotectonic analysis and GNSS observations for assessment of relative tectonic activity in Alaknanda basin of Garhwal Himalaya, India. *Geomorphology*, 301, 108–120.

Strahler, A. N. (1952). "Hypsometric (area–altitude) analysis of erosional topography." *Geological Society of America Bulletin*, 63, 1117–1142.

7 Methods to Quantify Salinity Problems Using Remote Sensing Technologies

Sanju Kumari, Shubhi Khare, Deepak Kumar, and Priyadarshi Maurya

7.1 INTRODUCTION

As per the global status of soil affected by Salt, published by Food and Agriculture Organization (FAO), it has been observed that salt-affected soils cover around 1,381 Mha, representing approximately 10.7% of the Earth's total land area. It also suggests that salinity impacts about 10% of irrigated and rainfed agricultural lands (FAO, 2024). Although the estimates are uncertain due to the scarcity of comprehensive data, estimates suggest that every year, nearly 10% of additional area is getting salinized, and by 2050, around 50% of the arable land would be salt-affected. In India, saline soils occupy 44% of the area, covering 12 states and one Union Territory, while sodic soils occupy 47% in 11 states (Kumar and Sharma, 2020).

Soil salination is the accumulation of water-soluble salts in the soil to a level that impacts agricultural production, environmental health, and economic welfare (Negacz et al., 2022). There are two main processes of soil salination. Primary process, also known as natural process, consists of weathering of soil minerals, saline groundwater intrusion with water table fluctuations, and secondary process occurring mainly due to agronomic practices such as high volume of fertilizer and pesticide application, saline groundwater intrusion with water table fluctuations, irrigation with saline water sourced from bore, recycled or waste waters, dumping of industrial and municipal wastes into soils (Rengasamy, 2016). A soil is considered saline when the electrical conductivity (EC) of a saturated soil paste is greater than 4 dS / m. Both World Reference Base (WRB) and Soil Taxonomy, international soil categorization systems, have elevated the minimum EC criteria for recognizing a soil horizon as saline. The WRB considers 15 dS / m as the standard value for defining a saline horizon, while Soil Taxonomy sets the limit at 30 dS / m (Canfora et al., 2014). Apart from EC and pH, Exchangeable Sodium Percentage (ESP) is also an important parameter in the determination of salinity. Based on EC, pH, and ESP, soil is classified into three main categories: saline soil, Sodic soil, and saline-Sodic soil. Saline soils have high levels of soluble salts, which reduce their fertility. Alkali soils are classified based on their productivity, which is affected by exchangeable sodium (Minhas and Sharma, 2003).

DOI: 10.1201/9781003596080-7

Saline–alkali soils possess both an excess of soluble salts and a high concentration of exchangeable sodium (Chhabra, 2004). The parameters along with their indicating values for assessment of salinity in soil are given in Table 7.1.

Building on this understanding, salinity arises primarily from unsustainable irrigation practices, such as excessive fertilizers, irrigation with saline water, and inadequate drainage systems to remove surplus water. These factors lead to the accumulation of salts in the soil, gradually reducing its fertility and making it less productive for agriculture. Furthermore, insufficient drainage results in waterlogging, which accelerates the salinization process by preventing the natural leaching of salts. Over time, this degrades soil health and causes irreversible groundwater contamination, as dissolved salts and harmful chemicals infiltrate underground water sources. This deterioration in water quality poses significant challenges for irrigation, drinking water supply, and overall ecosystem sustainability (Mohnavelu et al., 2021). Water evaporates in its pure form, leaving salts and other dissolved substances, which gradually accumulate in the soil. As water is lost through evapotranspiration (ET), the salt concentration in the remaining soil solution can increase significantly, often reaching four to ten times the original concentration in the irrigation water within just three to seven days after irrigation. As each irrigation cycle introduces additional salts, the total amount of salt added to the soil depends on the volume of water applied and the initial salt concentration of the irrigation water. Over time, without effective management practices such as leaching, these salts continue to build up in the soil, further reducing its productivity (Carter, 1975).

The rising salinity levels have significant consequences, particularly for agriculture, water resources, and the environment. Soil salinity significantly impacts soil structure, influencing the complex processes of aggregate formation and breakdown. Aggregate stability serves as a key indicator of soil structure. As the salt concentration in irrigation water rises, soil aggregates deteriorate due to swelling and the dispersion of clay particles, which occurs as the interlayer space between adjacent diffuse double layers expands (Tang et al., 2020). The physical properties of soil are severely impacted by high salt concentrations, leading to the deterioration of soil structure. This structural decline reduces soil hydraulic conductivity (K) and infiltration rates, limiting the soil's ability to absorb and retain water. Consequently, excess water accumulates on the surface, intensifying surface runoff and accelerating soil erosion. Over time, this erosion strips away the nutrient-rich topsoil, diminishing soil fertility and contributing to land degradation.

Soil porosity, texture, and mineral composition are crucial in determining soil hydrological properties, including water retention, permeability, and drainage.

TABLE 7.1
Categorization of Salt-Affected Soil (Qadir et al., 2000)

S. No	Parameter	Saline Soil	Sodic Soil	Saline-Sodic Soil
1	pH	< 8.5	>8.5	>8.5
2	EC	> 4dS / m	< 4dS / m	> 4 dS / m
3	ESP	<15	>15	>15

These properties are also influenced by the accumulation of salts on the soil surface, which can hinder water movement and impact soil fertility. A higher capillary fringe increases salt deposition on the surface, intensifying soil salinity and negatively affecting plant growth and overall soil health (Otlewska et al., 2020). The growth and development of plants are severely impacted by the low osmotic potential of the soil solution and nutritional imbalances caused by excessive salt accumulation. High soil salinity levels hinder seed germination and seedling growth by creating osmotic stress, which restricts water uptake, and exposing plants to specific ion toxicity that disrupts cellular functions. Additionally, salt stress interferes with essential metabolic processes, weakening plant vitality and significantly reducing productivity. Over time, these adverse effects compromise crop yields and threaten agricultural sustainability (Safdar et al., 2019).

Saline soil can be reclaimed by minimizing salt concentration, decreasing pH value, and improving soil structure. Soil desalinization can be achieved by either harvesting the aboveground portions of salt-tolerant plants, which absorb and store salts, or by facilitating the leaching process, where excess salts are washed from the upper soil layers beyond the root zone (Shankar and Evelin, 2019). Gypsum and sulfuric acid have long been recognized for their benefits in reclaiming sodic soils (Mace et al., 1999). Gypsum remains the most widely used amendment for reclaiming saline-sodic and sodic soils, owing to its cost-effectiveness, widespread availability, and ease of application (Chi et al., 2012). Pure gypsum was applied at 50% of the soil gypsum requirement at the time of planting rice and *Sesbania*, whereas 95% pure sulfuric acid was added at 50% soil gypsum requirement as one-third applications by mixing with the first three irrigations (Ahmad et al., 2013). In addition to chemical amendments, biological approaches such as the cultivation of halophytic plant species have shown promise in reclaiming saline soils. Another effective method for reclaiming saline soils is leaching, which involves the downward movement of water through the soil profile to dissolve and flush out soluble salts. Saline soils can be reclaimed by leaching with good-quality water, aiming to reduce salinity in the top 45–60 cm of soil below the crop-specific threshold. In the Delta area of Utah, saline–alkali soils have been successfully reclaimed using 4 feet of water. Applying water over crop requirements supports effective leaching, while seasonal precipitation, especially in fall and spring, also helps wash salts from the soil and dilute those present in irrigation water (Khoshgoftarmanesh et al., 2003).

7.2 ROLE OF REMOTE SENSING IN THE MANAGEMENT OF SALINE SOIL

Numerous remote sensing datasets and methodologies have been extensively applied to detect and monitor salt-affected regions, achieving varying levels of effectiveness. Frequently utilized data sources for such applications include aerial imagery, video recordings, thermal infrared scans, multispectral, and microwave imagery in the visible and infrared spectral ranges (Didi et al., 2019). Compared to non-salinized agricultural land, saline soils exhibit higher reflectance in the visible and near-infrared (NIR) spectral regions; the greater the salinity level, the stronger the reflectance observed in these wavelengths (Li et al., 2010). The mapping of soil salinity at a

TABLE 7.2
Overview of Satellites, Their Sensors, and Corresponding Bands

S. No.	Satellite	Sensor	Band
1	Landsat (4, 5, 7, 8, 9)	OLI, TM, ETM+, MSS	Visible, NIR, SWIR
2	Sentinel-2A/2B	MSI	VNIR, SWIR
3	MODIS (on Terra and Aqua)	MODIS	Visible, IR
4	IRS Resourcesat-1 and -2	LISS III	Visible, NIR, SWIR
5	IRS-1B	LISS I, LISS II	Visible, NIR
6	ASTER (on Terra)	ASTER	VNIR, SWIR, TIR
7	SPOT	SPOT XS	Visible, NIR
8	IKNOS	Optical	Panchromatic, Visible, NIR
9	RapidEye	Multispectral Push Broom	Visible, NIR

global level is done by integrating soil data with remote sensing technologies. This approach allows for the estimation of salinity-affected regions worldwide, which are believed to cover approximately 1 billion ha (Tola et al., 2024). A wide range of satellites, sensors, and spectral bands have been employed to monitor and map soil salinity at various scales effectively. To detect salinity-related features, these remote sensing platforms utilize different wavelengths across the electromagnetic spectrum, particularly in the visible, NIR, shortwave infrared (SWIR), and microwave regions. The satellites used for the measurement of salinity are listed in Table 7.2.

7.3 INDICES USED IN MONITORING SOIL SALINITY

7.3.1 Salinity Indices

Salinity indices (SIs) are spectral indicators designed to detect and assess the presence of salts in soil using remote sensing data. These indices typically involve mathematical combinations or ratios of reflectance values from specific spectral bands, most notably the third spectral band relative to others to enhance the spectral response associated with salt-affected areas (Rajitha et al., 2007). The various SIs are shown in Table 7.3.

7.3.2 Vegetation Indices

Normalized Difference Vegetation Index (NDVI): NDVI is functionally equivalent to the ratio obtained by subtracting NIR from the red (R) band and dividing that difference by the sum of the two bands. Furthermore, NDVI serves as an indicator of vegetation greenness or photosynthetic efficiency. A higher NDVI value suggests that the vegetation cover is absorbing a greater amount of incoming radiation. The index values typically range from −1 to +1. It is mathematically expressed as (Morshed et al., 2016; Ramos et al., 2020; Ijaz et al., 2020):

TABLE 7.3
Various SIs Used in Remote Sensing Technology for Studying the Salinity in Soil Are Explained

S. No.	Indices	Formula	References
1	SI	$(B \times R)^{0.5}$	Mirzaee et al. (2024);
2	SI1	$(G \times R)^{0.5}$	Wang et al. (2022), Golestani et al. (2023)
3	SI2	$(G^2 + R^2 + NIR^2)^{0.5}$	Wang et al. (2022)
4	SI3	$(G^2 + R^2)^{0.5}$	Golestani et al. (2023), Li et al. (2022)
5	SI4	B / R	Golestani et al. (2023), Wang et al. (2022)
6	SI5	$(B - R)/(B + R)$	Mirzaee et al. (2024)
7	SI6	$(G \times R)/B$	Barreto et al. (2023), Wang et al. (2022)
8	SI7	$(B \times R)/G$	Barreto et al. (2023), Mirzaee et al. (2024)
9	SI8	$(R \times NIR)/G$	Mirzaee et al. (2024), Cui et al. (2023)
10	SI9	$SWIR1 / NIR$	Li et al. (2022), Mirzaee et al. (2024)
11	SI10	$\frac{SWIR1 - SWIR2}{SWIR1 + SWIR2}$	Wang et al. (2022), Bouaziz et al. (2011)
12	SI11	$SWIR1 / SWIR2$	Bouaziz et al. (2011), Mirzaee et al. (2024), Wang et al. (2022)
13	Normalized Difference Salinity Index (NDSI)	$(R - NIR)/(R + NIR)$	

B: blue band, G: green band, R: red band, NIR: near-infrared band, SWIR: shortwave infrared band.

$$NDVI = (NIR - R)/(NIR + R) \tag{7.1}$$

Soil Adjusted Vegetation Index (SAVI): SAVI is a vegetation index calculated using the reflectance values from the NIR) band, red (R) band, and a constant L, which varies with location. SAVI is employed to adjust the Normalized Difference Vegetation Index (NDVI) for the impact of soil brightness, particularly in regions with low vegetation cover, and expressed as (Ijaz et al., 2020; Satir and Berberoglu, 2016):

$$SAVI = ((NIR - R) \times (1 + L))/(NIR + R + L) \tag{7.2}$$

Enhanced Vegetation Index (EVI): EVI, like the NDVI, is utilized to measure the greenness of vegetation. However, EVI compensates for certain atmospheric disturbances and background noise from the canopy, making it more responsive in regions with dense vegetation, and is expressed as (Peng et al., 2019; Cui et al., 2023)

$$\text{EVI} = 2.5\times\left(\left(NIR - RED\right)/\left(NIR + C1\right)\times\left(RED - C2\right)\times\left(BLUE + L\right)\right) \quad (7.3)$$

where C1, C2, and L are values that serve as correction factors to adjust the impact of atmospheric conditions.

Canopy Response Salinity Index (CRSI): CRSI is used to detect the sudden shift in reflectance between the red (R) and NIR bands. This index was developed based on observational data and is not directly linked to any specific physiological characteristics of plants, but rather reflects general plant health and is expressed as (Peng et al., 2019; Ramos et al., 2020):

$$\text{CRSI} = \left(\left(\text{NIR}\times\text{R}\right)-\left(\text{G}\times\text{B}\right)\right)/\left(\text{NIR}\times\text{R}\right)+\left(\text{G}\times\text{B}\right) \quad (7.4)$$

7.4 RECENT APPLICATIONS OF REMOTE SENSING IN SALINITY

Sethi et al. (2006) conducted a study to identify the most effective sequence for detecting and mapping the extent of land affected by secondary salinization in Shorapur Taluka, within the Upper Krishna Irrigation Project in Gulbarga District. To assess and map salt-affected soils, they utilized IRS LISS-III imagery with a combination of green, red, and NIR bands. Auger samples from two depths, 0–15 cm and 15–30 cm, and Profile samples were collected. For base mapping, Survey of India toposheets (56 D/7, 10, 11, 14, and 15) from March 1997, at a 1:50,000 scale, were used to conduct a ground survey. The soil samples' pH and EC were measured in real time in the field to determine ground points, which were then marked on topographical sheets and later transferred onto satellite imagery. The soil auger and profile samples were processed and analyzed in the laboratory to determine their physicochemical characteristics. The study found that reflectance could not be directly correlated with EC, and EC could not be incorporated as a direct indicator in the mapping and inventory of salt-affected soils. The findings also revealed that the estimated salt-affected area was 35–40% larger than previously reported.

Wu et al. (2008) explored remote sensing images to track changes in soil salinity over time in the Hetao Irrigation District, Inner Mongolia, China. Their study aimed to link these changes to land and water management practices and derive lessons for salinity control. The LANDSAT satellite was selected as the primary satellite. Similar satellites, such as IRS (Indian Remote Sensing Satellite) and CBERS (China–Brazil Earth Resources Satellite), were also used to compensate for the unavailability of a suitable LANDSAT satellite. Using an unsupervised method, the corresponding cropped areas were identified from NDVI values. The results suggested that the overall accuracy of remote sensing in detecting soil salinity was 90.2%, and in detecting cropped areas, it was 98%. Abdelfattah et al. (2009) conducted a study to develop a model that integrates remote sensing data with GIS techniques to assess, characterize, and map the state and behavior of soil salinity to map soil salinity using in the coastal area of Abu Dhabi Emirate in 2000 and 2002. Two datasets were taken from the multi-temporal Landsat-7 Enhanced Thematic Mapper (ETM) satellite. Four hundred and three soil samples were collected and analyzed in the laboratory by using the weighted average method at a depth of 0–50 cm. Then it was plotted on the map and interpolated to produce a salinity map with different classes.

Wu et al. (2014) conducted research on diachronic salinity mapping and analysis using multi-temporal remote sensing, focusing on a pilot site in the Dujaila area of Central Iraq. Atmospherically corrected Landsat, Satellite pour l'Observation de la Terre (SPOT), and RapidEye satellites were used for the study. Field surveys and soil sampling were conducted in 2011–2012, and a remote sensing dataset was created using satellite images from 1988–1993, 1998–2002, and 2009–2012. A new method called the multiyear maxima-based modeling approach was used to develop salinity models, as single-date images often give inaccurate salinity information. The results showed that the salinity models could predict salinity with 82.57% accuracy. Zhou et al. (2015) aimed to identify controllable risk factors, propose response measures, and design management strategies for salt-affected soils in the Yinchuan Plain (YCP) in a semiarid region of northwest China. To achieve these objectives, they integrated the Spatial Autoregressive (SAR) model, Multi-Attribute Decision Making (MADM), and the Analytic Hierarchy Process (AHP). The study results indicated that the spatial autocorrelation of soil salinization and its risk factors gradually weakened as distance increased in the YCP. Furthermore, a comparison of model fit demonstrated that the SAR model performed better than the Operational Linescan System (OLS) model.

Datta et al. (2016) highlight the importance of characterizing and classifying soil resources for sustainable land use planning and the effectiveness of reclamation measures in salt-affected soils in Rewari district, Haryana. For this study, 77 soil samples were collected from four depths: 0–15 cm, 15–30 cm, 30–60 cm, and 60–90 cm. Ground truth of satellite images was performed by using geo-referenced soil samples, which were visually interpreted for salt-affected soils and waterlogged areas using ISODATA and ERDAS IMAGINE. Supervised classification was performed for the entire Rewari district. Of the total 1,594 km^2 area, approximately 295.2 km^2 was classified as moderately salt-affected, 29.06 km^2 as severely salt-affected, and 116.95 km^2 as highly saline. IRS Resources at LISS-III satellite data from March 2012 and February 2014 were used for this analysis. Physicochemical analysis revealed that alkalinity and salinity increased with depth, with most soils being more affected by alkalinity than salinity. Among the analyzed cations and anions, sodium (Na^+) and chloride (Cl^-) were the dominant ions. Soil organic carbon (SOC) concentration was very low, with a mean value of 0.27%. Soluble sodium concentration in the soil extract showed a significant positive correlation with EC ($R^2 = 0.72$), while calcium (Ca^{2+}) and magnesium (Mg^{2+}) concentrations were also positively correlated with soil EC.

Whitney et al. (2018) tested MODIS satellite time series with vegetation indices to assess soil salinity in agricultural soil in the Western San Joaquin Valley (WSJV), CA, USA. The study aimed to analyze soil salinity using ground truth data and evaluate three key aspects: whether the MODIS Vegetation Index Time Series could effectively capture seasonal average values of the Vegetation Index, whether MODIS multiyear time series data would provide more accurate results for salinity assessment compared to single-season time series data, and whether the EVI had a stronger correlation with soil salinity in semiarid farmland compared to the NDVI. The study used surface reflectance data from MOD09A1 Version 6, collected between January 1, 2007 and December 31, 2013, totaling 322 data points at 8-day intervals. This surface reflectance data was used to calculate vegetation indices, which were then filtered using the Savitzky–Golay (SG) filter to smooth the calculated values. The Western San Joaquin Valley (WSJV) soil salinity dataset served as the ground truth, consisting of 41,779 electromagnetic induction readings and 267 salinity measurements from composite 0–1.2 m soil profiles. These data points were interpolated onto a 500 × 500 m MODIS grid, with 75 ground truth measurements recorded, of which 73 were used for further analysis. Multiyear remote sensing datasets provided stronger correlations with soil salinity than single-year data. Among vegetation indices, the correlation strength with salinity followed the order: CRSI > NDVI > EVI. However, the spatial resolution of MODIS was found unsuitable for salinity assessment in California farmland due to information loss caused by its coarse resolution, especially when compared to higher-resolution satellites such as Landsat and Sentinel.

Asfaw et al. (2018) mapped soil salinity using remote sensing and geographic information systems in one of the largest commercial farms of Ethiopia, that is, Wonji irrigation farm. Six selected remote sensing indices, such as SI, normalized difference salinity index (NDSI), brightness index (BI), NDVI, vegetation soil salinity index (VSSI), and soil adjusted vegetation index (SAVI), were used to discriminate and map salt-affected soils. SI is the ratio of the red band to the NIR (NIR) band, while NDSI is the ratio of the difference of the red to NIR divided by the summation of the two. Different remote sensing indices were calculated from the original bands of Landsat images. Statistical correlation between field measurements of EC and remote sensing spectral indices showed that the SI had the highest correlation with EC. The satellites used in this study included Landsat Thematic Mapper (TM) data of 2012, specifically from path 168 and row 054. Out of all indices tested, the SI was used to create enhanced images for soil salinity in this study, due to its highly significant correlation with EC. Gorji et al. (2017) generated soil salinity maps in the Tuz (Salt) Lake Region, Turkey, to track and monitor the changes in salt-affected areas in 1990, 2002, 2006, 2011, and 2015 using Landsat 8 data. Besides, this study utilized Coordination of Information on the Environment (CORINE) data to identify probable human-induced causes of soil salinization.

Didi et al. (2019) studied soil salinity mapping in the arid and semiarid zones of the Tadla plain in Morocco using optical remote sensing data and field measurements of the Sodium Absorption Ratio (SAR). They used a Landsat 8 satellite image from the Operational Land Imager (OLI) sensor, which has a spatial resolution of 30 m. Thirty samples were collected from different soil types affected by salinity in the irrigated Tadla perimeter. These samples were taken on the same date as the satellite image to measure EC. The EC values in the study area ranged from less than 2 dS/m (nonsaline soil) to more than 8 dS/m (highly saline soil), showing significant variation in salinity levels. The study focused on evaluating and validating salinity models. The selected model showed a strong correlation between the SI and the EC measured in the field. Hoa et al., (2019) researched on mapping of soil salinity intrusion in Ben Tre city located on the Mekong

River Delta of Southern Vietnam using the Sentinel-1 Synthetic Aperture Radar (SAR) C-band data combined with five state-of-the-art machine learning models, Multilayer Perceptron Neural Networks (MLP-NN), Radial Basis Function Neural Networks (RBF-NN), Gaussian Processes (GP), Support Vector Regression (SVR), and Random Forests (RFs). Sixty-three soil samples were collected during the field survey conducted from April 4 to 6, 2018, aligning with the Sentinel-1 SAR imagery. His results suggested that Sentinel-1 SAR data alone were generally effective for mapping soil salinity in arid and semiarid regions, but when combined with machine learning techniques; they can also be used for soil salinity mapping in tropical areas.

Khadim et al. (2019) conducted a study that combined remote sensing techniques to create spatiotemporal soil salinity maps, supported by an evaluation of existing research on vegetation salt tolerance and field assessments involving in situ EC measurements in the Everglades National Park (ENP). To generate soil salinity maps, a supervised machine learning technique, specifically the maximum likelihood classification algorithm, was applied to categorize the region into seven vegetation-based land cover types: mangrove forest, mangrove scrub, low-density forest, sawgrass, prairies and marshes, barren lands with woodland hammock, and water. This classification was carried out for 1996, 2000, 2006, 2010, and 2015, achieving accuracy rates between 82% and 94%, with the 2015 classification validated through ground truthing. Their research demonstrated that the soil salinity maps effectively illustrated the spatial distribution of non-, low-, and high-saline areas in the ENP. However, no significant temporal variations in salinity were observed throughout the study period.

Jamali et al. (2020) conducted a study to predict salinity expansion in Golpayegan, Isfahan, Iran, by estimating vegetation cover loss and wetland shrinkage using remote sensing and GIS techniques. Landsat satellite images from various sensors, including TM, Enhanced Thematic Mapper Plus (ETM+), and OLI, were analyzed. A total of 22 Landsat images from 1985 to 2016 were selected. The Fast Line-of-Sight Atmospheric Analysis of Spectral Hypercubes (FLAASH) model was applied to enhance the accuracy of atmospheric correction. To implement the model, salinity maps from 1985, 1994, 2008, and 2015 were introduced into the system, allowing for the analysis. This study employed six Soil SIs (SSI1, SSI2, SSI3, SSI4, SSI5, and SSI6) to identify salinity zones. Using the Cellular Automata (CA) Markov method, a forecast map for 2025 was created. The results indicated that among the different SIs, SSI6 was the most effective. The study revealed a progressive shift in land conditions, where areas initially classified as low salinity transitioned into medium- and high-salinity zones. Additionally, a negative correlation was observed between vegetation density and soil salinity, as wetland areas dried up and salinity increased, vegetation density declined. Mishra et al. (2020) carried out a study to assess the impact of sodic soil reclamation in the Indo-Gangetic plains of Uttar Pradesh, focusing on gypsum application, agronomic practices, and irrigation management to improve soil health and crop productivity. Soil characterization was conducted by measuring pH, EC, ESP, and gypsum requirement. Reclamation strategies included gypsum application (50% GR), organic amendments (press mud, farmyard manure), and improved irrigation techniques. Results showed a significant reduction in soil sodicity, with paddy yield increasing from 1.36 t/ha to 3.36 t/ha and wheat yield from 0.8 t/ha to 3.38 t/

ha. Adopting green manure and organic amendments enhanced soil fertility, while economic benefits were observed for 57% marginal, 36.4% small farmers, and 6.6% large farmers.

Kulmatov et al. (2021) conducted research on soil salinity assessment in the irrigated areas of Sirdarya, located on the left bank of the Syr Darya River in central Uzbekistan. The study analyzed raw data from traditional methods collected between 2000 and 2015 and from 2016 to 2019, using a combination of traditional techniques and GIS-based methods. Soil salinity maps of Sirdarya city were created to examine spatial and temporal changes using the inverse distance weighting (IDW) interpolation method, based on field surveys and laboratory studies. The results showed that the IDW interpolation method was highly effective and accurate in mapping changes in salt-affected irrigated areas. It was also found that local terrain conditions strongly influence soil salinity, while its dependence on climatic factors is weak. Naimi et al. (2021) investigated utilizing machine learning algorithms to predict spatial soil salinity (dS/m) by combining environmental covariates derived from remotely sensed (RS) data, a digital elevation model (DEM), and proximal sensing (PS) in southern Iran. They collected 300 surface soil samples and recorded spectral data using Sentinel-2, an electromagnetic induction instrument (EMI), and portable X-ray fluorescence (pXRF). The data was then analyzed using five machine learning methods: RF, k-nearest neighbor (kNN), support vector machines (SVM), partial least squares regression (PLSR), artificial neural networks (ANNs), and an ensemble of these models. Estimation of the EC of the saturated paste extract (ECe) was built based on three scenarios, including Scenario: Synthetic Soil Image (SySI) bands and SIs derived from it; Scenario: RS data, PS data, topographic attributes, and geology and geomorphology maps; and Scenario: the combination of Scenarios first and second. The results showed that the integrated approach was more reliable and provided more accurate soil salinity predictions than individual methods.

Narjary et al. (2022) conducted an investigation to test the hypothesis that an integrated approach of EMI, multivariate, and geostatistical techniques can provide rapid and reliable information on soil salinity at a spatiotemporal scale. The analysis of the EMI survey data revealed that soil EC showed a stronger correlation with the horizontal mode (EMH) of apparent soil EC compared to the vertical mode (EMV). Further analysis using Principal Component Analysis (PCA) and Pearson correlation showed that Na^+, Ca^{2+}, Mg^{2+}, and Cl^- are the key salt ions influencing apparent soil EC. Roohi et al. (2021) studied land subsidence changes from 2014 to 2020 using Sentinel-1 satellite imagery and SNAP software. They calculated soil EC, surface moisture content, and moisture at a depth of 1–2 m using Sentinel-2 imagery, Landsat 8 satellite data, and Famine Early Warning Systems Network (FEWS NET) Land Data Assimilation System (FLDAS) data on the Google Earth Engine (GEE) platform. The results showed that increasing land subsidence has expanded the area affected by soil salinity and increased its intensity. Consequently, moisture levels have decreased at different depths in recent years. Zhang et al. (2022) studied how soil salinity affects SOC stock over time on a large spatial scale. To analyze this, they used extreme gradient boosting models (XGBoost) along with field measurements from 433 sites and 41 environmental factors (both static and time-varying). Their goal was to quantify the impact of salinity on SOC changes in a typical inland river basin in China from the 1990s to the 2020s. The results showed that the XGBoost models accurately predicted soil EC and SOC stock at a depth of 0–20 cm. The models effectively captured the spatial and temporal patterns of soil EC and SOC stock.

Hossen et al. (2022) conducted research to determine soil salinity levels in different areas of Barguna Sadar Upazila, located in the coastal region of Bangladesh's Barguna District. The study used two types of datasets: Landsat 8 OLI satellite data and ground truth data from a printed map. The methodology involved using the Landsat 8 OLI dataset along with nine soil SIs to develop a soil salinity map. Soil salinity maps from the Soil Resource Development Institute (SRDI) were used. The results of the linear regression analysis revealed a very weak correlation between the satellite-based SI maps and the ground truth EC values. Ali (2023) focused on deriving soil salinity using spectral indices obtained from various bands of the Sentinel-2 satellite to effectively delineate management zones (MZs) within the study area. The study is located in the east of the Bitter Lake of the Suez Canal and inside the Sinai Peninsula, Egypt. For this study, four spectral SIs were utilized using blue, green, red, and NIR bands in different combinations. Through ground truthing, the relationship between soil EC and SIs is best described by exponential regression. Their study found that the Salinity Index 2 (SI2) and Salinity Index 4 (SI4) performed better than other SIs.

Gojiya et al. (2023) demonstrated the significance of monitoring and assessing soil salinity, especially in regions where traditional irrigation practices and inadequate drainage systems worsen the problem. Furthermore, they highlighted the importance of satellite-based technologies for the spatial and temporal mapping of soil salinity, as these technologies provided cost-effective, rapid, and efficient sources of both qualitative and quantitative spatial information. The results indicated that no single method was consistently the best for soil salinity mapping. The effectiveness of an approach depended on the study area's characteristics and data availability. Vegetation indices performed well in densely vegetated regions, while soil SIs were more suitable for bare or sparsely vegetated soils. Multispectral sensors, particularly Landsat, Sentinel, MODIS, IRS, and ASTER, have proven valuable for salinity detection and monitoring, with Landsat being the most widely used due to its extensive temporal coverage and global accessibility. Metternicht and Zinck (2003) reviewed different sensors, such as aerial photos, satellite and airborne multispectral sensors, microwave sensors, video imagery, airborne geophysics, hyperspectral sensors, and electromagnetic induction meters, for identifying and mapping salt-affected areas. They highlighted challenges in using remote sensing for this purpose, including variations in salt reflectance, uneven salt distribution, seasonal salinity changes, interference from vegetation, and difficulty distinguishing salts from other surfaces.

Das et al. (2023) aimed to predict soil salinity in five sites across India using data from the Airborne Visible–Infrared Imaging Spectrometer—Next Generation (AVIRIS-NG). The study focused on low to moderately salt-affected cropland soils. Soil EC was estimated using machine learning models, including RF, gradient boosting machines (GBM), and deep learning (DL). Consequently, a new hyperspectral soil SI, based on Shannon entropy and selected absorption features, was introduced. This index performed better than commonly used remote sensing-based SIs. It showed a strong correlation with measured EC (r = 0.68) and machine learning-predicted EC (r = 0.78), both statistically significant at the 1% level.

Sapkota et al. (2024) aimed to evaluate the effectiveness of PlanetScope and UAV-based NDVI imagery in identifying alfalfa MZs in areas with varying rootzone soil salinity. The study took place in an alfalfa field in Imperial Valley, California, where rootzone soil salinity was measured using electromagnetic induction (EMI) technology and deep soil sampling. MZs were created based on an ECe map derived from apparent electrical conductivity (ECa) data. NDVI images from PlanetScope and a UAV were used to define remote sensing-based MZs. Soil samples collected at different depths were analyzed in the lab for physicochemical properties. Soil salinity ranged from 2.2 to 13.4 dS m^{-1}, with moderate variability (CV = 37.7%). ECe-based MZs accounted for about 83% of the field's variability and showed clear differences in soil properties such as ECa, ECe, water content, and concentrations of Mg^{2+}, B, Ca^{2+}, Na^{+}, and Cl^{-}.

Wang et al. (2024) found that global soil salinity assessments are limited by the coarse spatial resolution of available datasets (>250 m) and semiqualitative classification methods (five ranks). They proposed a framework to estimate global soil salt content at a finer resolution of 10 m across five climate regions to overcome these limitations. This framework integrates Sentinel-1/2 remote sensing images, climate data, parent material, terrain data, and machine learning techniques. In hyper-arid and arid regions, models were developed using Sentinel-2 and other geospatial data. In semi-arid, dry sub-humid, and humid regions, models performed best using a combination of Sentinel-1, Sentinel-2, and other geospatial data. The Global Aridity Index and Potential Evapotranspiration (AI-ET) Climate Database was used to analyze global climate conditions at a 1 km scale. Climate classification was based on the annual average aridity index (AI; precipitation/ET) from 1970 to 2000, following the United Nations Environment Programme (UNEP) guidelines.

Deshpande et al. (2025) conducted research in the Balod district of Chhattisgarh, India. The objectives of this research were to estimate soil salinity using existing approaches and to develop a new soil SI for the study area. We collected two types of data: (1) remotely sensed data and (2) in situ EC data. The remotely sensed data included USGS Landsat 9 Level 2, Collection 2, Tier 1 images for the entire study site. For in situ EC data, 69 samples were collected. Both datasets were gathered during April–May 2024. It is observed that existing spectral indices had a weak correlation with field EC values. Therefore, a new index was developed by combining NIR, red, and SWIR-1 surface reflectance bands using linear regression analysis. Ahmad et al. (2017) carried out a study on 1 February 1997 and 7 February 2012 to map change detection in sodic lands in Raebareli district, India. Three spectral indices were used to identify soil salinity: the Soil Brightness Index (SBI), the NDVI, and the SAVI using IRS-IC LISS III and IRS-P6 LISS III satellites. It is observed that in 1997, sodic land covered 45,286.19 ha (10.14% of the total area), but by 2012, it had decreased by 12,650.56 ha, reducing the sodic land area to 32,635.63 ha (7.31% of the total area). This suggests successful land reclamation or management efforts over time.

7.5 SALINITY ANALYSIS OF MATHURA REGION: A CASE STUDY

Mathura, located in Uttar Pradesh, is situated on the banks of the Yamuna River at 27°41 ′ North latitude and 77°41′ East longitude, with the Yamuna sub-basin serving as the single drainage area, as shown in Figure 7.1. This region features two distinct climatic zones: sub-humid upstream of Mathura and semi-arid downstream. While both surface and groundwater are used for irrigation upstream, groundwater is predominantly utilized downstream, resulting in higher soil and groundwater salinity in the downstream areas. Mathura's geographical area spans 3.32 million ha, with 3.28 million ha cultivated and 3.11 million ha irrigated. The soil types are predominantly alluvial, sandy, and clay-loam, supporting major crops such as paddy, bajra, arhar, maize, cotton, jowar, wheat, barley, and mustard. The semi-arid climate experiences a rainy season from July to September with an average rainfall of approximately 593 mm, and temperatures ranging from 22°C to 45°C in summer and 14°C to 32°C in winter.

Mathura is important for religion and history, as it brings a lot of tourists and helps the economy, though increasing tourism leads to greater wastewater discharge, which contributes to salinity. The district's population, 25.47 lakhs in 2011, was projected to rise to 27.51 lakhs by 2015–16 and 30.29 lakhs by 2020 (as per census data assumptions), and this growth has increased demand for industries and agriculture. Agriculture, including dairy and horticulture, forms a significant part of the economy, but new farming techniques and expanded use of pesticides and fertilizers substantially impact groundwater quality. Excessive use of fertilizers and irrigation has increased soil salinity in the region. Furthermore, industrial activities significantly contribute to salinity; refineries release sulfur dioxide and nitrogen dioxide, which form acids that contribute to soil and water acidification and increase dissolved salt concentrations. Similarly, wastewater from textile industries, particularly those using dyes and chemicals, contains various salts and chemicals that can seep into the soil, leading to groundwater contamination and elevated salt concentrations.

For this study, Landsat-8 images from the OLI/TIRS C2 L2 sensor, with a 5% cloud mask applied to ensure atmospheric correction before processing in GEE, for 2015 and 2020, were downloaded from the Earth Engine Data Catalogue and processed. The Mathura area was then demarcated, and the SSI6 and SSI7 indices were applied based on the 17 bands used during processing. Subsequently, a supervised learning approach was employed by adding training points to generate soil salinity maps for 2015 and 2020. The resulting salinity maps were then imported into QGIS for further analysis. In QGIS, the percentage of salinity change between 2015 and 2020 was displayed on a map. Salinity levels in Mathura were categorized into three classes: low, moderate, and high. For quantitative analysis, these categories were assigned numerical values of 0, 1, and 2, respectively. Finally, the processed salinity map from QGIS was then extracted. A flowchart depicting the process used in the salinity assessment is shown in Figure 7.2. Also, the average annual rainfall for the study period, that is, from 2015 to 2020, has been analyzed for the Mathura region and is presented in Table 7.4.

From Figure7.3, it is observed that the percentage change in salinity from 2015 to 2020, as measured by SSI7, is −8.64%, +22.61%, and −13.95% corresponding to

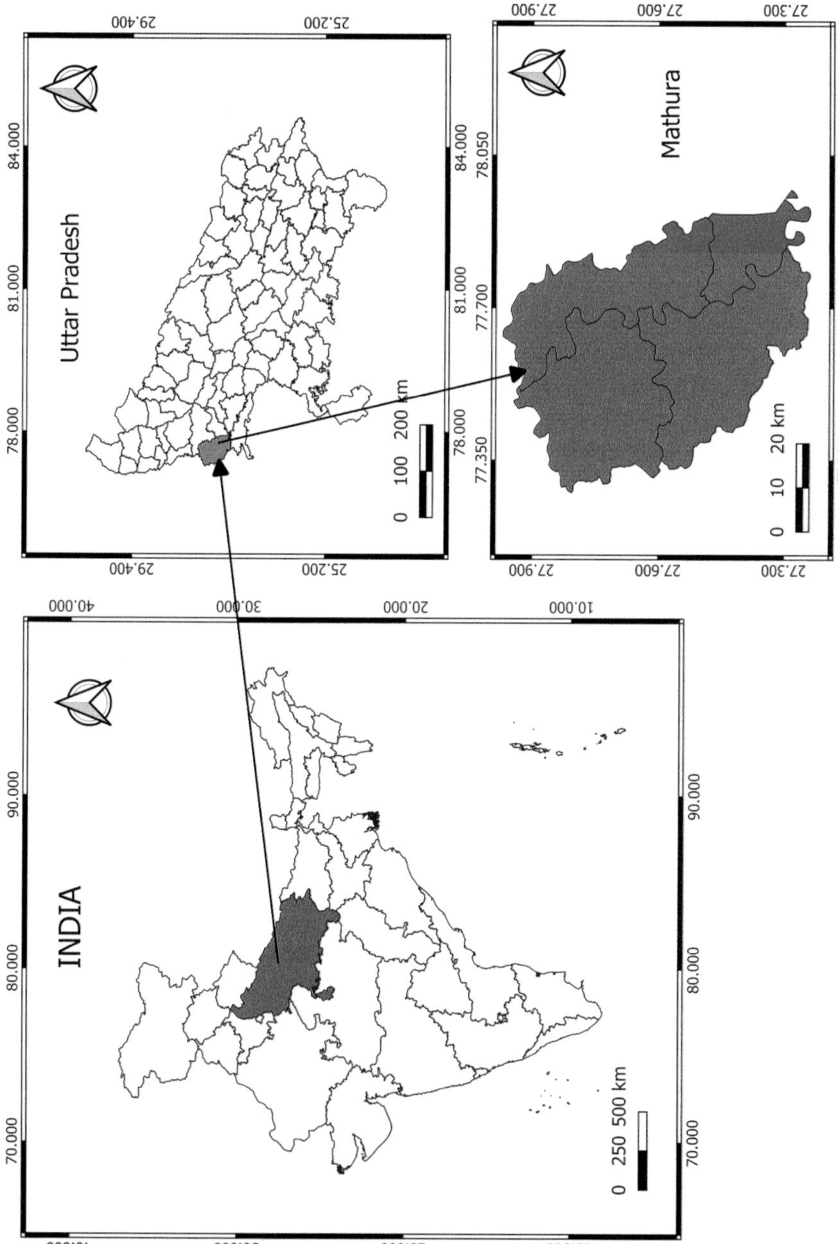

FIGURE 7.1 Map of Mathura district of Uttar Pradesh.

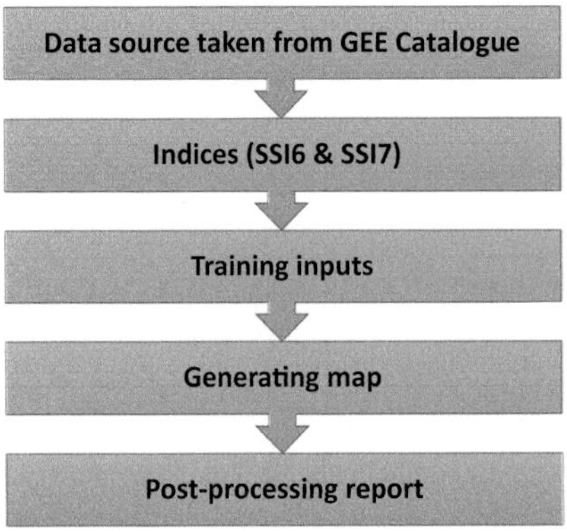

FIGURE 7.2 Process of data acquisition and processing for salinity assessment.

TABLE 7.4
Average Rainfall Data in Mathura from 2015 to 2020

Sr. No.	Year	Average Rainfall (mm)
1	2015	336.4
2	2016	315.8
3	2017	263.8
4	2018	803.5
5	2019	439.3
6	2020	424.8

Source: Uttar Pradesh Climate Change Authority (UPCCA), Department of Environment, Forest and Climate Change, Government of Uttar Pradesh.

low, medium, and high categories (Figure 7.4). Similarly, for the same period, salinity measured by SSI6 is +6.61%, +5.06%, and −11.67%, depicted by Figure 7.5. These results, shown in Figure 7.3, highlight that high salinity levels decreased in 2020, while moderate salinity levels increased. This change is likely due to natural leaching, driven by the very high average rainfall in 2018, and continued higher average rainfall after 2018 compared to 2015, 2016, and 2017. This is shown in Figure 7.6 of average rainfall variation by year.

7.6 CONCLUSION

This study demonstrates that remote sensing technologies enable rapid and efficient detection of soil salinity. Landsat and Sentinel are most commonly utilized among the various satellites due to their high-resolution multispectral data, which is highly effective in identifying salinity patterns. In addition to remote sensing, laboratory-based soil testing remains crucial for validating satellite data and ensuring accurate monitoring of soil salinity. In India, regions such as Gujarat, Uttar Pradesh,

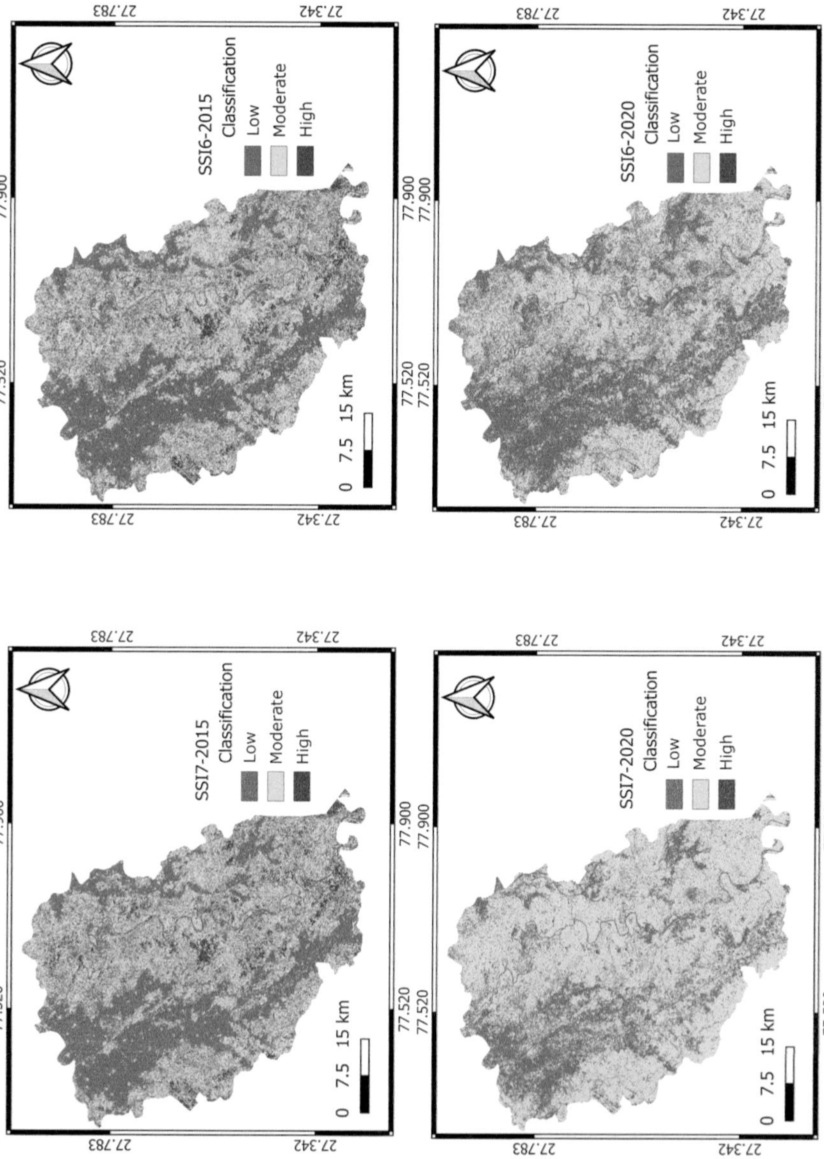

FIGURE 7.3 Shift in salinity from 2015 to 2020.

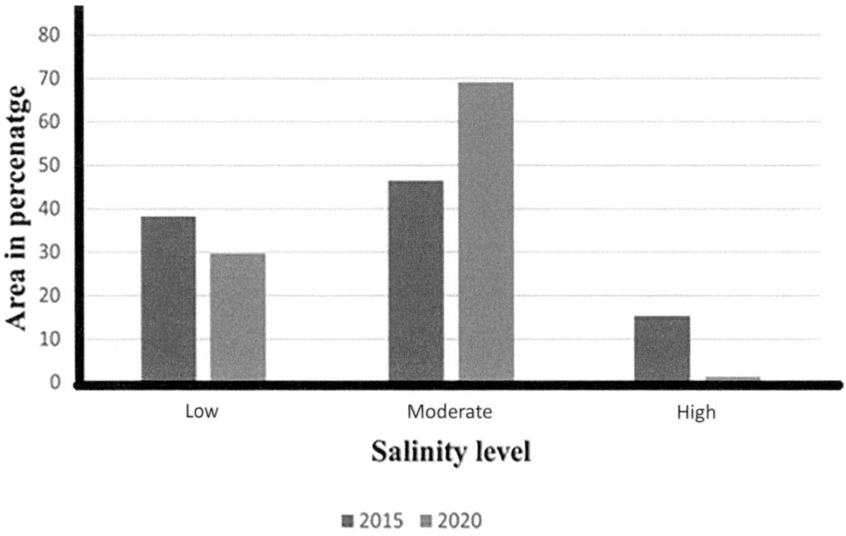

FIGURE 7.4 Changing trend of salinity from 2015 to 2020 using SSI6.

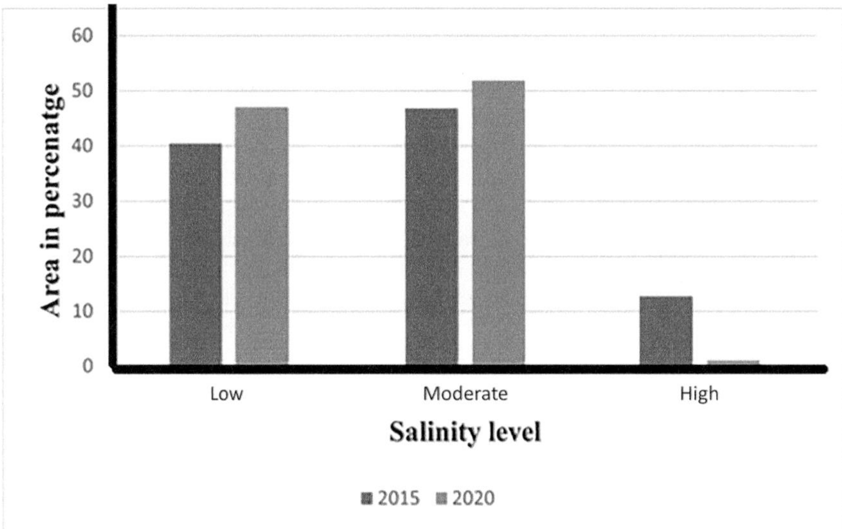

FIGURE 7.5 Changing trend of salinity from 2015 to 2020 using SSI7.

Maharashtra, West Bengal, Rajasthan, and Haryana are notably affected by salinization. Globally, countries like China, Bangladesh, Iran, Egypt, and Australia also experience significant soil salinity issues. A case study of Mathura found that the highly saline categorized area had decreased in salinity for a given period. During this period, the maximum average rainfall in 2018 was found to have leached out the

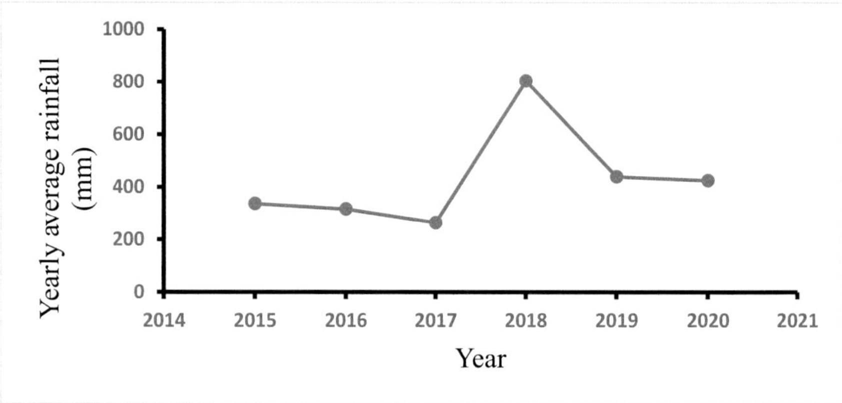

FIGURE 7.6 Average rainfall variation by year.

salt through a natural leaching process. After 2018, average rainfall did not significantly increase, but compared to the years prior to 2018, average rainfall considerably increased. However, there was an increase in salinity in the moderately saline-categorized area. The moderately saline-categorized area needs to be considered to reduce salinity there. There is a need to control the increasing population, intensive agricultural practices, exponential industrial and tourism growth, and urbanization, which are the main reasons for increasing salinity in Mathura.

REFERENCES

Abdelfattah, M. A., Shahid, S. A., and Othman, Y. R. (2009). Soil salinity mapping model developed using RS and GIS—a case study from Abu Dhabi, United Arab Emirates. *European Journal of Scientific Research*, 26(3), 342–351.

Ahmad, A., Upadhyay, R. K., Lal, B., and Singh, D. (2017). Change Detection of sodic land in Raebareli District using remote sensing and GIS techniques. In *Environmental Pollution: Select Proceedings of ICWEES-2016* (pp. 487–498). Singapore: Springer Singapore.

Ahmad, S., Ahmad, A., Ali, H., Hussain, A., Garcia y Garcia, A., Khan, M. A., Zia-Ul-Haq, M., Hasanuzzaman, M. and Hoogenboom, G. (2013). Application of the CSM-CERES-Rice model for evaluation of plant density and irrigation management of transplanted rice for an irrigated semiarid environment. *Irrigation Science*, 31(3), 491–506.

Ali, A. M. (2023). Detection of the spatial variability of soil salinity east of the Bitter Lake of the Suez Canal, Egypt, as a basis for management zones using Sentinel-2 multispectral imagery. *Communications in Soil Science and Plant Analysis*, 54(14), 1989–2001.

Allbed, A., and Kumar, L. (2013). Soil salinity mapping and monitoring in arid and semi-arid regions using remote sensing technology: A review. *Advances in Remote Sensing*, 2(4), 373–385.

Asfaw, E., Suryabhagavan, K. V., and Argaw, M. (2018). Soil salinity modeling and mapping using remote sensing and GIS: The case of Wonji sugar cane irrigation farm, Ethiopia. *Journal of the Saudi Society of Agricultural Sciences*, 17(3), 250–258.

Barreto, A. C., Neto, M. F., de Oliveira, R. P., Moreira, L. C. J., de Medeiros, J. F., and da Silva Sá, F. V. (2023). Comparative analysis of spectral indexes for soil salinity mapping in irrigated areas in a semi-arid region, Brazil. *Journal of Arid Environments*, 209, 104888.

Bouaziz, M., Matschullat, J., and Gloaguen, R. (2011). Improved remote sensing detection of soil salinity from a semi-arid climate in Northeast Brazil. Comptes Rendus. *Géoscience*, 343(11–12), 795–803.

Canfora, L., Bacci, G., Pinzari, F., Lo Papa, G., Dazzi, C., and Benedetti, A. (2014). Salinity and bacterial diversity: To what extent does the concentration of salt affect the bacterial community in a saline soil? *PLoS One*, 9(9), e106662.

Carter, D. L. (1975). Problems of salinity in agriculture. In *Plants in Saline Environments* (pp. 25–35). Berlin, Heidelberg: Springer Berlin Heidelberg.

Chhabra, R. (2004). Classification of salt-affected soils. *Arid Land Research and Management*, 19(1), 61–79.

Chi, C. M., Zhao, C. W., Sun, X., and Wang, Z. C. (2012). Reclamation of saline-sodic soil properties and improvement of rice (Oriza sativa L.) growth and yield using desulfurized gypsum in the west of Songnen Plain, northeast China. *Geoderma*, 187, 24–30.

Cui, X., Han, W., Zhang, H., Dong, Y., Ma, W., Zhai, X., ... and Li, G. (2023). Estimating and mapping the dynamics of soil salinity under different crop types using Sentinel-2 satellite imagery. *Geoderma*, 440, 116738.

Das, A., Bhattacharya, B. K., Setia, R., Jayasree, G., and Das, B. S. (2023). A novel method for detecting soil salinity using AVIRIS-NG imaging spectroscopy and ensemble machine learning. *ISPRS Journal of Photogrammetry and Remote Sensing*, 200, 191–212.

Datta, A., Sethi, M., Basak, N., Yadav, A. K., Chinchmalatpure, A. R., Khurana, M. L., and Barman, A. (2016). Salt-affected soils of Rewari District, Haryana: Distribution and characteristics. *Journal of Soil Salinity and Water Quality*, 8(2), 144–152.

Deshpande, V. P., Ahmad, I., and Singh, C. K. (2025). Development of novel soil salinity spectral index using remotely sensed data: A case study on Balod district, Chhattisgarh, India. *Journal of Landscape Ecology*, 18(2), 62–81.

Didi, S., Housni, F. E., Bracamontes del Toro, H., and Najine, A. (2019). Mapping of soil salinity using the Landsat 8 image and direct field measurements: A case study of the tadla plain, Morocco. *Journal of the Indian Society of Remote Sensing*, 47, 1235–1243.

FAO. 2024. Global status of salt-affected soils – Main report. Food and Agriculture Organization of the United Nations, Rome.

Gojiya, K. M., Rank, H. D., Chauhan, P. M., Patel, D. V., Satasiya, R. M., and Prajapati, G. V. (2023). Remote sensing and GIS applications in soil salinity analysis: A comprehensive review. *International Journal of Environment and Climate Change*, 13, 2149–2161.

Golestani, M., Ghahfarokhi, Z. M., Esfandiarpour-Boroujeni, I., and Shirani, H. (2023). Evaluating the spatiotemporal variations of soil salinity in Sirjan Playa, Iran using Sentinel-2A and Landsat-8 OLI imagery. *Catena*, 231, 107375.

Gorji, T., Sertel, E., and Tanik, A. (2017). Monitoring soil salinity via remote sensing technology under data scarce conditions: A case study from Turkey. *Ecological indicators*, 74, 384–391.

Hoa, P. V., Giang, N. V., Binh, N. A., Hai, L. V. H., Pham, T. D., Hasanlou, M., and Tien Bui, D. (2019). Soil salinity mapping using SAR sentinel-1 data and advanced machine learning algorithms: A case study at Ben Tre Province of the Mekong River Delta (Vietnam). *Remote Sensing*, 11(2), 128.

Hossen, B., Yabar, H., and Faruque, M. J. (2022). Exploring the potential of soil salinity assessment through remote sensing and GIS: Case study in the coastal rural areas of Bangladesh. *Land*, 11(10), 1784.

Ijaz, M., Ahmad, H. R., Bibi, S., Ayub, M. A., and Khalid, S. (2020). Soil salinity detection and monitoring using Landsat data: A case study from Kot Addu, Pakistan. *Arabian Journal of Geosciences*, 13, 1–9.

Jamali, A. A., Naeeni, M. A. M., and Zarei, G. (2020). Assessing the expansion of saline lands through vegetation and wetland loss using remote sensing and GIS. *Remote Sensing Applications: Society and Environment*, 20, 100428.

Khadim, F. K., Su, H., Xu, L., and Tian, J. (2019). Soil salinity mapping in Everglades National Park using remote sensing techniques and vegetation salt tolerance. *Physics and Chemistry of the Earth, Parts A/B/C*, 110, 31–50.

Khoshgoftarmanesh, A. H., Shariatmadari, H., and Vakil, R. (2003). Reclamation of saline soils by leaching and barley production. *Communications in Soil Science and Plant Analysis*, 34(19-20), 2875–2883.

Kulmatov, R., Khasanov, S., Odilov, S., and Li, F. (2021). Assessment of the space-time dynamics of soil salinity in irrigated areas under climate change: A case study in Sirdarya province, Uzbekistan. *Water, Air, & Soil Pollution*, 232, 1–13.

Kumar, P., and Sharma, P. K. (2020). Soil salinity and food security in India. *Frontiers in Sustainable Food Systems*, 4, 533781.

Li, J., Wang, C., Zhang, J. H., Cai, J. M., Cao, Y. P., and Sun, X. J. (2010). Hydrogen-rich saline improves memory function in a rat model of amyloid-beta-induced Alzheimer's disease by reduction of oxidative stress. *Brain Research*, 1328, 152–161.

Li, Y., Chang, C., Wang, Z., and Zhao, G. (2022). Remote sensing prediction and characteristic analysis of cultivated land salinization in different seasons and multiple soil layers in the coastal area. *International Journal of Applied Earth Observation and Geoinformation*, 111, 102838.

Mace, J. E., Amrhein, C., and Oster, J. D. (1999). Comparison of gypsum and sulfuric acid for sodic soil reclamation. *Arid Soil Research and Rehabilitation*, 13(2), 171–188.

Metternicht, G. I., and Zinck, J. A. (2003). Remote sensing of soil salinity: Potentials and constraints. *Remote Sensing of Environment*, 85(1), 1–20.

Minhas, P. S., and Sharma, O. P. (2003). Management of soil salinity and alkalinity problems in India. *Journal of Crop Production*, 7(1-2), 181–230.

Mirzaee, S., Nafchi, A. M., Ostovari, Y., Seifi, M., Ghorbani-Dashtaki, S., Khodaverdiloo, H., ... and Raei, B. (2024). Monitoring and assessment of spatiotemporal soil salinization in the Lake Urmia region. *Environmental Monitoring and Assessment*, 196(10), 958.

Mishra, V. K., Jha, S. K., Arora, S., Singh, Y. P., Damodaran, T., Verma, C. L., ... and Mishra, A. K. (2020). Sodic land reclamation for livelihood security in Central Indo-Gangetic Plains of Uttar Pradesh. *Journal of Soil and Water Conservation*, 19(2), 135–141.

Mohanavelu, A., Naganna, S. R., and Al-Ansari, N. (2021). Irrigation induced salinity and sodicity hazards on soil and groundwater: An overview of its causes, impacts and mitigation strategies. *Agriculture*, 11(10), 983.

Morshed, M. M., Islam, M. T., and Jamil, R. (2016). Soil salinity detection from satellite image analysis: An integrated approach of salinity indices and field data. *Environmental Monitoring and Assessment*, 188, 1–10.

Naimi, S., Ayoubi, S., Zeraatpisheh, M., and Dematte, J. A. M. (2021). Ground observations and environmental covariates integration for mapping of soil salinity: A machine learning-based approach. *Remote Sensing*, 13(23), 4825.

Narjary, B., Kumar, S., Meena, M., Kamra, S. K., and Sharma, D. K. (2022). Spatio-temporal mapping and analysis of soil salinity: An integrated approach through Electromagnetic induction (EMI), multivariate and geostatistical techniques. *Geocarto International*, 37(25), 8602–8623.

Negacz, K., Malek, Ž., de Vos, A., and Vellinga, P. (2022). Saline soils worldwide: Identifying the most promising areas for saline agriculture. *Journal of Arid Environments*, 203, 104775.

Otlewska, A., Migliore, M., Dybka-Stępień, K., Manfredini, A., Struszczyk-Świta, K., Napoli, R., ... and Pinzari, F. (2020). When salt meddles between plant, soil, and microorganisms. *Frontiers in Plant Science*, 11, 553087.

Peng, J., Biswas, A., Jiang, Q., Zhao, R., Hu, J., Hu, B., and Shi, Z. (2019). Estimating soil salinity from remote sensing and terrain data in southern Xinjiang Province, China. *Geoderma*, 337, 1309–1319.

Rajitha, K., Mukherjee, C. K., and Chandran, R. V. (2007). Applications of remote sensing and GIS for sustainable management of shrimp culture in India. *Aquacultural Engineering*, 36(1), 1–17.

Ramos, T. B., Castanheira, N., Oliveira, A. R., Paz, A. M., Darouich, H., Simionesei, L., ... and Gonçalves, M. C. (2020). Soil salinity assessment using vegetation indices derived from Sentinel-2 multispectral data. Application to Lezíria Grande, Portugal. *Agricultural Water Management*, 241, 106387.

Rengasamy, P. (2016). Soil salinization. In *Oxford Research Encyclopedia of Environmental Science*. Oxford: Oxford University Press, 2016.

Roohi, M., Faeli, M., Irani, M., and Shamsaei, E. (2021). Calculation of land subsidence and changes in soil moisture and salinity using remote sensing techniques. *Environmental Earth Sciences*, 80(12), 423.

Safdar, H., Amin, A., Shafiq, Y., Ali, A., Yasin, R., Shoukat, A., ... and Sarwar, M. I. (2019). A review: Impact of salinity on plant growth. *Natural Sciences*, 17(1), 34–40.

Sapkota, A., Verdi, A., Scudiero, E., and Montazar, A. (2024). Assessing the effectiveness of satellite and UAV-based remote sensing for delineating alfalfa management zones under heterogeneous rootzone soil salinity. *Smart Agricultural Technology*, 9, 100583.

Satir, O., and Berberoglu, S. (2016). Crop yield prediction under soil salinity using satellite derived vegetation indices. *Field Crops Research*, 192, 134–143.

Sethi, M., Dasog, G. S., Van Lieshout, A., and Salimath, S. B. (2006). Salinity appraisal using IRS images in Shorapur Taluka, upper Krishna irrigation project, phase I, Gulbarga District, Karnataka, India. *International Journal of Remote Sensing*, 27(14), 2917–2926.

Shankar, V., and Evelin, H. (2019). Strategies for reclamation of saline soils. In *Microorganisms in Saline Environments: Strategies and Functions* (pp. 439–449). Cham: Springer International Publishing.

Tang, S., She, D., and Wang, H. (2020). Effect of salinity on soil structure and soil hydraulic characteristics. *Canadian Journal of Soil Science*, 101(1), 62–73.

Tola, D., Satgé, F., Pillco Zolá, R., Sainz, H., Condori, B., Miranda, R., ... and Espinoza-Villar, R. (2024). Soil salinity mapping of plowed agriculture lands combining Radar Sentinel-1 and Optical Sentinel-2 with topographic data in machine learning models. *Remote Sensing*, 16(18), 3456.

Wang, N., Chen, S., Huang, J., Frappart, F., Taghizadeh, R., Zhang, X., ... and Shi, Z. (2024). Global soil salinity estimation at 10 m using multi-source remote sensing. *Journal of Remote Sensing*, 4, 0130.

Wang, N., Peng, J., Chen, S., Huang, J., Li, H., Biswas, A., ... and Shi, Z. (2022). Improving remote sensing of salinity on topsoil with crop residues using novel indices of optical and microwave bands. *Geoderma*, 422, 115935.

Whitney, K., Scudiero, E., El-Askary, H. M., Skaggs, T. H., Allali, M., and Corwin, D. L. (2018). Validating the use of MODIS time series for salinity assessment over agricultural soils in California, USA. *Ecological Indicators*, 93, 889–898.

Wu, J., Vincent, B., Yang, J., Bouarfa, S., and Vidal, A. (2008). Remote sensing monitoring of changes in soil salinity: A case study in Inner Mongolia, China. *Sensors*, 8(11), 7035–7049.

Wu, W., Mhaimeed, A. S., Al-Shafie, W. M., Ziadat, F., Dhehibi, B., Nangia, V., and De Pauw, E. (2014). Mapping soil salinity changes using remote sensing in Central Iraq. *Geoderma Regional*, 2, 21–31.

Zhang, W., Zhang, W., Liu, Y., Zhang, J., Yang, L., Wang, Z., ... and Yin, Z. (2022). The role of soil salinization in shaping the spatio-temporal patterns of soil organic carbon stock. *Remote Sensing*, 14(13), 3204.

Zhou, D., Xu, J., Wang, L., Lin, Z., and Liu, L. (2015). Identifying and managing risk factors for salt-affected soils: A case study in a semi-arid region in China. *Environmental Monitoring and Assessment*, 187, 1–16.

8 Dynamic Assessment of Soil Erosion Risk Using Hydrological Model and Geospatial Approach

A Case Study from Chamoli, Uttarakhand

Sandeep Kunwar, Harish Khali, Akash Deep, Nitin Mishra, and Amit Kumar Sharma

8.1 INTRODUCTION

Soil degradation has had a considerable influence on natural resources and agricultural productivity around the world in recent decades (Koirala et al., 2019). In mountains, it places considerable concerns, including strong rainfall and surface water flow on exposed terrain, which contributes to land degradation. It is a natural phenomenon that results in the gradual degradation of the topsoil due to the erosive forces of wind and water (Foster et al., 2003). According to Thompson et al. (2001), soil erosion is caused by the depletion of the rich soil layer and the accumulation of sediments. It was noticed that an estimated 75 billion tonnes of arable land are predicted to disappear each year worldwide. According to Gobinath et al. (2022), Asia, Africa, and South America frequently experience 28–40 tonnes per hectare rate of soil erosion annually (Ferreira et al., 2018). They elaborated in their article that topsoil erosion brought on by water runoff is the most common kind of soil deterioration. Across many countries, the topsoil loss due to erosion severely reduces agricultural output. The uppermost layer of soil experiences notable productivity decline, particularly in the most fertile regions prone to erosion. Recent studies indicate significant disruptions in the movement and settling of soil due to changes in nutrient and carbon cycling. Soil erosion can cause a loss of 75–85% of the soil's carbon content, which releases carbon into the atmosphere (Conoscenti et al., 2014). By understanding the Upper Alaknanda watersheds, their morphological and topographical characters can help users to enable sustainable land use to reduce the erosion. Soil properties in terms of their physiochemical signatures can be used for effective planning in the

DOI: 10.1201/9781003596080-8

watershed management. Soil erosion is believed to be influenced by various factors, including changes in ground cover, soil composition, infrastructure, topographical gradients, mapping units, and agricultural techniques (Jalayer et al., 2022). Globally, Revised Universal Soil Loss Equation (RUSLE) is the most widely used method for estimating long-term rates of inter-rill and rill erosion in field- or farm-sized units (Goudie, 2023). Sediment deposition on rivers also damages reservoirs and dams, increasing maintenance costs and eventually rendering them useless (Chalise et al., 2019). Numerous studies have been conducted to better understand this condition, and their conclusions have helped to regulate soil erosion and encourage ecological restoration (Šurda et al., 2007). Even though erosion modeling is crucial for inaccessible mountainous areas, the mountain region remains highly susceptible to resource degradation due to landslides, soil erosion on steep slopes, and deforestation. Deforestation, decimated local vegetation, and access to overgrazing are all believed to stem from soil erosion in hilly regions, which is attributed to the depletion of topsoil and alterations in land structure (Foster et al., 2003). It is not feasible to extrapolate linear soil loss due to the dynamic landscape system, encompassing soil composition, vegetation patterns, field surveys, geomorphological features, land usage, and slope variations. Soil erosion processes give rise to distinctive soil features such as rills, gullies, and internal landscape erosion. These features are shaped by both the inherent characteristics and structures of the landscape, which act as keys and disruptors in the erosion process. The topographic elevation profile serves as a vital component for soil erosion modeling (SEM), derived through the radar-based observation and optical measurements (remote sensing data). Remote sensing (RS) plays a pivotal role in furnishing precise and timely ground and soil information. Consequently, consistent access to standardized data over large areas can significantly improve regional erosion assessment. Mapping land use and land cover (LULC) more accurately is made possible by the use of multi-temporal satellite data, which allows us to record recurrent fluctuations and provide important information on seasonal variations in land use (Bharadwaj et al., 2022). Spatially derived soil categorization includes the combined effect of soil characteristics, lithology, topography, vegetation, and climate. These are useful resources for evaluating changes in soil erodibility, which is a crucial component of erosion modeling (Weng et al., 2024). The primary objective of this study was to assess soil erosion through the utilization of RS and geographic information system (GIS) methodologies, while soil mapping predominantly relies on optical satellite imagery to delineate soil patterns. The present study examined soil erosion patterns in the Chamoli District using the RUSLE model coupled with GIS. The framework offers guidance to decision-makers in planning erosion control activities, bridging the gap in information within a Chamoli District.

8.2 STUDY AREA

The study region (Chamoli District) is located in the north of the state of Uttarakhand approximately between 30.3° to 31.0° North and 79.2° to 80.2° East (Figure 8.1). The region has a rich network of streams/rivers, with notable ones including the Alaknanda, Dhauliganga, and Nandakini rivers, among others. Geologically, the major proportion of the region belongs to the lesser and central crystalline. Topographically, the

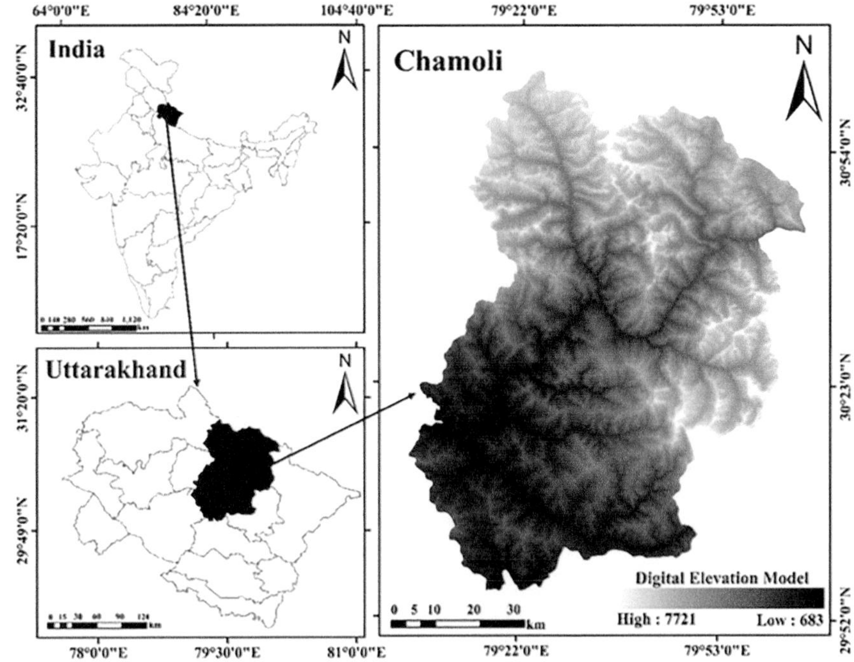

FIGURE 8.1 Study area map.

vertical profile of the area ranges between 683 and 7721 msl. The district experiences a diverse range of rainfall patterns, influenced by its varied topography (1235 mm yr^{-1}). The rainfall plays a crucial role in shaping the local environment and sustaining agricultural activities and has a direct influence on the soil. In the district, precipitation occurs mainly during the monsoon season, from June to September, at varying intensities. Similarly, the average temperature goes up to 22°C in the district, defining semi-arid to semi-humid conditions. Sometimes regional factors play an important role in uneven precipitation and trigger unexpected events like landslides, soil erosion, subsidence, etc. In Chamoli District, soil erosion is a significant concern among the environmentalists due to its mountainous terrain and its proneness to heavy rain and landslides. Additionally, soil erosion in the area has been increased by deforestation, inappropriate land use, and infrastructure expansion, especially during the monsoon season. Khali et al. (2023) contend that soil erosion, deforestation, and terrace farming have made the land more vulnerable in recent years. Anthropogenic activities exacerbate/accelerate the modification of the land's cover and usage and are a key additional element causing the increasing erosion rates here (Šurda et al., 2007; Kumar et al., 2021).

8.3 DATASET

The source of data used for estimating soil erosion has been shown in Table 8.1. As shown in Table 8.1, DEM, LULC, soil, and rainfall data have been collected from

TABLE 8.1
Datasets and Source Information in This Study

S. No.	Dataset	Data Source	Site Description
1	DEM	Shuttle Radar Topographic Mission (SRTM) USGS	https://earthexplorer.usgs.gov/
2	LULC	Environmental Systems Research Institute, Inc. (ESRI)	https://livingatlas.arcgis.com/landcover/
3	Soil Data	National Bureau of Soil Survey and Landuse Planning (NBSS & LUP)	https://icar-nbsslup.org.in
4	Rainfall Data	IMD (India Meteorological Department, India)	https://dsp.imdpune.gov.in/

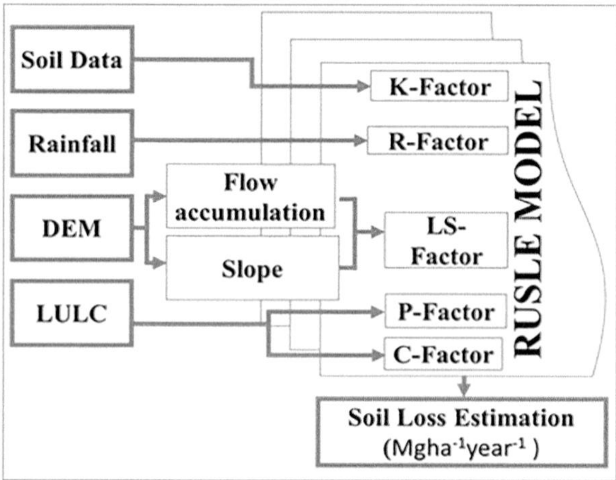

FIGURE 8.2 Methodology used in the study.

Earth Explorer, Living Atlas, ICAR and India Meteorological Department (IMD) Pune, respectively. Details have been provided in Table 8.1.

8.4 METHODOLOGY

In this study, various datasets (soil, rainfall, LULC and slope length) (Table 8.1) were used for RUSLE modeling with GIS integration. It determines how soil deterioration occurs on steep terrain. In essence, Equation 8.1 determines the net annual soil loss in the natural conditions. It aids in determining erosion risk areas and further management strategies for such places. In the last decades, development in mapping approaches and systematic upgradation has made the entire process so suitable for the global scientific community (Khali et al., 2023). With the period, its ability to make more accurate observations has improved with advances in GIS and RS technologies. The framework for calculating the RUSLE model is shown in Figure 8.2 and is represented by Equation 8.1:

$$A = R \times K \times LS \times P \times C. \tag{8.1}$$

Here, '*A*' denotes soil loss (unit in mg ha^{-1} yr^{-1}), '*R*' defines the rainfall erosivity factor (unit in mm ha^{-1} yr^{-1}), '*K*' is the soil erodibility factor (unit in mg ha^{-1} yr^{-1}), '*LS*' is the slope length and slope steepness factor (unit is dimensionless), '*C*' represents the land management component (unit is dimensionless), and '*P*' is the conservation practice factor (dimensionless).

8.5 RESULTS AND DISCUSSION

8.5.1 Model Input Processing and Factor Generation

8.5.1.1 Rainfall Erosivity (*R*) Factor

According to Koirala et al. (2019), it indicates the amount of heavy rainfall and its contribution to soil erosion at a certain location (Equation 8.2):

$$R = 38.5 + 0.35P, \tag{8.2}$$

where *R* is the rainfall erosivity factor (mm ha^{-1} h^{-1} yr^{-1}), and *P* is the mean annual rainfall (mm).

It calculates the effect of rainfall volume and intensity on runoff. IMD rainfall data was used in this study to create an erosion factor based on rainfall. Using integrated calculations to get the *R* factor as suggested by Natarajan (2023), this map displays the average annual precipitation throughout the district.

8.5.1.2 Soil Erodibility (*K*) Factor

It indicates its susceptibility to erosion in typical circumstances, including the ease of sediment displacement and water runoff in relation to rainfall. The traditional measurement is the unit plot, measuring 22.6 m in length with a 9% gradient, which serves as the standard condition. It undergoes continuous flow and is tilled both downhill and uphill (Getu et al., 2022). It derived from soil roughness data, which depicts the rate of soil loss per unit function of rainfall, indicating the erosivity of runoff (*R*). These values are determined according to the classification made by Gobinath et al. (2022). A soil map is presented in Figure 8.3.

It reveals a relatively uniform susceptibility to soil erosion influenced by precipitation and runoff, with values ranging from 0.3 to 0.02. The consistency of the soil is a result of its uniform composition in terms of particle size, where more compaction is associated with higher values. Additionally, soils with greater porosity exhibit enhanced water infiltration, thereby mitigating erodibility. The values for this factor specifically for Chamoli District are shown in Table 8.2.

8.5.1.3 Slope Length Steepness (*LS*) Factor

There is a direct correlation between '*K*' and '*LS*', by comparing the slope with a 9% grade and a length of 22.6 m. An *LS* and steepness (*S*) determine its erosion potential. Using the parameter (*L*), one can measure how much erosion is caused by the steepness of the terrain. To define overland flow, the distance between those points

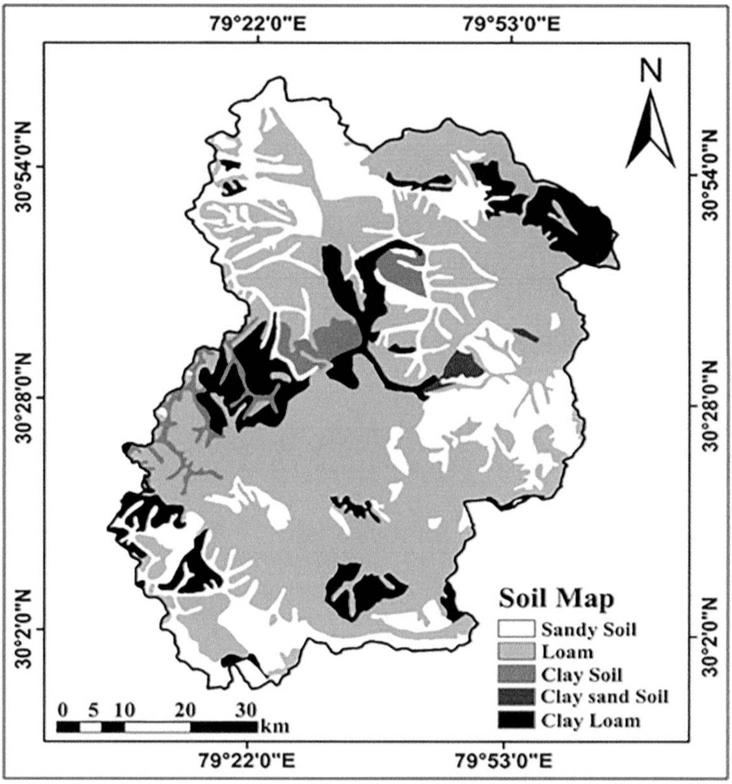

FIGURE 8.3 Soil map of the study area.

TABLE 8.2
Factor K Values According to the Soil Texture

S. No.	Soil Texture	K factor
1	Loam	0.3
2	Clay soil	0.22
3	Sandy soil	0.02
4	Clay sand soil	0.14
5	Clay loam	0.3

must be determined from the place where deposition or runoff enters the channel. As the angle of inclination increases, the propensity of erosion also increases. The slope steepness parameter (*S*) represents how steep the slope is in relation to weathering. Slope steepness is more important than slope length in determining how much soil is carried away. Greater slope steepness results in increased erosion. The slope distribution map of the study area is shown in Figure 8.4.

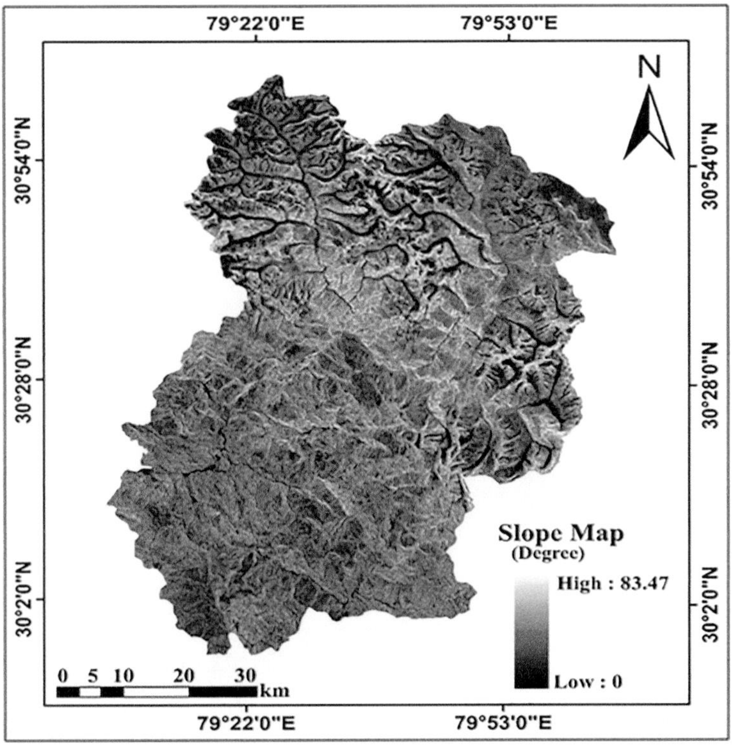

FIGURE 8.4 Slope distribution in the study area.

The equation by Moore and Burch (1986) is expressed as follows:

$$LS = \left(A_s / 22.13\right)^{0.4} \left(s / 0.0896\right)^{1.3} \tag{8.3}$$

where A_s is the area of the catchment, and s is the slope (in degree).

8.5.1.4 Conservation Practice Factor (*P*) Factor

It represents the relative reduction in soil erosion. It is determined by particular land management practices compared to conventional straight-row upslope and downslope tillage. It includes changes in hydrological patterns, precipitation flow concentration, flow velocity, and the hydraulic forces that affect soil owing to runoff (Gobinath et al., 2022). Essentially, the '*P*' factor accounts for control practices that diminish the erosion potential of runoff by influencing factors such as drainage patterns, runoff concentration, runoff velocity, and the hydraulic forces exerted by runoff on the soil. The LULC map of Chamoli District is shown in Figure 8.5.

Table 8.3 presents the land cover type average values for '*P*' parameters. The '*P*' factor is assigned a value between '0' and '1', where '0' indicates effective support practices and a value approaching '1' signifies inadequate support practices.

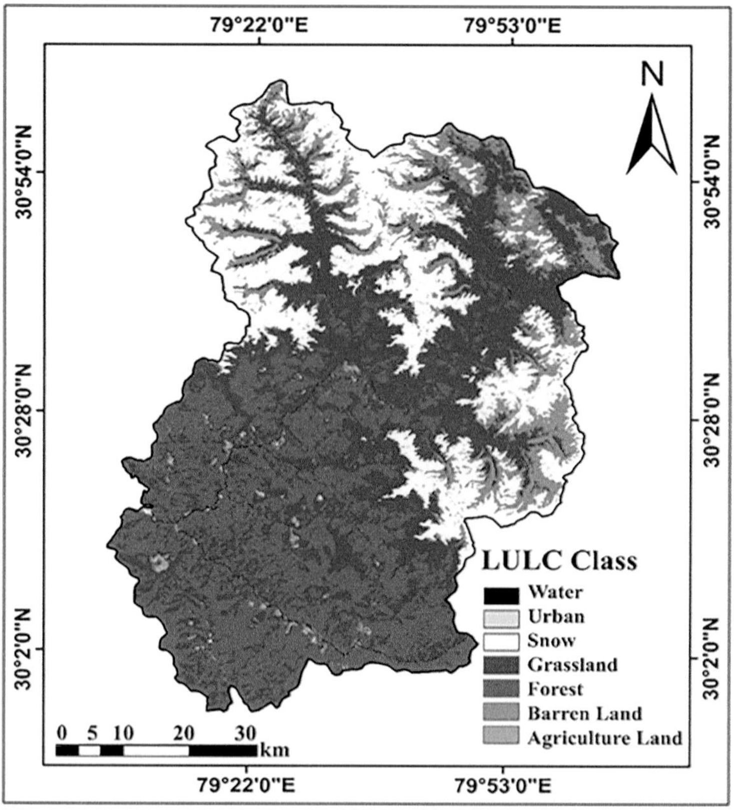

FIGURE 8.5 LULC map of Chamoli District.

TABLE 8.3
Different Land Cover Types and P Factor for Study Area

S. no.	1	2	3	4	5	6	7
LULC classes	Barren land	Snow cover	Grassland	Water bodies	Agricultural land	Built-up	Forest area
P values	1	0	0.5	0	0.5	1	1

Source: USDA Handbook No. 282 (1981).

8.5.1.5 Cover-Management (C) Factor

According to Chalise et al. (2019), the '*C*' component illustrates the influence of various practices, such as cropping, on erosion rates. Due to its careful tracking of plant development and rainfall dynamics, it is the most spatiotemporally sensitive. To evaluate how well crop and soil management strategies mitigate soil loss in comparison, one useful metric is the '*C*' factor. The '*C*' factor sheds light on how

TABLE 8.4
'C' Factor Values for Different Classes

S. No.	LULC Classes	C Values
1	Barren land	1.00
2	Snow cover	0.00
3	Grassland	0.01
4	Water bodies	0.00
5	Agricultural land	0.40
6	Built-up	0.00
7	Forest area	0.03

Source: USDA (1972).

conservation measures affect the average yearly soil loss and its temporal distribution throughout different activities including building, crop rotations, and other management schemes (Joshi et al., 2023). Slope steepness and length variables along with vegetation cover have a major influence on soil loss sensitivity. It is dimensionless (Gobinath et al., 2022) and ranges from '0' to '1'. Seven different land use categories were employed in this study, which are shown in LULC map (Figure 8.5) that was obtained from the Environmental Systems Research Institute, Inc. (ESRI) to prepare a '*C*' factor map (Table 8.4). Each land use has a '*C*' value provided by the reference source: USDA (1972), which ranged from '0' to '1'. A lower '*C*' value denotes no loss, while a larger '*C*' value denotes significant and exposed risks of soil loss.

All the factors for SEM are shown in Figure 8.6.

8.5.1.6 Soil Loss Estimation of Chamoli District

In this study, based on the requirements, various data inputs were prepared and processed to produce factors (*R*, *K*, *LS*, *P*, and *C*) and the final output as a soil erosion map for the study area. Raster mapping and sorting of the dataset was performed in the geospatial environment to build the RUSLE relationship among the factors for the Chamoli District (Table 8.5; Figure 8.7).

Based on the findings, it is observed that soil erosion rates vary across different severity levels (depending upon the topographic slope and elevation range), posing significant concerns for land management and environmental stability.

The data categorizes erosion rates into six groups based on severity. The lowest erosion rates, ranging from 0 to 5 Mg ha^{-1} yr^{-1}, cover the largest area, representing 21.48% of the district. This level is classified as 'Very Low (VL)' severity. Moving up the severity scale, areas experiencing erosion rates between 5 and 10 Mg ha^{-1} yr^{-1} encompass 7.42% of the district, categorized as 'Low (L)'. The severity escalates as erosion rates increase, with 16.64% of the area facing 'Moderate (M)' erosion rates of 10–20 Mg ha^{-1} yr^{-1}, followed by 'High (H)' severity 20–40 Mg ha^{-1} yr^{-1} covering 18.8% of the district. Additionally, 'Very High (VH)' erosion rates of 40–80 Mg ha^{-1} yr^{-1} affect 19.16% of the area, while 'Severe (S)' erosion rates of >80 Mg ha^{-1} yr^{-1} are present in 16.49% of the district. It is noticed that the values of the *R* factor and *K* soil

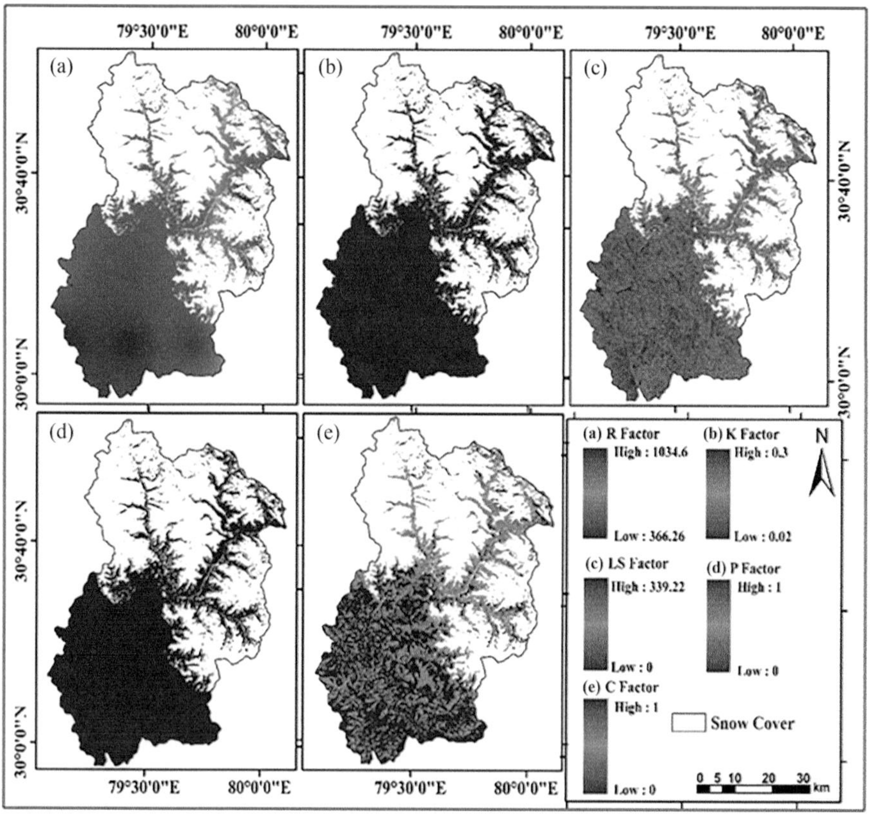

FIGURE 8.6 Factors maps for the SEM : (a) 'R' factor map, (b) 'K' factor map, (c) 'LS' factor, (d) 'P' factor, and (e) 'C' factor.

TABLE 8.5
Soil Loss Class Distribution in Chamoli District

S. No.	Rate of Erosion Mg ha^{-1} yr^{-1}	Area (sq km)	Area (%Distribution)	Severity
1	0–5	1170.39	21.48	Very Low
2	5–10	404.14	7.42	Low
3	10–20	907.18	16.64	Moderate
4	20–40	1024.94	18.8	High
5	40–80	1044.43	19.16	Very High
6	>80	898.71	16.49	Severe

erodibility factor vary from 366.26 to 1034.6 mm ha^{-1} yr^{-1} and 0–0.3 (Figure 8.6a and b, respectively). A K value of 0.3 indicates the surface as a soil is relatively resistant to erosion, whereas higher K values are determined to be less prone to erosion. It reflects the surface properties to make them more stable against the force of water.

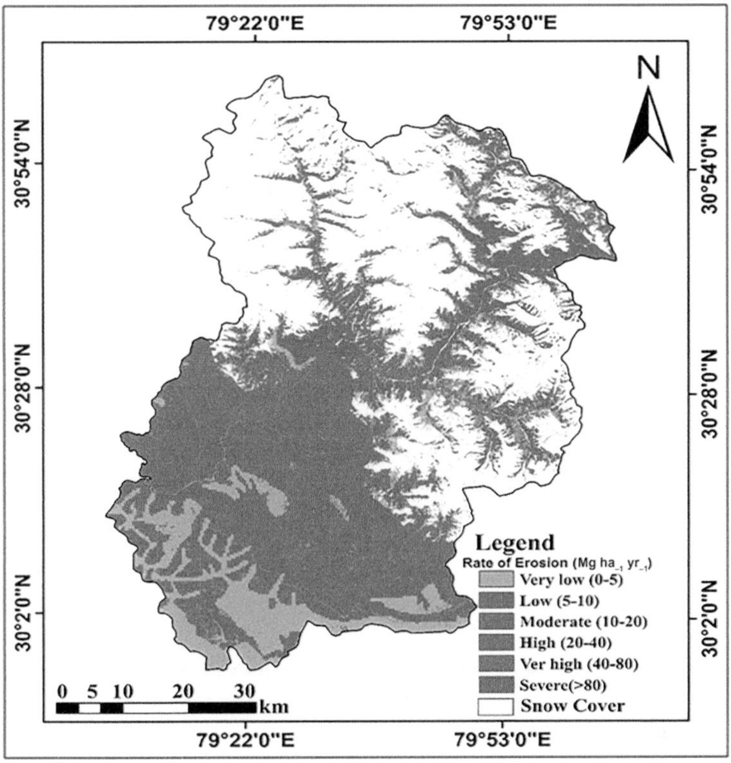

FIGURE 8.7 Soil erosion map of Chamoli District.

The surface soil with lower '*K*' was determined to be susceptible to erosion because of poor structure, lack of organic matter content, and high sand content. According to Figure 8.6c, *LS* factor values vary from 0 to 339.22. It indicates closure of the value to zero, lower erosion potential, and higher ranges (>200), indicating VH erosion potential. It demands necessary, immediate, and comprehensive erosion control strategies. Throughout the region, the *P* factor value varied between '0' and '1' (Figure 8.6d). '0' represents the area where there is no or less erosion control and practices are being implemented. These places experience higher erosion significance, whereas a *P* factor close to '1' denotes that the erosional practices are fully controlled and effectively prevented. As shown in Figure 8.6e, the *C* factor value ranged from '0' to '1'. Taking into account the erosive soils over time, the rate on slopes can be predicted using RUSLE, an *R* factor that forecasts soil loss under similar topography and weather circumstances. Figure 8.7 shows the annual soil erosion map of the study area.

Numerous studies have shown that the Chamoli area is vulnerable to the risks associated with weathering patterns and soil erosion because of a number of important elements, including excess annual precipitation, the texture and steep slopes of the soil/surface, land cover, and the efficacy of soil conservation measures along slopes. The model's strength enables its capacity to forecast soil loss with less information using the availability of spaceborne and GIS data and is extremely helpful in those areas where the information is scarce. The results focusing on ground observation and their accuracy steps closer to challenging in the high hill regions due to variable topography and changing conditioning factors. Consequently, it is impossible to monitor heavy precipitation occurrences with enough accuracy, which greatly affects soil erosion. Based on the spaceborne data and their broad area synopticity these days, it is becoming easy and cost-effective to produce such outputs.

8.6 CONCLUSIONS

Measuring the potential loss of soil in a given region is a necessary step in controlling surface soil erosion. In the present study, total loss was predicted and potential locations were mapped using the adopted (RUSLE) methodology using spaceborne datasets and their modules in GIS environment. It takes many elements such as rainfall, soil type, topography, land use, and land cover for evaluation. For conservation efforts and plant regeneration, this study shows how these methods may be affected in identifying soil erosion-prone regions. Here, significant regional differences in erosion risk were observed and are shown by the soil erosion rate, which is classified into six groups according to severity. Based on the outcomes, the regions have been categorized into a six-pointer scale ranging from Very Low to Severe, representing 21.48%, 7.42%, 16.64%, 18.8%, 19.16%, and 16.49% for VL, L, M, H, and VH, respectively. Their pattern in the study area implies that high-elevation and high-precipitation locations are especially vulnerable to soil erosion. It indicates the importance of understanding local environmental factors for further erosion control strategies in the future. Forecast levels of severity are an important source of information for planning and conservation efforts, giving decision-makers what they need to prioritize the use of resources and help. To prevent environmental deterioration and preserve agricultural output, areas with high to severe soil erosion require extra care and preventive efforts. To improve its precision and efficacy, it also emphasizes the necessity of continual improvement and verification at the same time with ground observation. To increase the model dependability and suitability for use in a variety of landscapes, future research should concentrate on adding new data sources, adjusting model parameters, and independently verifying forecasts. By supporting sustainable farming practices, afforestation, and the use of conservation techniques like terracing, efforts are made to lessen erosion and save the soil from further deterioration.

REFERENCES

Bhardwaj, R. K., Sharma, S., and Kumar, D. (2022). Impact of LULC dynamics on evapotranspiration using GIS: Case study of Uttarakhand. *Journal of Materials and Environmental Science*, 13, 631–639.

Chalise, D., Kumar, L., Spalevic, V., and Skataric, G. (2019). Estimation of sediment yield and maximum outflow using the IntErO model in the Sarada river basin of Nepal. *Water*, 11(5), 952.

Conoscenti, C., Angileri, S., Cappadonia, C., Rotigliano, E., Agnesi, V., and Märker, M. (2014). Gully erosion susceptibility assessment by means of GIS-based logistic regression: A case of Sicily (Italy). *Geomorphology*, 204, 399–411.

Ferreira, C. S., Walsh, R. P., and Ferreira, A. J. (2018). Degradation in urban areas. *Current Opinion in Environmental Science & Health*, 5, 19–25.

Foster, G. R., Yoder, D. G., Weesies, G. A., McCool, D. K., McGregor, K. C., and Bingner, R. L. (2003). *Users Guide: Revised Universal Soil Loss Equation*. USDA-Agricultural Research Service, Washington, DC, 76.

Getu, L. A., Nagy, A., and Addis, H. K. (2022). Soil loss estimation and severity mapping using the RUSLE model and GIS in Megech watershed, Ethiopia. *Environmental Challenges*, 8, 100560.

Gobinath, R., Ganapathy, G. P., Gayathiri, E., Salunkhe, A. A., and Pourghasemi, H. R. (2022). Ecoengineering practices for soil degradation protection of vulnerable hill slopes. *In Computers in Earth and Environmental Sciences* (pp. 255–270). Elsevier, The Netherlands.

Goudie, A. (2023). Water erosion and mass movements. *In Landscapes of the Anthropocene with Google Earth* (pp. 191–219). Cham: Springer Nature Switzerland.

Jalayer, S., Sharifi, A., Abbasi-Moghadam, D., Tariq, A., and Qin, S. (2022). Modeling and predicting land use land cover spatiotemporal changes: A case study in chalus watershed, Iran. *IEEE Journal of Selected Topics in Applied Earth Observations and Remote Sensing*, 15, 5496–5513.

Joshi, P., Adhikari, R., Bhandari, R., Shrestha, B., Shrestha, N., Chhetri, S., Sharma, S., and Routh, J. (2023). Himalayan watersheds in Nepal record high soil erosion rates estimated using the RUSLE model and experimental erosion plots. *Heliyon*, 9(5), e15800, 1–14.

Khali, H., Rawat, K. S., and Chatterjee, R. S. (2023). Dynamic association of slope movements in the Uttarakhand Himalaya: A critical review on the landslide susceptibility assessment. *Geomatics, Natural Hazards and Risk*, 14(1), 2273214.

Koirala, P., Thakuri, S., Joshi, S., and Chauhan, R. (2019). Estimation of soil erosion in Nepal using a RUSLE modeling and geospatial tool. *Geosciences*, 9(4), 147.

Kumar, D., Thakur, S., Bhardwaj, R. K., and Sharma, S. (2021). Anthropogenic activity-induced water quality degradation in Girital Lake, Uttarakhand. *International Research Journal of Engineering and Technology*, 9(2), 1083–1087.

Moore, I. D., and Burch, G. J. (1986). Physical basis of the length-slope factor in the universal soil loss equation. *Soil Science Society of America Journal*, 50(5), 1294–1298.

Natarajan, S. (2023). Prediction of recently occurred soil erosion by integrating revised universal soil loss equation (RUSLE) with geo-spatial techniques: A case study on Pettimudi hills, Kerala-India, Research Square, 1–29, https://doi.org/10.21203/rs.3.rs-2827820/v1.

Šurda, P., Šimonides, I., and Antal, J. (2007). A determination of area of potential erosion by geographic information systems. *Journal of Environmental Engineering and Landscape Management*, 15(3), 144–152.

Thompson, J. A., Bell, J. C., and Butler, C. A. (2001). Digital elevation model resolution: effects on terrain attribute calculation and quantitative soil-landscape modeling. *Geoderma*, 100(1-2), 67–89.

Weng, X., Zhu, J., Wang, D., Zhong, M., Luo, M., Mei, Y., and Tang, G. (2024). Assessing rainfall erosivity changes over China through a Bayesian averaged ensemble of high-resolution climate models. *Environmental Research Communications*, 6(3), 035025.

9 Study of Rainfall Variability Using Multi-Scale Entropy for Mandi and Shimla Regions of Himachal Pradesh, India

Deepak Kumar, Amneesh Singh Saini, Kritika Singh, and Ankit Bajaj

9.1 INTRODUCTION

Rainfall is an integral part of the hydrological cycle, and understanding the spatio-temporal variability of rainfall is important for proper water resources planning and management, flood and drought mitigation, soil erosion control, and other problems. According to the World Meteorological Organization, floods were the most frequent extreme phenomenon that occurred in the 2001–2010 decade. This phenomenon affected several regions of the world, producing hundreds or even thousands of victims. In 2016, almost 23.5 million people were displaced because of natural disasters related to extreme meteorological events, mainly storms and floods, in the Asia-Pacific region. Higher living standards, technology, global warming, deforestation, and many other issues have contributed to the changes in rainfall intensity, spatial and temporal distribution, and average rainfall amount at various places across the globe. Rainfall-influenced hazards are becoming more intense and frequent in recent decades and are also affecting agriculture and the water management sector. With increasing concerns about the intensification of climate change, it is essential to understand the nature and variability of rainfall and its changing patterns at different spatiotemporal scales. In this view, a quantitative assessment of the spatiotemporal variability at different time scales using the entropic measures can effectively help get a deeper knowledge about the rainfall magnitude and its variability (Borwein et al., 2014; Poveda & Salas, 2015; Silva et al., 2017; Li et al., 2016; Zhao et al., 2016). Azman et al. (2017) studied rainfall variability on the East Coast of Peninsular Malaysia from 1960 to 2012. Entropy of stable rainfall and apportionment entropy were used to describe the rainfall variability. Findings confirmed that the apportionment entropy values indicated that the stations on the east coastline were more variable

 DOI: 10.1201/9781003596080-9

as compared to other stations inland for Kelantan. Ilunga (2017) worked on the mean annual runoff for the upper catchment in South Africa. Shannon entropy has been used in this study to assess the mean annual rainfall of the region. Roushangar et al. (2018) worked on water resource planning and management of the regional region of Iran. Data was taken by 31 rain gauges from 1960 to 2010. This study explored the ability of a proposed methodology based on continuous wavelet transform (CWT) and multi-scale entropy to analyze complexity and regionalize monthly precipitation data. The application of this method gave promising results for the monthly data of Iran. Tongal and Sivakumar (2019) studied in the island state of Tasmania in Australia. Entropy theory provided a framework to analyze the spatiotemporal variability in stream flows. Shannon entropy was used in this study to assess the space–time variability of season. Results suggested that the spatial variability of uncertainty in stream flow was associated with the spatial pattern of rainfall in Tasmania. Xavier et al. (2019) studied the complex characteristics of precipitation series in Paraiba state, Brazil. Data was taken from 69 meteorological stations. Sample entropy was used to find the dynamic change of precipitation series, providing a new way to investigate the complexity of hydrological series. The result confirmed that the intricacies of monthly average precipitation have differences in the macro-regions. Garrido-Arévalo et al. (2020) studied information entropy theory to assess rainfall station distribution in Colombia. Information entropy was used in the study to find rainfall variability to design the rainfall station distribution system. Saha and Chattopadhyay (2020) studied the maximization of Shannon entropy using probability distribution in the Himalayas region of India from 1844 to 2005. Results shown by the study demonstrated how entropy was affected by the fluctuation of mean rainfall on a seasonal scale and a yearly scale. The probability distribution curve was also generated by the rainfall series. Silva et al. (2021) studied sample entropy in the state of Pernambuco, Northeastern Brazil. Using 133-gauge stations in the state of Pernambuco, Northeastern Brazil, in an area of 1,542,000 km^2 for the period of 50 years from 1962 to 2012. Sample entropy was used in this study to describe the regularity of temporal series. A study confirmed that using modified multi-scale entropy, the regularity of temporal series on different temporal scales was suitable for short time series. Ghorbani et al. (2021) studied the spatiotemporal variability of rainfall patterns in North Algeria for water resource management. Tsallis entropy, pattern recognition, and precipitation concentration index (PCI) have been analyzed for a period of 33 years from 1980 to 2013. Mann–Kendall (MK) trend tests and structural tests were carried out to find rainfall trends. The results showed a significant correlation between El Nino Modoki EMI and the Southern Oscillation Index (SOI) with the rainfall pattern of North Algeria. Junior et al. (2021) studied monthly rainfall temporal series at 69 meteorological stations in Paraiba state, Brazil, recorded from 1962 to 2012. The methodology used in this study was modified multi-scale entropy, which was suitable for short time series and for analyzing rainfall regularity. The result shown by the study was to establish the use of multi-scale entropy with methods from information entropy in climatological studies. Ray and Chattopadhya (2021) studied a univariate analysis of air temperature and rainfall over northern India. In this study, they test the relative contributions of meteorological phenomena. Maximization of Shannon entropy was used in this study. Sílva et al. (2021) studied the Modified Multi-Scale Entropy (MMSE) method of Paraiba state in northeastern Brazil. Data was taken from 1962 to 2012 from 69 meteorological stations in Paraiba state. This work contributed to establishing its use in climatological studies. The MMSE method is suitable for short time series. Singh et al. (2023) have discussed the variability of rainfall in southern Uttarakhand, India by using entropy theory. In general, variability is defined as the quality of unevenness and lack of uniformity over different intervals, and entropic measures have been used to measure it (Singh and Kumar, 2021). One of the main advantages of entropy-based measures is that they do not make any prior assumptions on the probability distribution or statistical properties of the data, and thus, they are applicable to any type of data (Singh and Kumar, 2020). Marginal entropy (ME) is applied at different timescales, that is, annual, seasonal, and monthly, to assess the interannual variability associated with the individual timescales. So, the present analysis will notably add knowledge to the field of spatiotemporal rainfall variability.

9.2 STUDY AREA DESCRIPTION

Himachal Pradesh is a northern Indian state. The climate varies from hot and sub-humid tropical (450–900 m) in the southern low tracts to warm and temperate (900–1,800 m), cool and temperate (1,900–2,400 m), and cold glacial and alpine (2,400–4,800 m) in the northern and eastern high elevated mountain ranges. Figure 9.1 shows the geographical location of the study area. In the present study, the Shimla and Mandi districts of Himachal Pradesh have been selected.

The Mandi district in Himachal Pradesh lies in the central part of the state, spanning the mid-Himalayan zone. It features a diverse topography with deep valleys, high mountain ranges, and the Beas River traversing through it. Elevations range from

FIGURE 9.1 Location map of study area (Mandi and Shimla).

around 500 m to over 3,000 m above sea level. The region has varied terrain, including forested hills, fertile plains, and rugged slopes. Mandi experiences a temperate climate with distinct seasons. Average annual rainfall ranges between 1,200 mm and 2,000 mm, primarily during the monsoon months of June to September, supporting rich agriculture and lush vegetation across the district.

Shimla district, located in the south-central part of Himachal Pradesh, is known for its mountainous terrain, forested landscapes, and picturesque valleys. It lies in the lower and middle Himalayan ranges, with elevations ranging from about 300 m in the southern lowlands to over 4,500 m in the high mountain areas. The district is characterized by steep slopes, deep gorges, and river valleys formed by rivers such as the Sutlej, Giri, and Pabbar. The terrain supports dense forests of deodar, pine, and oak, along with fertile land in valleys used for agriculture and horticulture, especially apple cultivation. Shimla experiences a temperate climate with cool summers and cold, snowy winters at higher altitudes. The district receives an average annual rainfall of about 1,000–1,500 mm, mainly during the monsoon season (June to September), with occasional winter precipitation in the form of snow in the upper regions, contributing to the region's scenic beauty and water availability.

For the present study, rainfall data have been used. Rainfall data from 1980 to 2021 on a monthly scale has been taken from the Water Resources Information System (WRIS) website. Further, the time series data has been sorted into four seasons, namely, January–February (Winter Season, abbreviated as WS), March–June (Pre-Monsoon season, abbreviated as Pre-M), July–September (Monsoon season), October–December (Post-Monsoon season, abbreviated as Post-M). In this study, QGIS has been used for geo-referencing AME of both the districts for the time period 1980–2021.

9.3 ENTROPY THEORY

In the present study, ME has been used. ME is defined as the average information content of a random variable X with probability distribution $P(x)$ to measure the uncertainty. It is used to investigate the variability linked with monthly, seasonal, and annual time series. The mathematical expression for evaluating ME is given below (Singh and Kumar, 2024):

$$ME = -\sum_{i}^{n} \frac{r_i}{R} log_2 \left[\frac{r_i}{R} \right] \tag{9.1}$$

where r_i denotes the rainfall amount for the ith year and R denotes the total rainfall for the period taken, here 1980–2021.

9.4 RESULTS AND DISCUSSION

In the present study, the rainfall variability of Mandi and Shimla in Himachal Pradesh has been studied using ME, whose results have been presented and discussed here.

9.4.1 Entropy Analysis of Mandi

ME quantifies the uncertainty or variability associated with a single hydrological variable, like stream flow or precipitation, when considered independently. Table 9.1 and Figure 9.2 show the statistical parameters analysis for the ME of the Mandi district. From this table, it can be depicted that for the period 1980–2000, for the monsoon season, the number of years having ME higher than AME was maximum as compared to other selected seasons of the year. In addition, the standard deviation value of Post-M was highest as compared to other seasons. Further, ME shows that in the period of 2001–2021, the number of years above AME decreased.

Further, from Table 9.1, it can be observed that out of 41 years (1980–2021), 23 years in the WS have exceeded the value of average ME. The range of ME for the WS from 1980 to 2021 was 0.049–0.197, with a standard deviation value of 0.037 and a median value of 0.129. In the Pre-M period, from 1980 to 2021, the value of average ME exceeded in 21 years, with a median value of 0.127. The range of ME from 1980 to 2021 in Pre-M periods was 0.043–0.190, and the standard deviation for the Pre-M period was observed to be 0.030. From 1980 to 2021, there were 19 years in the monsoon season, when the value of average ME exceeded the ME range of 0.089–0.205. The standard deviation value of the monsoon season for the study period was 0.024, and the median value for the same season and time period was 0.125. In the Post-M season, during the study period (1980–2021), out of 41 years, there were 16 years in which the value of average maximum entropy exceeded. The range of ME for the study period in the Post-M season varied from 0.011 to 0.296, with a standard deviation value of 0.069 and a median value of 0.111.

TABLE 9.1
Statistical Performance of ME for the Mandi District

Year		1980–2000	-2001–2021	1980–2021
WS	No. of years exceeding the average ME	12	11	23
	Median	0.129	0.129	0.129
	Standard deviation	0.028	0.045	0.037
	Range	0.075–0.196	0.049–0.197	0.049–0.197
Pre-M	No. of years exceeding average ME	12	9	21
	Median	0.145	0.120	0.127
	Standard deviation	0.032	0.026	0.030
	Range	0.084–0.190	0.043–0.152	0.043–0.190
Monsoon	No. of years exceeding average ME	13	6	19
	Median	0.135	0.121	0.125
	Standard deviation	0.028	0.021	0.024
	Range	0.089–0.205	0.091–0.174	0.089–0.205
Post-M	No. of years exceeding average ME	10	6	16
	Median	0.117	0.108	0.111
	Standard deviation	0.080	0.052	0.069
	Range	0.019–0.296	0.011–0.213	0.011–0.296

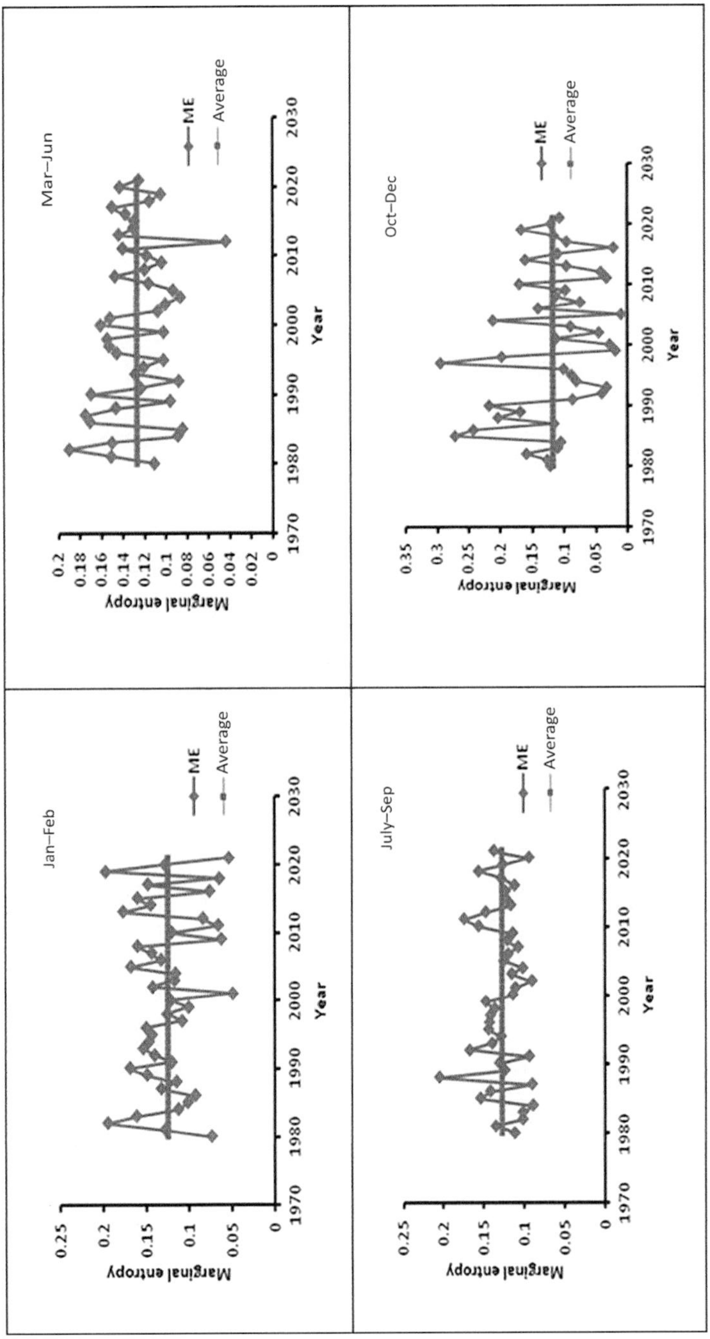

FIGURE 9.2 ME of the Mandi district of Himachal Pradesh.

Figure 9.2 shows the variation of maximum entropy from 1980 to 2021 for the Mandi district. There are four graphs in Figure 9.2 for four seasons: WS (January–February), Pre-M season (March–June), monsoon season (July–September), and Post-M season (October–December), showing the variations in ME.

9.4.2 Entropy Analysis of Shimla

ME provides a numerical value representing the degree of uncertainty or randomness associated with a specific hydrological variable. It is often used to analyze time series data (e.g., monthly, seasonal, or yearly precipitation) to understand how the variability of a hydrological process changes over time. Table 9.2 and Figure 9.3 show the statistical parameters analysis for the ME of the Shimla district. From this table, it could be depicted that for the period 1980–2000, except for the monsoon season, the number of years having ME higher than AME was maximum as compared to other selected seasons of the year. In addition, the standard deviation value of Post-M was highest as compared to other seasons. Further, ME showed that in the period of 2001–2021, the number of years above AME has decreased. From Table 9.2, it was evident that out of the 41 years of the study period, 20 years in the WS exceeded the value of average ME, while the range of ME for the season during the study period was 0.027–0.259. The median and standard deviation values for 1980–2021 in the WS for the Shimla district were 0.121 and 0.048, respectively. In the Pre-M season, the maximum number of years, that is, 23 years out of 41 years of the study period, crossed the value of average ME. The range of ME for the Pre-M season was 0.048–0.222, and the values of median and standard deviation observed were 0.128 and 0.038, respectively. For the monsoon season, the range of ME was 0.074–0.189. There were

TABLE 9.2
Statistical Performance of ME for Shimla District

Year		1980–2000	2001–2021	1980–2021
WS	No. of years exceeding average ME	12	8	20
	Median	0.129	0.099	0.121
	Standard deviation	0.039	0.054	0.048
	Range	0.074–0.259	0.027–0.232	0.027–0.259
Pre-M	No. of years exceeding average ME	12	11	23
	Median	0.141	0.128	0.128
	Standard deviation	0.036	0.038	0.038
	Range	0.069–0.222	0.048–0.195	0.048–0.222
Monsoon	No. of years exceeding average ME	11	8	19
	Median	0.131	0.123	0.127
	Standard deviation	0.026	0.028	0.027
	Range	0.077–0.189	0.074–0.189	0.074–0.189
Post-M	No. of years exceeding average ME	12	7	19
	Median	0.130	0.090	0.097
	Standard deviation	0.079	0.052	0.071
	Range	0.022–0.301	0.011–0.200	0.011–0.301

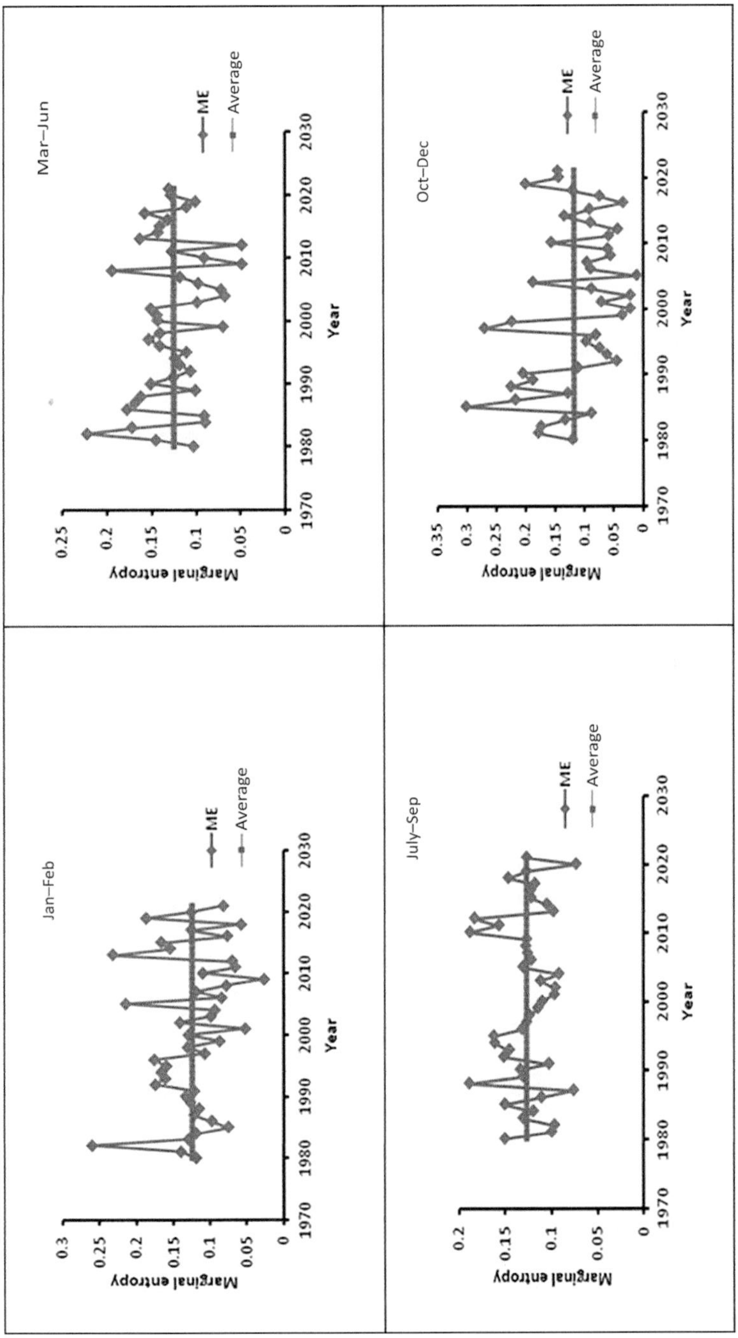

FIGURE 9.3 ME of Shimla district of Himachal Pradesh.

19 years in which the value of average ME was exceeded in the monsoon season in the Shimla district from 1980 to 2021. The median value for the monsoon season was 0.127, and the standard deviation value was observed to be 0.027. Maximum standard deviation was observed in the Post-M season in the Shimla district from 1980 to 2021, with a value of 0.097, and the median for the same was 0.097. There were 19 years out of 41 years of the study period in which the value of average ME exceeded.

In Figure 9.3, the trend of ME in the Shimla district from 1980 to 2021 has been shown. There are four graphs in the figure showing the variation in ME for four seasons, that is, WS (January–February), Pre-M (March–June), monsoon (July–September), and Post-M (October–December), respectively.

9.5 CONCLUSION

ME provides a statistical measure of how unpredictable the rainfall data is, without considering dependencies with other variables. In practical terms, higher ME indicates greater variability and less predictability in rainfall patterns, while lower entropy suggests more consistent or uniform rainfall behavior. In the present study, ME has been used to observe variability in the rainfall patterns of the Mandi and Shimla districts of Himachal Pradesh, India. Both Mandi and Shimla are tourist places in India and attract thousands of visitors worldwide. Global warming is affecting natural processes on Earth. Rainfall is one such natural process. ME in the study area was found to be higher than average ME in at least 50% of the total years of the study period.

REFERENCES

Azman, M. A. Z., Zakaria, R., Satari, S. Z., and Radi, N. F. A. (2017, May). Entropy of stable seasonal rainfall distribution in Kelantan. In *AIP Conference Proceedings* (Vol. 1842, No. 1, p. 020002). AIP Publishing LLC, Kuala Lumpur, Malaysia.

Borwein, J., Howlett, P., and Piantadosi, J. (2014). Modelling and simulation of seasonal rainfall using the principle of maximum entropy. *Entropy*, 16(2), 747–769.

Garrido-Arévalo, A. R., Agudelo-Otálora, L. M., Obregón-Neira, N., Garrido-Arévalo, V., Quiñones-Bolaños, E. E., Naraei, P., ... and Bustillo-Lecompte, C. F. (2020). Application of artificial neural network and information entropy theory to assess rainfall station distribution: A case study from Colombia. *Water*, 12(7), 1973.

Ghorbani, M. A., Kahya, E., Roshni, T., Kashani, M. H., Malik, A., and Heddam, S. (2021). Entropy analysis and pattern recognition in rainfall data, north Algeria. *Theoretical and Applied Climatology*, 144(1), 317–326.

Ilunga, M. (2017). Assessing catchment resilience using entropy associated with mean annual runoff for the Upper Vaal catchment in South Africa. *Entropy*, 19(5), 147.

Júnior, S. F. A. X., Xavier, É. F. M., Jale, J. S., Stosic, T., and dos Santos, C. A. C. (2021). Multiscale entropy analysis of monthly rainfall time series in Paraíba, Brazil. *Chaos, Solitons & Fractals*, 151, 111296.

Li, C. Q., Li, Y. W., Wang, P., and Shen, J. (2016). Natural runoff prediction of the yellow river in the future under climate change. In *International Conference on Advanced Education and Management Engineering* (pp. 177–179), Bangkok, Thailand.

Poveda, G., and Salas, H. D. (2015). Statistical scaling, Shannon entropy, and Generalized space-time q-entropy of rainfall fields in tropical South America. *Chaos: An Interdisciplinary Journal of Nonlinear Science*, 25(7), 075409.

Ray, S. N., and Chattopadhyay, S. (2021). Analyzing surface air temperature and rainfall in univariate framework, quantifying uncertainty through Shannon entropy and prediction through artificial neural network. *Earth Science Informatics*, 14(1), 485–503.

Roushangar, K., Alizadeh, F., and Adamowski, J. (2018). Exploring the effects of climatic variables on monthly precipitation variation using a continuous wavelet-based multiscale entropy approach. *Environmental Research*, 165, 176–192.

Saha, S., and Chattopadhyay, S. (2020). Exploring of the summer monsoon rainfall around the Himalayas in time domain through maximization of Shannon entropy. *Theoretical & Applied Climatology*, 141, 133–141.

Silva, A. S. A., Menezes, R. S. C., Rosso, O. A., Stosic, B., and Stosic, T. (2021). Complexity entropy-analysis of monthly rainfall time series in northeastern Brazil. *Chaos, Solitons & Fractals*, 143, 110623, 1–12.

Silva, A. S. A. D., Barreto, I. D. D. C., Cunha-Filho, M., Menezes, R. S. C., Stosic, B., and Stosic, T. (2021). Multiscale complexity analysis of rainfall in Northeast Brazil. *Water*, 13(22), 3213.

Silva, V. D. P. R. D., Belo Filho, A. F., Singh, V. P., Almeida, R. S. R., Silva, B. B. D., de Sousa, I. F., and Holanda, R. M. D. (2017). Entropy theory for analysing water resources in northeastern region of Brazil. *Hydrological Sciences Journal*, 62(7), 1029–1038.

Singh, S., and Kumar, A. (2024). Understanding the intricacies of rainfall dynamics using entropy measures. *Journal of Water and Climate Change*, 15(9), 4817–4839.

Singh, S., and Kumar, D. (2020). Entropy analysis of seasonal groundwater level fluctuations for Delhi and Haridwar regions of India using grace satellite data. *Journal of Water Engineering and Management*, 1(3), 27–33.

Singh, S., and Kumar, D. (2021). Investigation of rainfall variability of the southern part of Uttarakhand using entropy theory. *Indian Journal of Soil Conservation*, 49(2), 83–88.

Singh, S., Kumar, D., Kumar, A., and Kuriqi, A. (2023). Entropy-based assessment of climate dynamics with varying elevations for hilly areas of Uttarakhand, India. *Sustainable Water Resources Management*, 9(4), 130.

Tongal, H., and Sivakumar, B. (2019). Entropy analysis for spatiotemporal variability of seasonal, low, and high streamflows. *Stochastic Environmental Research and Risk Assessment*, 33(1), 303–320.

Xavier, S. F. A., da Silva Jale, J., Stosic, T., dos Santos, C. A. C., and Singh, V. P. (2019). An application of sample entropy to precipitation in Paraíba State, Brazil. *Theoretical and Applied Climatology*, 136(1), 429–440.

Zhao, J., Wang, D., Yang, H., and Sivapalan, M. (2016). Unifying catchment water balance models for different time scales through the maximum entropy production principle. *Water Resources Research*, 52(9), 7503–7512.

10 Ecosystem Restoration of Lakes through Biomanipulation and Aeration

Ashutosh Mishra, A. P. Sharma, and I. J. Singh

10.1 INTRODUCTION

Lakes are dynamic and complex ecosystems with many species of animals and plants interacting with each other and their environment. These are valued as water sources and for fishing, recreation, and tourism. Lakes and their surroundings are unique assets and valuable ecosystems of society and nature, which have social, cultural, and aesthetic value. Lakes in mountainous terrain are considered an extreme environment because they are small and sensitive ecosystems with rapid flushing rates (Vreca and Muri, 2006). Every lake is a specific body of water that indicates many of the characteristics of the surrounding watershed and the climate, as well as the shape and volume of the lake basin. Indian lakes can be classified geographically into Himalayan, Peninsular, and Coastal lakes, limnologically into fresh water and brackish water lakes (Reddy and Char, 2006). The undesirable anthropogenic activities and unscientific utilization of resources from the lakes have caused adverse environmental changes, leading to environmental deterioration and threatening the biodiversity sustained in them. Increase in population, urbanization, industrialization, and injudicious agricultural practices have brought about a large number of changes in the aquatic environment of lakes and further aggravated the awful situation, which is already under stress due to deterioration of water quality by natural and anthropogenic phenomena. Illegal construction, litter, discharge of domestic waste, and recreational use of lake water are major concerns for sedimentation and eutrophication of the lake water (Purushothaman and Chakrapani, 2012).

Healthy lakes have a natural lake restoration capacity to cleanse themselves. Each lake is an individual ecosystem with a food chain of organisms that assimilate the incoming nutrients (Gophen, 1990). The food chain moves nutrients up from the simplest single-celled bacteria to fish (Carpenter and Kitchell, 1996). This natural lake restoration system works very well to improve lake water quality, keeping the ecosystem in balance until excessive nutrient inflow overwhelms the ability of the ecosystem to assimilate the nutrients (Breukelaar et al, 1994). The natural assimilation of nutrients in the lake ecosystems begins to break down when these natural processes are limited by low oxygen levels. Aerobic organisms are much more

 DOI: 10.1201/9781003596080-10

efficient at digesting organic material than anaerobic organisms. Aerobic organisms feed on organic material contained in the sediments and assimilate these nutrients into increased body mass and reproduction. Aerobically assimilated nutrients become part of the food chain, rather than being recycled within the water column, as they are when anaerobic conditions exist. Maintaining the aerobic environment will also reduce or prevent the accumulation of organic sediments. Aerobic conditions at the lake bottom benefit all aspects of the aquatic environment, resulting in true lake restoration (Hansson et al, 1998).

The Kumaon region of Uttarakhand has many famous natural lakes such as Nainital, Naukuchiatal, Khurpatal, Bhimtal, Sattal, etc. The water quality of Kumaon lakes has deteriorated, and most of the lakes have become eutrophicated (Mishra et al, 2017). The stability of the ecosystem of a lake is dependent on the food chains present therein. In the previous decade, lakes of the Kumaon region were dominated by spurious blooms of algae and profuse growth of aquatic weeds (Sharma et al, 1982). The important zooplankton, viz. rotifers and cladocerans, were rarely observed. The fish fauna was represented by weed fishes in the littoral region, bighead carp, mahseer and grass carp in the limnetic region, common carp, and some catfish in the benthic region. The weed fishes feed on zooplankton, and common carp causes the release of phosphate and nitrogen compounds into the lake due to its habit of stirring the sediments.

Many steps have been taken to control the eutrophication of these lakes. The removal of hypolimnetic water, which is deficient in dissolved oxygen (DO) but has a high concentration of nutrients, was done to reduce nutrient concentration in the lake. A sewerage system has also been installed to check and divert the inflowing nutrients. Afforestation in the catchment area is being done to reduce the input of nutrients into the lake. The other measures to reduce external and internal load of nutrients are desilting of sediments, aeration of the whole lake, and withdrawal of hypolimnetic water of the lake. The synergistic effect of all these corrective measures will be able to solve some problems of lake eutrophication and pollution. Among them, the biomanipulation of food chain spectrum and aeration of hypolimnetic waters is most effective in the restoration of the lakes (Jeppesen et al., 2007).

10.2 BIOMANIPULATION AND AERATION FOR ECOSYSTEM RESTORATION

The use of biomanipulation as a tool for water management started in the 1970s in small lakes, and because of its success, it was also applied in large lakes in the 1990s. Shapiro and Wright originally defined biomanipulation as "The deliberate exploitation of the interactions between the components of the aquatic ecosystem to reduce the algal biomass." Biomanipulation is a kind of biological engineering that reconstructs the ecosystem by cascading trophic interactions (Drenner and Hambright, 1999; Kasprzak et al., 2002; Jeppesen et al., 2007). The biomanipulation strategy allows for the zooplankton community to be dominated by large-sized species, which feed mostly by sight (Ramachandra and Solanki, 2007). It was known that planktivorous fish affect the zooplankton community and indirectly influence the phytoplankton community (Garcia et al., 2002). Biomanipulation is an efficient practical tool in water management, bringing a water body more quickly into the alternative stable state. Considering the most important factors in lakes and reservoirs, nutrients, fish, and water depth are the only variables that can be controlled by the water managers. Biomanipulation with regard to fish always involves methods to remove fish. Fish may be removed by nets, or poison, or by stocking predatory fish (Amarsinghe, 1998). These techniques may be combined, but the extent and the period of time in which they are applied may differ considerably (Amarsinghe, 1998).

The aeration technology is also useful in restoring the ecosystem of lakes. Naturally, most lakes undergo two stratification periods annually. Without aeration, many lakes will develop an oxygen-deprived hypolimnion throughout the summer (Smirnoff and Stewart, 1987). The same lakes may also run the risk of oxygen in the ecosystem being depleted to the extreme of experiencing episodes of fish kills. By disrupting the stratification, the lake ecosystem is supplied with adequate oxygen, preventing the suffocating effect that would have otherwise occurred. Aeration is also an important tool for reducing the accumulation of organic muck that builds upon the lake bottom. Under low oxygen conditions, decomposition takes place at a much slower rate than under oxygenated conditions. Consequently, organic material is decomposed at a very slow rate, in fact, usually slower than the rate at which new organic material is being created.

10.3 IMPACT OF BIOMANIPULATION AND AERATION ON THE LAKE ECOSYSTEM

The balanced ecosystem of a lake has two main functions, *viz.*, production and decomposition. These are essential for the equilibrium and sustainability. Primary productivity (growth of phytoplankton) is the main and most basic food source in most of the freshwater lakes. Phytoplankton growth requires inorganic nutrients, adequate light, and temperature to meet the primary producer's needs. Productivity from lake to lake can range from low to high, and can be considered "healthy" at any level depending on the context. There is high productivity of primary producers in the Nainital and Naukuchiatal lakes. However, overproduction can lead to premature lake aging and often is accompanied by mostly undesirable characteristics and a decline in water quality. A highly productive lake is known as eutrophic, and an extremely productive one is called hyper-eutrophic (Nainital and Naukuchiatal are eutrophic lakes). Nutrient cycling is the process that recycles, through decomposition, what has not been used or has been discarded by lake organisms. The bacterial community contributes to system "cleanup" and provides essential nutrients (mainly phosphorus and nitrogen) for the phytoplankton. Manipulation of the food web, incorporating removal of unwanted fishes and the stocking of environment-friendly planktivorous fishes, helps in restoring the ecosystem of the lake, improving the water transparency, nutrient load, and rehabilitation of its pride mahseer fishery. Our study gives additional support to the view that fish may have a major influence on plankton community structure and biomass.

10.3.1 Lake Nainital

The impact of biomanipulation and aeration activities has been studies (through 30 aeration discs fixed at the bottom and connected with two 30 HP compressors supplying oxygen at 8.5 bar pressure) started in 2007 for restoration of ecosystem of lake Nainital (area 48.76 ha) situated at 29°24′ N latitude, 79°28′ E longitude, at 1935 m above sea level altitude, on trophic structure and water quality for five years was assessed. Due to increased urbanization, sewage disposal, and other anthropogenic activities, Lake Nainital became eutrophicated and was dominated by blooms of planktonic algae and unwanted fishes. Its commendable mahseer fishery has almost vanished. To restore the ecosystem of lake, manipulation of the trophic structure, removal of unwanted biomass especially fishes, aeration and at the same time stocking of environment-friendly fish species like mahseer and silver carp, was done. The above techniques were applied after preparing eco-database of Lake Nainital initially for 1 year by adopting standard methods. To control the population of unwanted fishes and blooms of algae, a large population of existing unwanted fishes {*Puntius* spp, *Gambusia* spp (about 9,212 kg) and *Hypophthalmichthys nobilis* (about 7,042 kg)} was removed, and a large number of advanced fingerlings of the selected eco-friendly candidate fish species {*Tor putitora* (58,000) and *Hypophthalmichthys molitrix* (108,000)} were stocked phase-wise. During initial harvesting (2007), the catch composition of *Puntius* spp. and *Gambusia* spp was 87%, while it decreased to 7% at the end (2012). The composition of food fish has increased from 12% to 89%. During the study period, 29 species of phytoplankton, 15 species of zooplankton, and 15 species of fish were recorded from the lake. Before the start of this work, the physicochemical condition of the lake was not satisfactory. There was wide variation in the physico-chemical parameters at initial stage of the work, *viz* Secchi disc transparency (35.0–178.0 cm), TDS (160.0–270.0 mg/l), DO (3.8–8.4 mg/l), pH (6.9–7.9), total alkalinity (134.0–230.0 mg/l), nitrate (0.3–2.0 mg/l), phosphate (0.08–0.19 mg/l), but now these parameters are showing narrow variation (Mishra et al, 2017). Due to the impact of regulation of fish species and aeration in the lake, the average plankton population in terms of phytoplankton and zooplankton has changed from 28.486×10^5 to 27.52×10^5 cells/l and 1962–2085 individuals/l, respectively, with dominant species of *Microcystis, Melosira, Cyclops* while the rarely observed organisms during initial stage of the work viz. *Daphnia, rotifers, Chironomus, Atherix, Leptocella, Baetis, etc.,* are now showing their presence in significant amounts. Earlier mixing of epilimnetic and hypolimnetic water occurred regularly in winters, causing mass mortality of fishes, but this has not been observed for the past 18 years after the start of biomanipulation and aeration activities (Figures 10.1 to 10.7).

10.3.2 Lake Naukuchiatal

The impact of biomanipulation and aeration (through 39 aeration discs fixed at the bottom and connected with three 30 HP compressors supplying oxygen at 8.5 bar pressure) started in 2015 for restoration of ecosystem of an eutriphied Lake Naukuchiatal (area 30.6 ha) situated at 29°19′ N latitude, 79°37′ E longitude and

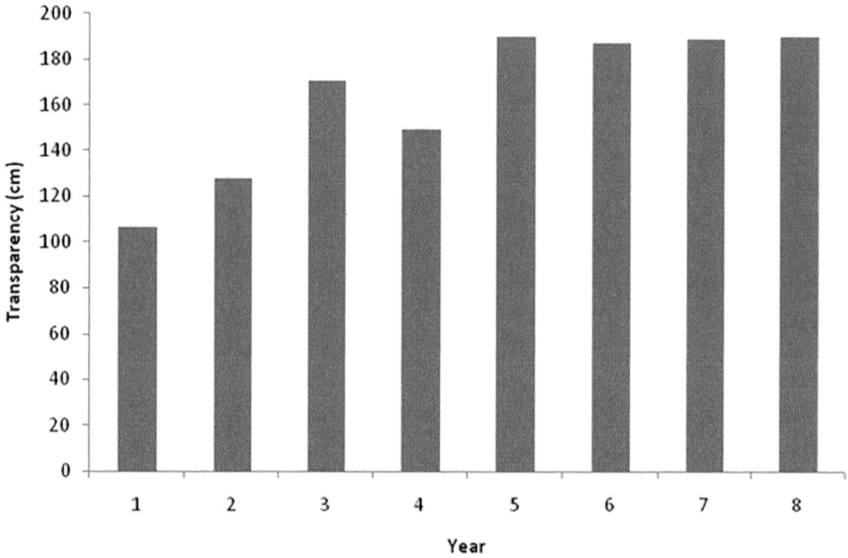

FIGURE 10.1 Variation in water transparency.

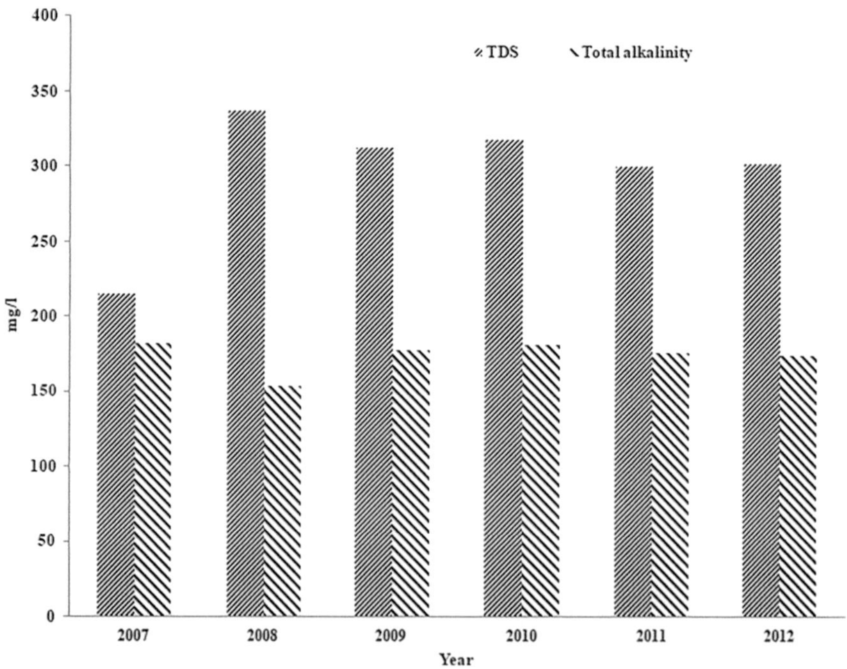

FIGURE 10.2 Variation in TDS and total alkalinity.

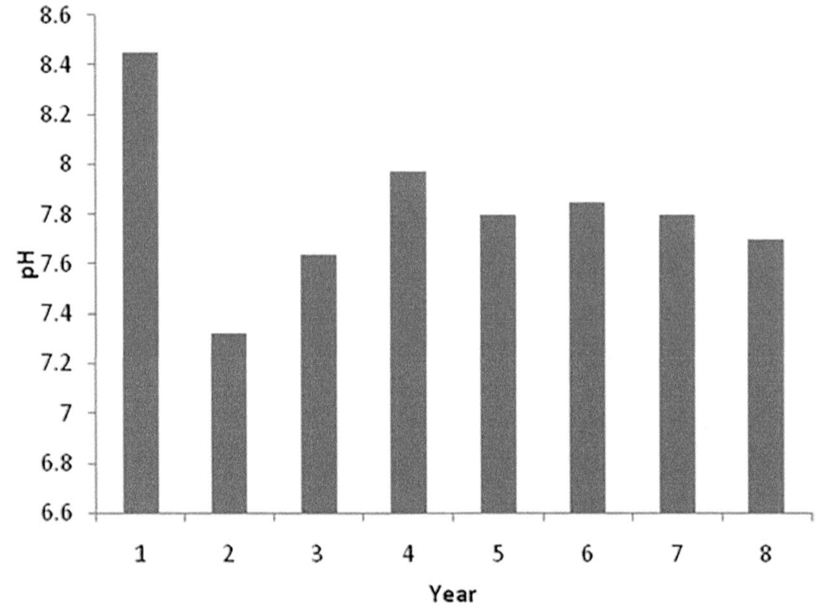

FIGURE 10.3 Variation in water pH.

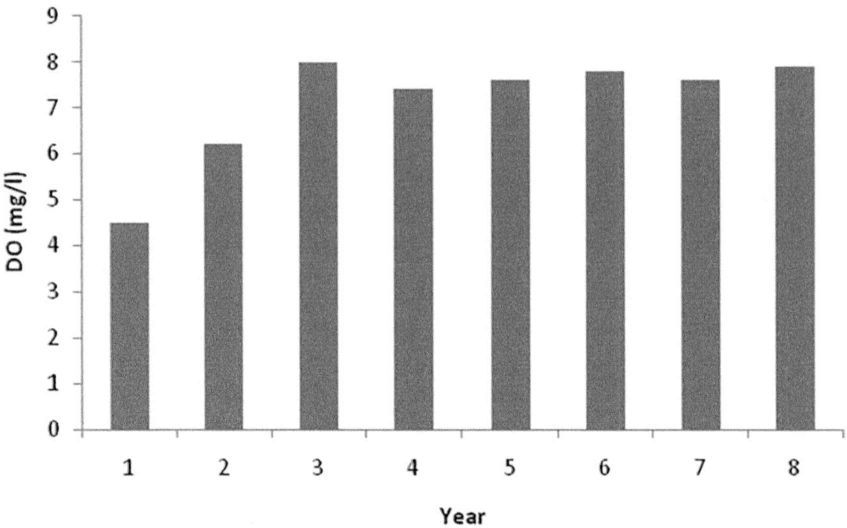

FIGURE 10.4 Variation in DO.

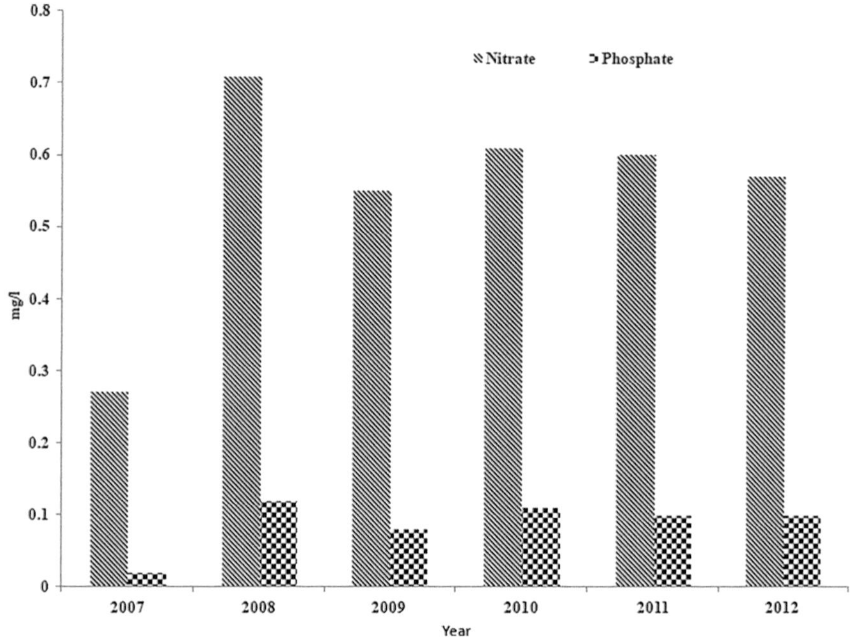

FIGURE 10.5 Variation in nitrate and phosphate concentration.

1,220 m ASL altitude, on trophic structure. Its water quality was assessed for 4 years. To restore the ecosystem of the lake, manipulation of the trophic structure through releasing fingerlings of eco-friendly fishes, viz., golden mahseer (14,700), silver carp (56,750), Indian major carps (32,750), etc., and aeration was done. To prepare the eco database, water quality parameters (water temperature, transparency, pH, DO, nitrate, phosphate, phytoplankton, zooplankton, and macrophytes) were analyzed weekly. There was wide variation in physico-chemical parameters at initial stage of the work *viz* water temperature (15.33–24.09°C) Secchi disc transparency (113.0–239.75 cm), DO (0.09–11.43 mg/l), pH (6.9–8.8), total alkalinity (88.0–115.50 mg/l), nitrate (0.4–2.18 mg/l), phosphate (0.01–1.53 mg/l) but now these parameters are showing stability. During the investigation period, a total of 20 species of phytoplankton (dominant ones *Spirogyra* sp. and *Nitzschia* sp.), 9 species of zooplankton (dominant ones *Brachionus* spp. and *Keratella* sp.), and 5 species of macrophytes (dominant one *Vallisneria* spp.) were recorded. At the beginning of the investigation, the lake was nutrient-rich, but due to proper aeration and biomanipulation, the ecosystem of the Lake is progressing toward restoration, and all the water quality parameters are under permissible limits. Earlier mixing of epilimnetic and hypolimnetic water occurred regularly in winters, causing mass mortality of fishes, but this has not been observed for the past 10 years after the start of work (Figures 10.8 to 10.14).

The values of recorded water quality parameters, viz., water temperature, water transparency, DO, pH, total alkalinity, nitrate nitrogen, and phosphate in these lakes indicate that their values are in a moderate range with fair variation. The values of

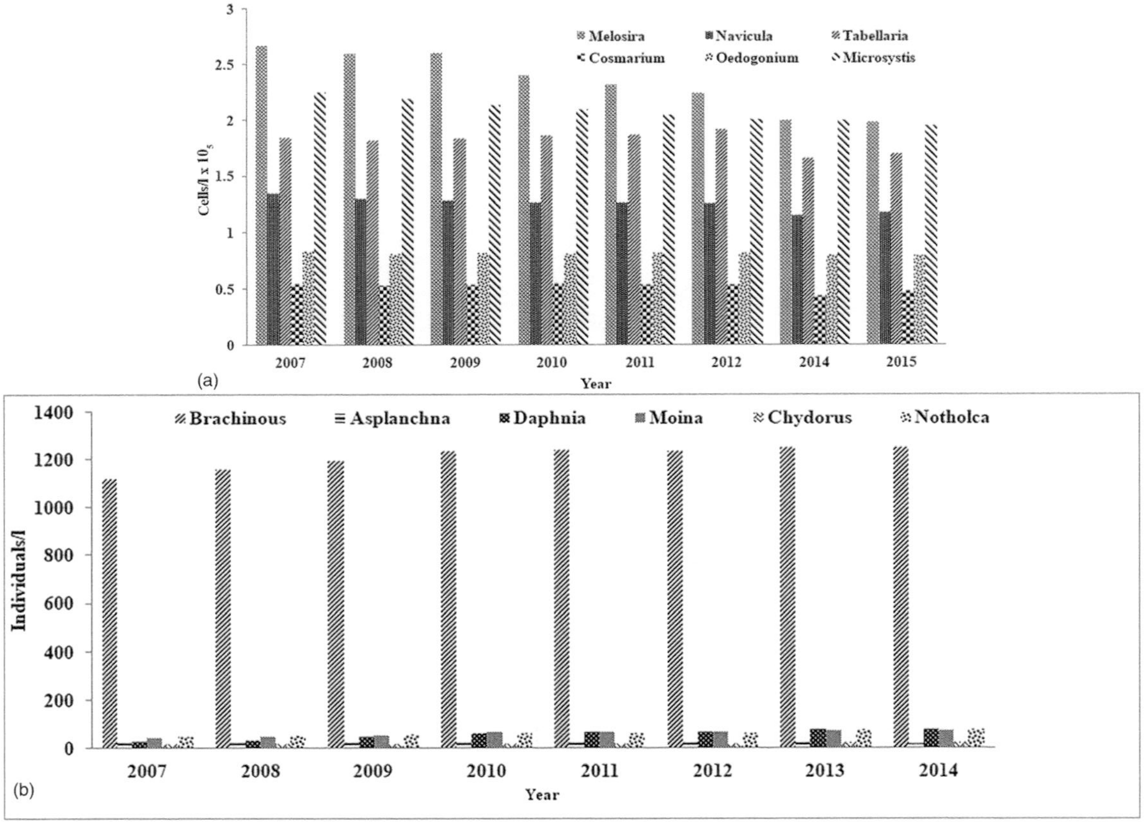

FIGURE 10.6 Variation in dominant species of (a) phytoplankton and (b) zooplankton.

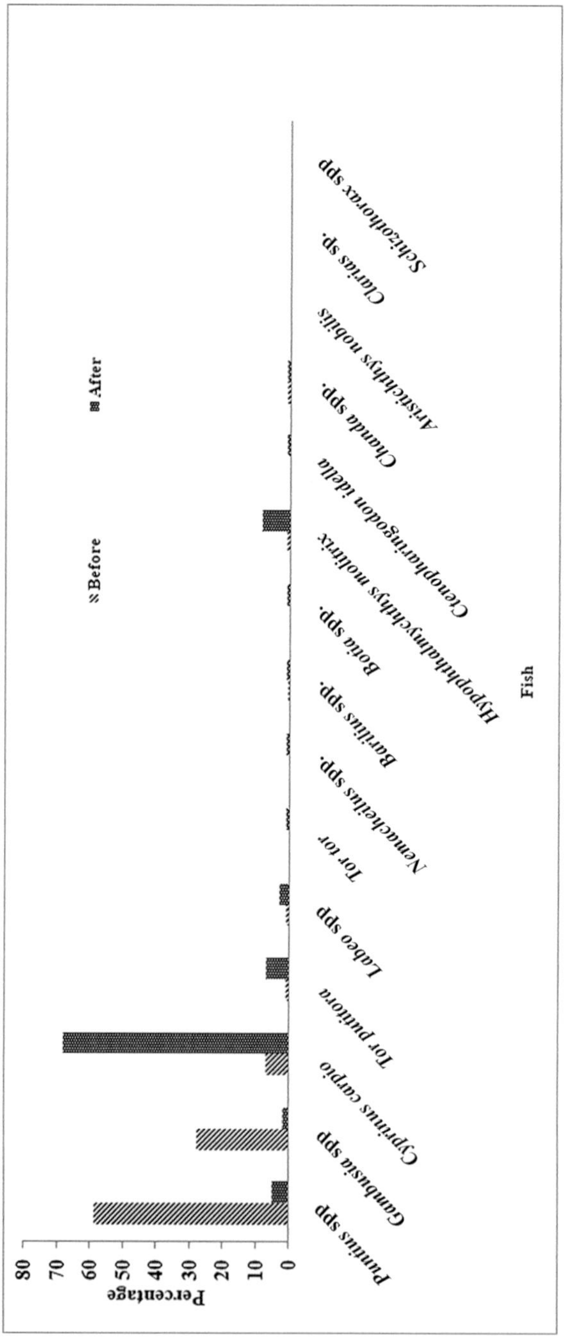

FIGURE 10.7 Composition of fishes in Lake Nainital.

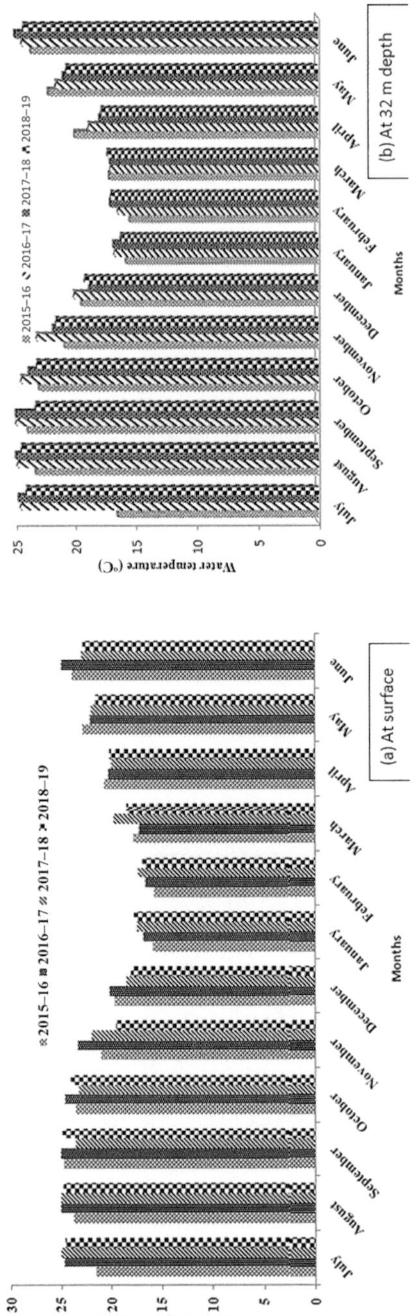

FIGURE 10.8 (a, b) Variation of lake water temperature (°C) at different depths.

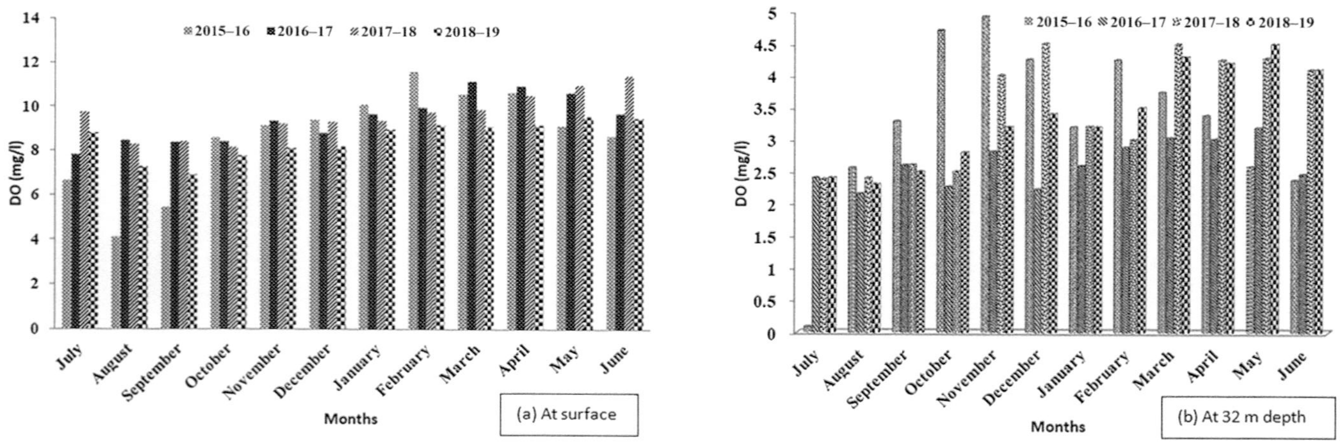

FIGURE 10.9 (a, b) Variation of DO (mg/l) at all the depths of the lake water.

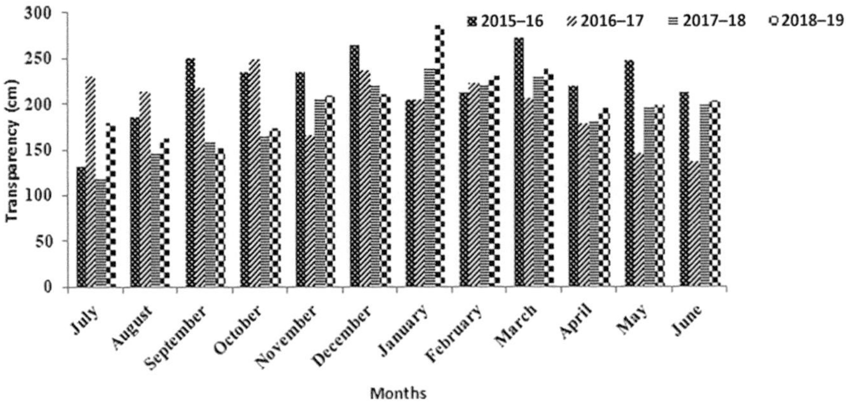

FIGURE 10.10 Variation of water transparency (cm) in lake water.

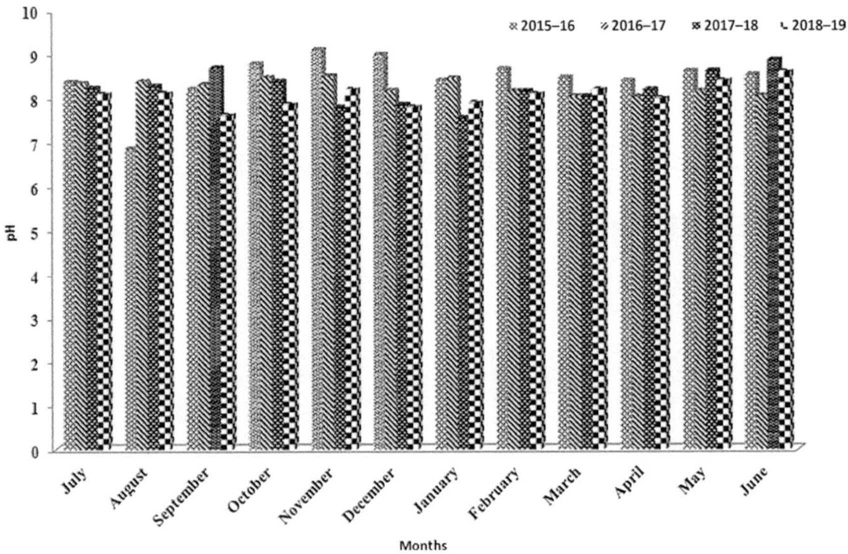

FIGURE 10.11 Variation in pH value of lake water.

water temperature and DO were uniformly distributed up to the deepest part. The introduction of planktivorous and omnivorous fish species controlled the blooms of coarse phytoplankton species and facilitated the growth of desired species, viz. *Nitzschia, Navicula, Cosmarium, Scenedesmus,* and others. Also, the quantity of the desired species of zooplankton, viz. *Brachionus, Daphnia, Chydorus, Bosmina*, etc., were higher than others. The study of fish composition shows that the percentage share of eco-friendly food fishes, viz. *Tor putitora, Hypophthalmichthys molitrix, Ctenopharyngodon idella*, Indian Major Carps, etc., were increased from 36.2 (at

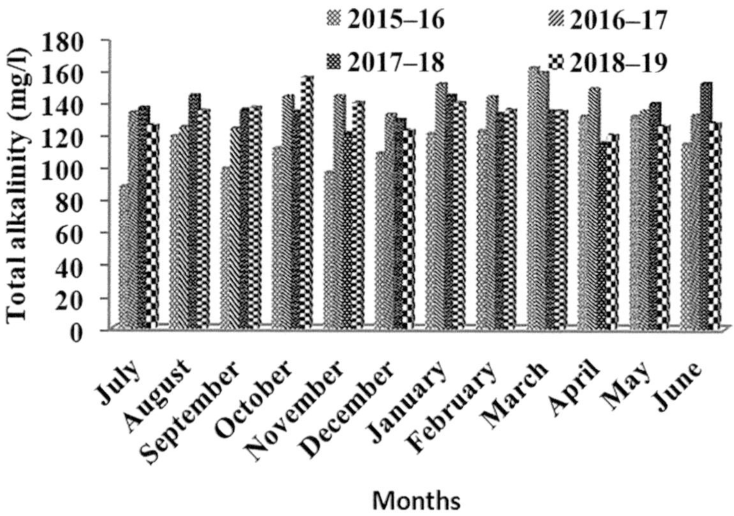

FIGURE 10.12 Variation in total alkalinity (mg/l) values of lake water.

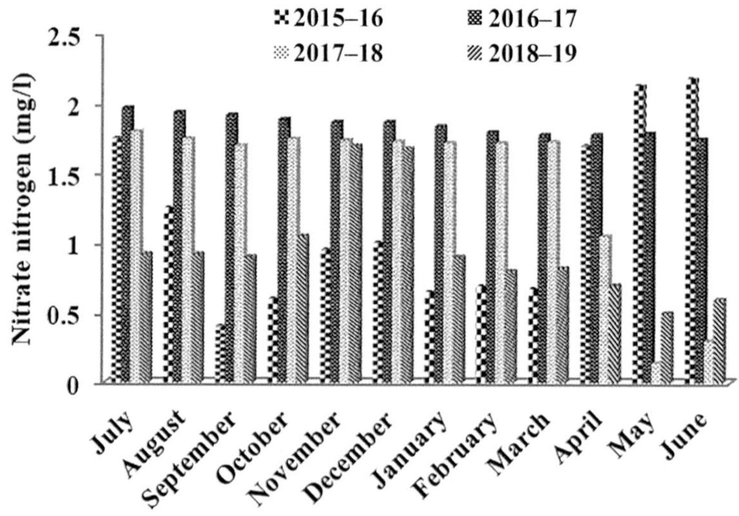

FIGURE 10.13 Variation of nitrate concentration (mg/l) in lake water.

start) to 52.69% (at end). The fish population of the lake shows that eco-friendly fish are in good percentage, growing well, and no fish mortality has occurred to date, which was a frequent phenomenon of this lake in the past. The result of the biomanipulation and artificial aeration activities shows that this lake is progressing toward its ecological restoration.

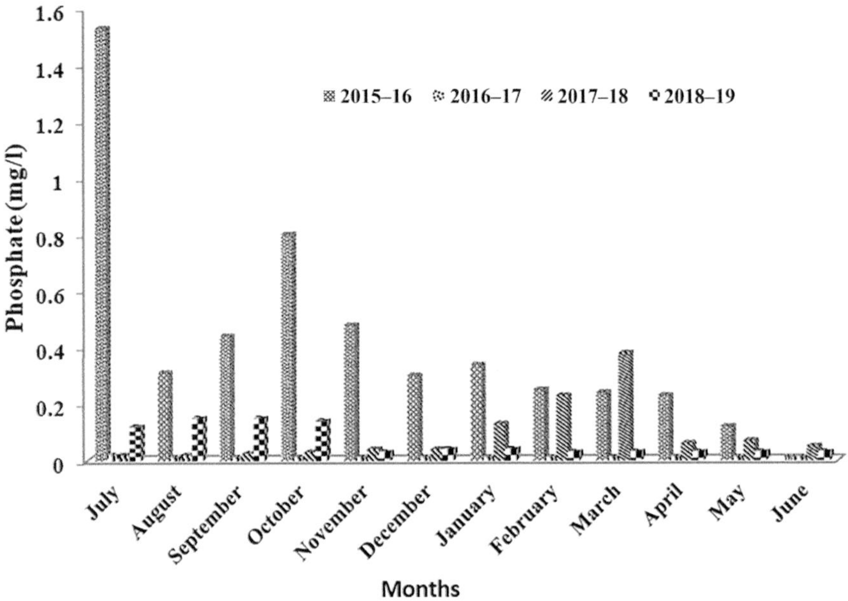

FIGURE 10.14 Variation of phosphate concentration (mg/l) in lake water.

10.4 CONCLUSIONS

Biomanipulation of the trophic structure, incorporating the removal of unwanted fish, stocking of environment-friendly planktivorous fish species like mahseer, silver carp, etc., and aeration of hypolimnetic water, has helped in restoring the ecosystem of lakes Nainital and Naukuchiatal, improving the water transparency, nutrient load, and rehabilitation of their prized mahseer fishery. These techniques can be regarded as an approach to water quality improvement that requires the combination of research and management. Our study gives additional support to the view that fish may have a major influence on plankton community structure and biomass. We thus conclude that the trophic efficiencies in the pelagic food web of lakes Nainital and Naukuchiatal are not unusually high. The crustacean zooplankton production is small, but the recorded fish yield is quite normal relative to the measured primary production of pelagic phytoplankton. The overall success of biomanipulation and aeration is debatable. Some sources proclaim the approach as successful, while others question its reliability. In our case, it is most successful in the form of increasing transparency, decreasing phytoplankton and unwanted fishes and increasing zooplankton and many food fishes. There is some agreement between fisheries and water quality management with respect to the goals of biomanipulation and ways to achieve these goals. Furthermore, sound scientific knowledge on the behavior of aquatic ecosystems can only be obtained by manipulating systems of adequate size over long periods.

REFERENCES

Amarsinghe, U.S. (1998). How effective are the stocking strategies for the management of reservoir fisheries of Srilanka? In: I.G. Cowx (ed.) *Stocking and Introduction of Fish* (pp. 422–436). Oxford: Fishing News Books, Blackwell Science.

Breukelaar, A. W., Lammens, E. H., Breteler, J. G. K., and Tatrai, I. (1994). Effects of benthivorous bream (Abramis brama) and carp (Cyprinus carpio) on sediment resuspension and concentrations of nutrients and chlorophyll a. *Freshwater Biology*, 32(1), 113–121.

Carpenter, S. R., and Kitchell, J. F. (Eds.). (1996). *The Trophic Cascade in Lakes*. USA: Cambridge University Press.

Drenner, R. W., and Hambright, K. D. (1999). Biomanipulation of fish assemblages as a lake restoration technique. *Archiv für Hydrobiologie*, 146(2), 129–165.

Garcia, P. R., Nandini, S., Sharma, S. S. S., Valderrama, E. R., Cuesta, I., and Hurtado, M. D. (2002). Seasonal variations of zooplankton abundance in the freshwater reservoir Valle de Bravo (Mexico). *Hydrobiologia,* 467, 99–108.

Gophen, M. (1990). Biomanipulation: Retrospective and future development. *Hydrobiologia,* 200/ 201, 1–11.

Hansson, L. A., Annadotter, H., Bergman, E., Hamrin, S. F., Jeppesen, E., Kairesalo, T., ... and Strand, J. (1998). Biomanipulation as an application of food-chain theory: Constraints, synthesis, and recommendations for temperate lakes. *Ecosystems*, 1, 558–574.

Jeppesen, E., Meerhoff, M., Jacobsen, B. A., Hansen, R. S., Sondergaard, M., Jensen, J.P., Lauridsen, T.L., Mazzeo, N., and Branco, C. W. C. (2007). Restoration of shallow lakes by nutrient control and biomanipulation- the successful strategy varies with lake size and climate. *Hydrobiologia,* 581, 269–285.

Kasprzak, P., Benndorf, J., Mehner, T., and Koschel, R. (2002). Biomanipulation of lake ecosystems: An introduction. *Freshwater Biology,* 47 (12), 2277–2281.

Mishra, A., Sharma, A. P., Singh, I. J., and Nainwal, G.C. (2017). Ecosystem restoration of lake Nainital through biomanipulation. *Environment & Ecology,* 35, 3231–3239.

Purushothaman, P., and Chakrapani, G. J. (2012). Trace metals biogeochemistry of Kumaon Himalayan Lakes, Uttarkhand, India. *Environmental Monitoring and Assessment,* 184, 2447–2965.

Ramachandra, T. V., and Solanki, M. (2007). Ecological assessment of lentic water bodies of Bangalore. *The Ministry of Science and Technology*, 25(96), 2.

Reddy, M. S., and Char, N. V. V. (2006). Management of lakes in India. *Lakes and Reservoirs,* 11(4), 227–237.

Sharma, A. P., Jaiswal, S., Nei, V., and Pant, M. C. (1982). Phytoplankton community analysis in lakes of Kumaun Himalaya. *Archiv für Hydrobiologie,* 93(2), 173–193.

Smirnoff, N., and Stewart, G. R. (1987). Nitrogen assimilation and zinc toxicity to zinc-tolerant and non-tolerant clones of *Deschampsia cespitosa* (L.) Beauv. *New Phytologist*, 107(4), 671–680.

Vreca, P., and Muri, G. (2006). Changes in accumulation of organic matter and stable carbon and nitrogen isotopes in sediments of two Slovenian maintain lakes (Lake Ledvica and Lake Planina) induced by eutrophication changes. *Limnology and Oceanography,* 51, 781–790.

11 A Performance Evaluation of K-Nearest Neighbor and Artificial Neural Networks for Predicting Total Dissolved Solids Concentration in Groundwater Sources

A. R. Malavika and V. Santhosh

11.1 INTRODUCTION

Groundwater is a vital renewable natural resource used for drinking water supply, agriculture, and industrial activities. In areas with little or no surface water, groundwater often represents the only freshwater source for millions of people all over the globe (Aranguren–Díaz et al., 2024). Rapid population growth and increased urbanization have been permanently increasing the demand for groundwater due to its other importance concerning water security (Zango et al., 2022). Apart from its availability, groundwater quality is equally important. Unsatisfactory quality of groundwater increases risks to human health, agriculture, and ecosystems. Groundwater quality assessment also considers Total Dissolved Solids (TDS), which vary according to the amount of dissolved inorganic salts and organic compounds therein (Barroso et al., 2023). Aside from the amount, the quality of groundwater is important in determining safe drinking water. Groundwater quality refers to the physical, chemical, and biological characteristics of water as regards its fitness for human consumption and use. An assessment of the quality is necessary as certain contaminants can cause considerable risk to human health and the environment (Mao et al., 2023; Kumar et al., 2021). Common indicators that signify groundwater quality are TDS, pH, hardness, and concentrations of various ions that have been measured in the field, such as nitrates, chlorides, and heavy metals. TDS, which refers to the sum of all dissolved substances in water, is one of the most frequently used parameters for assessing water

DOI: 10.1201/9781003596080-11

quality. The TDS level tends to sour the waters with undesirable taste, with health implications and many operational issues in the course of water treatment processes. For example, a TDS level greater than 500 mg/L without advanced treatment will adversely affect water quality and render it unsafe for consumption (BIS: Bureau of Indian Standards 10500, 2012). Quite high levels of TDS deter people from using a water supply because of poor taste and odor quality. Nevertheless, low TDS is desirable for most uses, such as drinking and irrigation. Excess TDS concentration exerts stress from osmotic pressure on plants, resulting in the failure of plants to absorb water. Salty soils develop a condition that does not permit enough aeration for continuous and sustainable plant growth (Adjovu et al., 2023). Monitoring groundwater quality while analyzing TDS is vital for sound management; it helps assess pollution sources and causal agents, water treatment planning, and compliance with established laws for drinking water quality. It is believed, therefore, that if solutions to these current concerns are taken up, the future generations would retain this precious resource and development toward sustainability assured. Recent studies focus on advancing using machine learning (ML) and artificial intelligence to predict and assess groundwater quality. Such methods provide researchers with the ability to analyze huge datasets to recognize potentials in groundwater quality patterns, unrecognizable through conventional analytical methods (Pourmorad et al., 2024; Singh et. al., 2024). Some of the ML models used include Artificial Neural Network (ANN), Support Vector Machine (SVM), and Random Forest (RF), which have shown promise in predicting water quality parameters like TDS based on hydrogeological and other environmental conditions (Haggerty et al., 2023). Models such as these have many advantages over traditional numerical ones in that they are more effective, user-friendly, cheaper, and save time. By contrast, numerical methods can require substantial computing power and detailed geological and hydrological data; however, ML and artificial intelligence techniques can still apply to incomplete or noisy datasets. This enables their use and application in real-world problems. ANNs have been one of the most used ML approaches because they model the complex nonlinear relationship among groundwater quality parameters (Khaki et al., 2015). For example, Asadollahfardi et al., (2012) reported on the successful application of ANNs in predicting TDS from the Talkheh Rud River. Their study revealed that the ANN model provided more accurate forecasts than those based on traditional methods, underlining the significance of being set for real-life water quality management contexts. Hussein et al.(2024) evaluated three ML models—XGBoost, Support Vector Regression (SVR), and K-Nearest Neighbor (KNN)—for predicting Irrigation Water Quality Index. All these results emphasize the potential of ML models as valuable tools for groundwater quality prediction, suggesting that these models can provide essential insights for informed decision-making in water management and resource allocation.

11.2 STUDY AREA

The Kadalundi River, which flows across the Malappuram and Palakkad districts in Kerala, emerges from the Western Ghats near the western edge of Silent Valley National Park and empties itself into the Arabian Sea later. The river flows proudly by the two main tributaries, Olipuzha and Veliyar, as shown in the study area in Figure 11.1. The Olipuzha takes its origin from Cherakkobban Mala (1160 m) and the Veliyar originates from the forest of the Erattakomban Mala (1190 m above Mean Sea Level (MSL). This rain-fed river flows through a diverse landscape, highlighting the distinct ecological and geological characteristics of the tropical mountain river system (District Survey Report of Minor Minerals (Except River Sand Malappuram District, 2016). The basin covers an area of approximately 1243 sq kilometers and is characterized by its complex topography, which includes rugged hills, plains, densely forested regions, and coastal areas. The area is primarily defined by Archean formations, in addition to the Laterite deposits and Quaternary sediments. The study area experiences four distinct seasons: the dry season from December to February; the hot season from March to May; the south-west monsoon from June to September; and the north-east monsoon from October to December. Most of the annual rainfall comes in the south-west monsoon, while a relatively lesser amount occurs in the north-east monsoon (District Survey Report of Minor Minerals (Except River Sand) Malappuram District, 2016).

11.3 MATERIALS AND METHODS

During the pre-monsoon season, 60 groundwater samples were collected from the study area. The principal analysis was done for parameters like pH, electrical conductivity (EC), TDS, salinity, total alkalinity (TA), total hardness (TH), sodium (Na^+), potassium (K^+), magnesium (Mg^{2+}), calcium (Ca^{2+}), chloride (Cl^-), nitrate (NO_3^-), sulfate (SO_4^{2-}), and fluoride (F^-).

To develop a solid model, the following preprocessing steps were adopted:

1. *Missing Values:* For addressing missing values, mean imputation was used for the quantitative parameters.
2. *Normalization:* Normalization was done so that all input variable features share a common scale using Min–Max scaling.
3. *Feature Selection:* Pearson correlation shown in Figure 11.2 was utilized to find parameters most correlated with TDS concentrations, therefore reducing overfitting risk.
4. *Dataset Splitting:* The dataset was stratified with respect to TDS concentration into training (80%) and test (20%) sets to maintain the TDS value distribution within training and test subsets.
5. *Modeling Techniques Used:* ANN and KNN models are applied here to predict the concentration levels of TDS.

11.3.1 Artificial Neural Networks

An ANN is a computer model inspired by the processing relationship of the human brain to learn complex patterns and relationships in data. An ANN is made up of an input layer, one or more hidden layers, and a visible layer (Manning et al., 2014). Each layer consists of neurons that are connected by weights, which change through

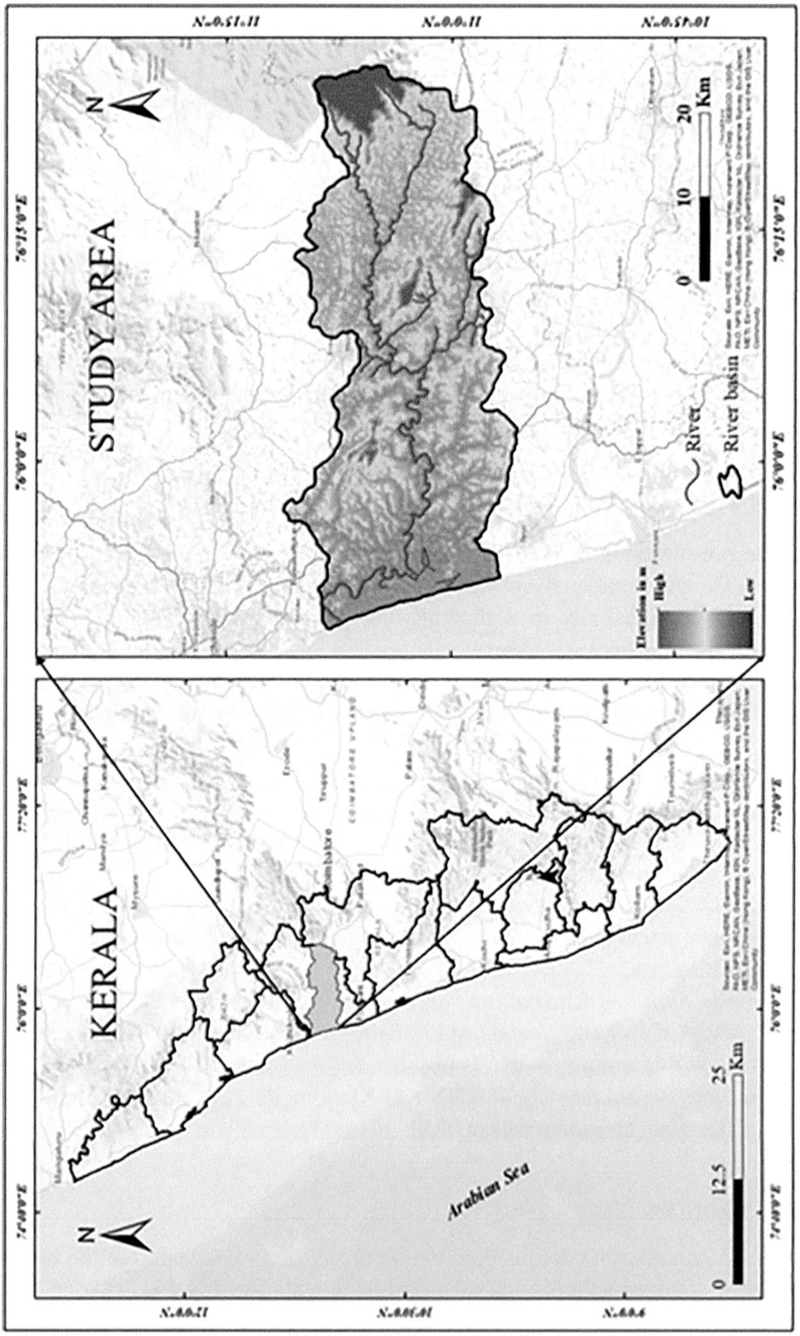

FIGURE 11.1 Study area location map.

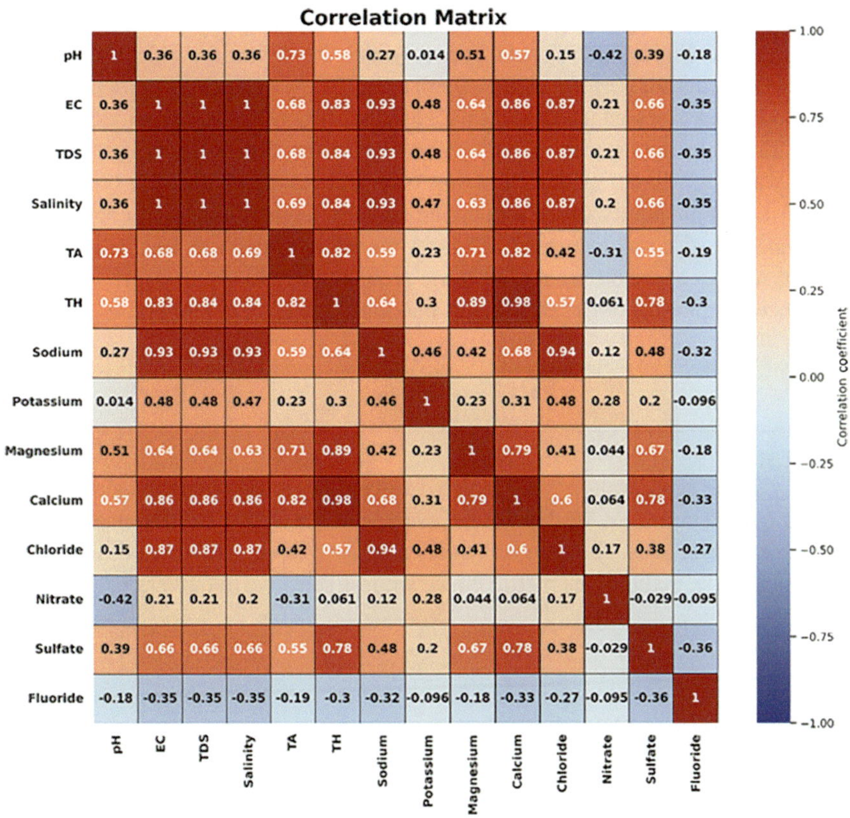

FIGURE 11.2 Pearson correlation matrix.

the learning process. Activation functions such as Rectified Linear Unit (ReLU) or Sigmoid introduce nonlinearity to the network when modeling complex relationships. To train the model, one minimizes the difference between the predicted and actual outputs matched, using common optimization algorithms such as Adam or Gradient Descent.

11.3.2 K-Nearest Neighbors

KNN is a relatively simple, non-parametric approach used in classification and regression tasks. It predicts the unknown value of a target variable by searching for k, the nearest data points in the feature space (Zamri et al., 2022). The similarity between two points is measured by some distance metric, for instance, Euclidean distance. In regression, the predicted value is usually the average or weighted average of the target values of the neighbors. KNN is particularly straightforward to implement and interpret; it works very well on short datasets. On increasing data sizes, it becomes computation-intensive and, therefore, sensitive to irrelevant or noisy attributes.

Prediction accuracy was evaluated using the Mean Absolute Error (MAE), Root Mean Squared Error (RMSE) and in a way that the lower the values from these sources, the better the performance is (Sohil et al., 2022). The generalization ability of each model was assessed using the test dataset, where predictions provided coverage greater than the training data. Comparison of each in terms of computational efficiency was made based on their training time and resources, thereby bringing out the real-life applicability of all models under large populations. The coefficient of determination (R^2) was also used to assess model performance. RMSE is the root mean of squares of the differences between predictions from actual errors, as squared differences emphasize larger errors, thus making it sensitive to outliers:

$$RMSE = \sqrt{\frac{1}{n}\sum_{i=1}^{n}\left(yi - \hat{y}_i\right)^2} \tag{11.1}$$

where, y_i is the actual value; $\hat{y}_i$ is predicted values; $\bar{y}$ is mean of actual value; n is the number of data points. MAE denotes the average of the absolute differences between predictions and actuals. Thus, one may very simply interpret a MAE value in terms of error magnitude, making this metric, however, less sensitive to outliers:

$$MAE = \frac{1}{n}\sum_{i=1}^{n}\left|yi - \hat{y}i\right|^2 \tag{11.2}$$

The R^2 metric, or the coefficient of determination, interprets how well a model explains variability in the target variable.

$$R^2 = 1 - \frac{\sum_{i=1}^{n}\left(yi - \hat{y}i\right)^2}{\sum_{i=1}^{n}\left(yi - \bar{y}\right)^2} \tag{11.3}$$

An R^2 value of 1 signifies a perfect prediction where a value of zero means that the model performs no better than the mean, and negative values signify an even worse performance than the mean.

ANNs are among the best in capturing nonlinear patterns, while KNN provides simplicity and interpretability. In their collective use, the trio of RMSE, MAE, and R^2 will furnish a clear assessment of the performance of a model, helping narrow down the most suitable approach for predicting TDS concentrations in groundwater.

Feature importance was analyzed using scatter plots on KNN, establishing the relationships between single features and TDS concentrations. Sensitivity analysis explored how varying the input features affected ANN predictions and, therefore, the relative importance of each feature for prediction tasks. Validation was done with the support of a holdout dataset of samples taken from 20 locations. The predicted TDS values were compared to values that were taken, with validation metrics calculated to assess the robustness of the different models. This comprehensive step for validation ensured that the models could perform reliably in new unobserved data, thus reinforcing their applicability in valuing groundwater quality management.

11.4 RESULTS AND DISCUSSION

Figure 11.3 lays out a comparison between the performance of KNN and ANN methods in predicting the concentration of TDS in groundwater, through a scatter plot of predicted values against actual values drawn. The diagram shows appropriate performance metrics, including the R^2 coefficient, followed by the MAE and mean squared error.

11.4.1 KNN Model Results

KNN appears to attain a slightly moderate level of accuracy, as is indicated by an R^2 value of 0.88, which further implies that the TDS concentration variability of 88% represents what is explained by the model. The model yielded an MAE of 19.39—the average absolute difference between TDS predictions and actual values—and an RMSE of 43.94, implying a relatively greater penalty on larger errors. The scatter plot in Figure 11.4 indicates that predicted values generally follow closely the actual values, as most points lay very close to the red dashed line marking the ideal y=x line. However, there are a number of data points that lie noticeably away from this line, which illustrates the point at which the model performed relatively poorly, especially in predicting higher concentrations of TDS in this instance.

11.4.2 ANN Model Results

The performance of the ANN model is shown in Figure 11.5 and indicates that ANN works better than the KNN model. A remarkable R^2 value of 0.99 indicates that the ANN has explained nearly all the variation (99%) in the TDS concentration, predicting almost perfectly. This MAE value of 7.73 indicates that there are fewer average errors in the model and this RMSE of 10.58 also implies that, compared to KNN, ANN disproportionately penalizes large errors, making them more sensitive to outliers than to smaller errors. The scatter plot signifies the almost perfect fit of the predicted values with the true values, as all the points nearly lie along the red dashed line, which is the trademark of the ANN model for perfectly modeling the relationships contained within the data.

11.4.3 Comparative Interpretation

In all respects, the performance benchmarks and scatter plots indicate that the ANN model outperforms KNN. The ANN shows far better quadratic coefficients, implying that it fits the data with significantly greater accuracy. Further, the MAE and RMSE values computed for ANN being much smaller show that an ANN model is capable of reducing both average and large-sized errors well. Thus, this implies that ANN models are better suited to handle complex, nonlinear functional relations existing within a given dataset. In contrast, KNN is often bound to close-proximity predictions, potentially suffering from generalization difficulties, especially in the faces of outliers and higher TDS concentrations. The ANN model is more accurate than any of the other models for predicting TDS concentrations in groundwater, thus aligning practically

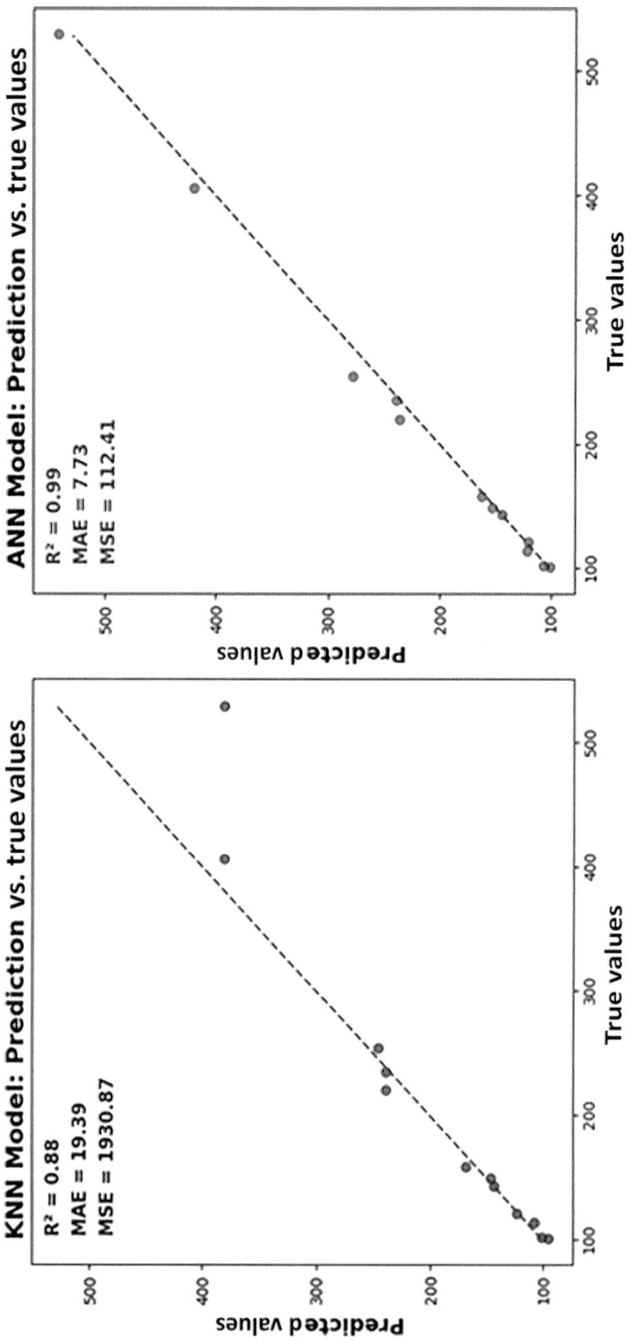

FIGURE 11.3 Comparison of performance of KNN and ANN models.

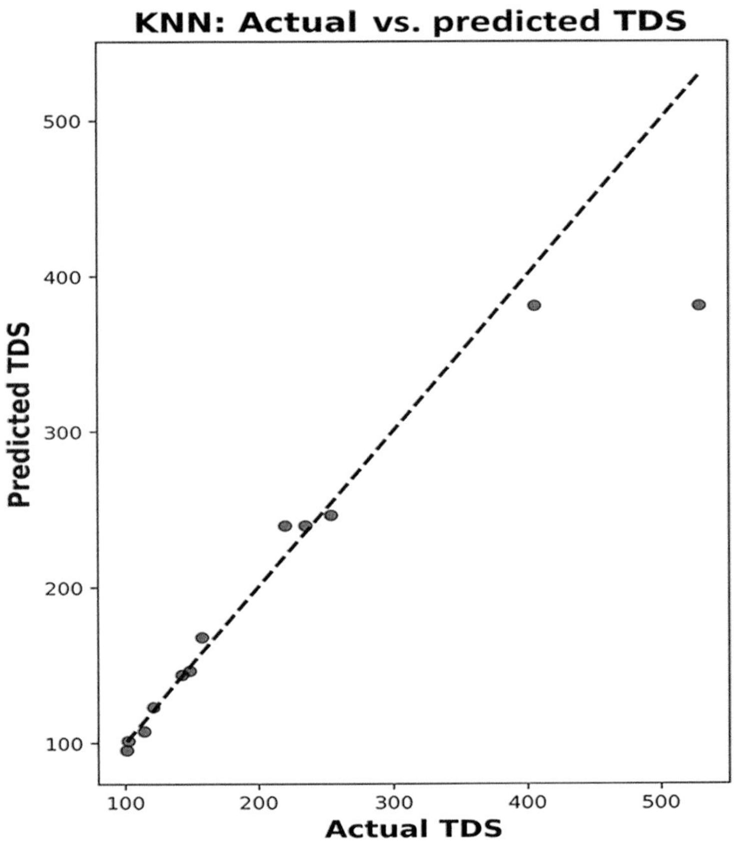

FIGURE 11.4 Scatter plot of KNN prediction model.

perfect with actual measurements, and showing human error scores relatively less, thus ANN is a very viable option for modeling complex relationships in water quality data, more so than KNN.

11.5 CONCLUSION

The study focused on employing ANN and KNN models for predicting the TDS concentrations in groundwater, utilizing pre-monsoon season data from 60 sampling locations in the study area. Extensive preprocessing steps were undertaken to clean the dataset for modeling. Missing ancillary data were added using mean imputation to ensure completeness of the data. Min–Max normalization was conducted to scale the input features uniformly and improve model efficiency and stability. Feature selection from among those most correlated with the TDS was carried out by means of Pearson correlation to shrink the dimension of the data and minimize the risk of model overfitting. This resulted in further splitting the dataset into training (80%) and test subsets (20%) by stratified sampling to best hold the

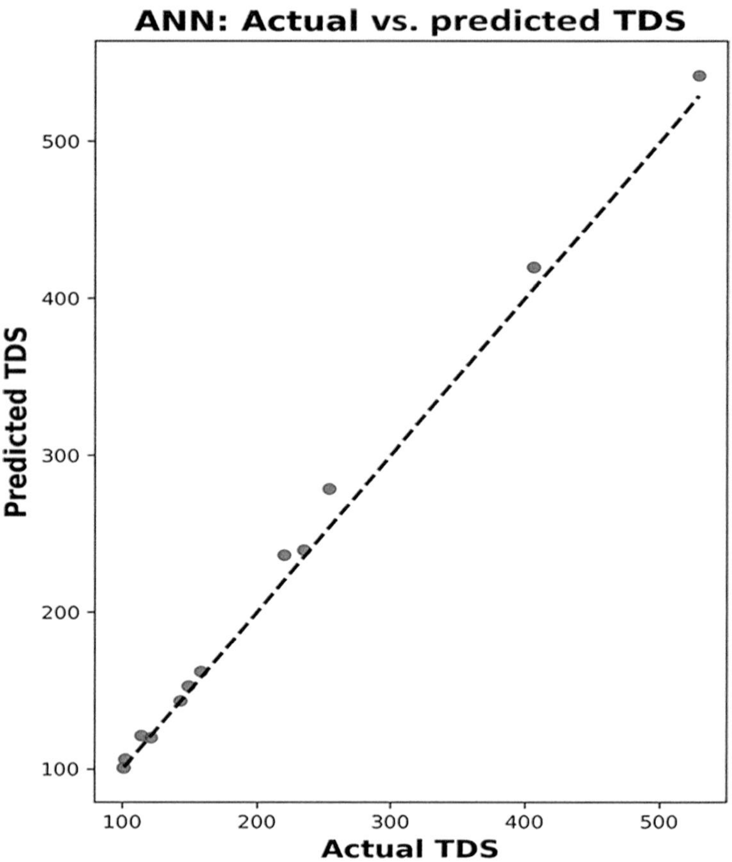

FIGURE 11.5 Scatter plot of ANN prediction model.

TDS distribution, giving the model a better basis for evaluation. ANN and KNN models were developed and compared based on their predictive accuracy, error metrics, and generalization ability. The ANN model explained 99% of the variability in TDS concentrations with $R^2 = 0.99$, while MAE was interpreted as 7.73 and RMSE as 10.58. The scatterplot of ANN closely shows that predicted values conforming to actual values cluster almost only along the ideal line. This supports the fact that ANNs can describe complex patterns present in the data with high reliability regarding groundwater quality predictions. KNN model, being easier to implement and computationally efficient, performed modestly. With an R^2 value of 0.88, KNN explained that 88% of variation of TDS concentrations was due to inputs. Its MAE was 19.39 and RMSE was 43.94, showing larger average errors and higher penalties for greater deviations from the actual versus ANN. From the scatter plot for KNN, almost all predictions, in general, aligned together with the actual values. Still, outliers and higher values on TDS concentration led to prominent deviations. This shows that KNN failed to generalize well, in particular regarding complicated relationships and the outliers present in the data.

This comparison showed that ANN was considerably more accurate, error-minimizing, and generalizable as compared with KNN. ANN is capable of modeling nonlinear interactions between variables and requires higher computational effort. KNN, being much simple and easy to implement, is more error-prone and less capable of handling highly complex datasets, particularly for outliers. Validation with a holdout dataset of 20 samples, with ANN predicting unseen data accurately throughout, reiterates its capability of practical application. Further, sensitivity analysis showed how the key features influenced the concentration of TDS, making the model interpretable and giving insights into these factors of groundwater quality. Although KNN is considered appropriate for small amounts of data or where computational efficiency is of utmost weight, its incapability to account for nonlinearity and make big errors leaves it less effective for complex prediction purposes. Therefore, in conclusion, the ANN model's superiority to KNN in predicting TDS levels in groundwater is proven. ANN's better accuracy, smaller error rates, and capability to represent complex relations well position it to be a boon for sustainable groundwater management. Such advanced ML methods like ANN provide decision-makers with reliable assessments of groundwater quality and act toward the effective management and protection of this vital resource.

REFERENCES

Adjovu, G. E., Stephen, H., James, D., and Ahmad, S. (2023). Measurement of total dissolved solids and total suspended solids in water systems: A review of the issues, conventional, and remote sensing Techniques. *Remote Sensing*, 15(14), 3534.

Aranguren-Díaz, Y., Galán-Freyle, N. J., Guerra, A., Manares-Romero, A., Pacheco-Londoño, L. C., Romero-Coronado, A., Vidal-Figueroa, N., and Machado-Sierra, E. (2024). Aquifers and groundwater: Challenges and opportunities in water resource management in Colombia. *Water*, 16(5), 685.

Asadollahfardi, G., Taklify, A., and Ghanbari, A. (2012). Application of artificial neural network to predict TDS in Talkheh Rud River. *Journal of Irrigation and Drainage Engineering*, 138(4), 363–370.

Barroso, A., Valente, T., Marinho Reis, A. P., and Antunes, I. M. H. R. (2023). A new acidity-based approach for estimating total dissolved solids in acidic mining influenced water. *Water*, 15(16), 2995.

BIS (Bureau of Indian Standards) 10500. (2012). New Delhi, India.

District Survey Report of Minor Minerals (Except River Sand) Malappuram District. (2016). Department of Mining and Geology, Govt. of Kerala.

Haggerty, R., Sun, J., Yu, H., and Li, Y. (2023). Application of machine learning in groundwater quality modeling—A comprehensive review. *Water Research*, 233, 119745.

Hussein, E. E., Derdour, A., Zerouali, B., Almaliki, A., Wong, Y. J., Ballesta-de Los Santos, M., Minh Ngoc, P., Hashim, M. A., and Elbeltagi, A. (2024). Groundwater quality assessment and irrigation water quality index prediction using machine learning algorithms. *Water*, 16(2), 264.

Khaki, M., Yusoff, I., and Islami, N. (2015). Application of the artificial neural network and neuro-fuzzy system for assessment of groundwater quality. *CLEAN—Soil, Air, Water*, 43(4), 551–560.

Kumar, D., Thakur, S., Bhardwaj, R. K., and Sharma, S. (2021). Anthropogenic activity induced water quality degradation in Girital Lake, Uttarakhand. *International Research Journal of Engineering and Technology*, 9, 1083–1087.

Manning, T., Sleator, R. D., and Walsh, P. (2014). Biologically inspired intelligent decision making: A commentary on the use of artificial neural networks in bioinformatics. *Bioengineered*, 5(2), 80–95.

Mao, X., Zhang, S., Wang, S., Li, T., Hu, S., and Zhou, X. (2023). Evaluation of human health risks associated with groundwater contamination and groundwater pollution prediction in a landfill and surrounding area in Kaifeng City, China. *Water*, 15(4), 723.

Pourmorad, S., Kabolizade, M., and Dimuccio, L. A. (2024). Artificial intelligence advancements for accurate groundwater level modelling: An updated synthesis and review. *Applied Sciences*, 14(16), 7358.

Singh, S., Kumar, D., Vishwakarma, D. K., Kumar, R., and Kushwaha, N. L. (2024). Seasonal rainfall pattern using coupled neural network-wavelet technique of southern Uttarakhand, India. *Theoretical and Applied Climatology*, 155(6), 5185–5201.

Sohil, F., Sohali, M. U., and Shabbir, J. (2022). An introduction to statistical learning with applications in R: By Gareth James, Daniela Witten, Trevor Hastie, and Robert Tibshirani, New York, Springer Science and Business Media, 2013, $41.98, eISBN: 978-1-4614-7137-7. *Statistical Theory and Related Fields*, 6(1), 87–87.

Zamri, N., Pairan, M. A., Azman, W. N. A. W., Abas, S. S., Abdullah, L., Naim, S., Tarmudi, Z., & Gao, M. (2022). River quality classification using different distances in k-nearest neighbors algorithm. *Procedia Computer Science*, 204, 180–186.

Zango, B.-S., Seidou, O., Sartaj, M., Nakhaei, N., & Stiles, K. (2022). Impacts of urbanization and climate change on water quantity and quality in the Carp River watershed. *Journal of Water and Climate Change*, 13(2), 786–816.

12 Remote Sensing and GIS Applications for Solid Waste Management in Hilly Terrains

Shiva Tiwari, Shubhi Khare, Deepak Kumar, and Victor Nyabuti Ongéra

12.1 INTRODUCTION

The term 'solid waste' is used to describe non-liquid waste materials arising from domestic, trade, commercial, agricultural, industrial activities, and public services (Aibor & Olorunda, 2006). Solid waste management (SWM) is a vital service for public health, environmental protection, and sustainable development. Hilly areas characterized by steep slopes, varying elevations, and dispersed settlements present specific challenges for effective SWM. Difficult terrain hinders waste collection, transportation, and disposal, often leading to uncontrolled dumping, pollution of water resources, and visual impairment. Traditional SWM approaches are often designed and unstable in these complex environments. Remote sensing (RS) and geographic information systems (GIS) provide powerful tools to address these challenges. RS provides an overview of the terrain, enabling the acquisition of valuable data on land use, topography, and waste accumulation. GIS integrates this data with other spatial information, facilitating analysis, modeling, and decision-making for optimized SWM strategies.

The world's cities produce approximately 1.3 billion tons of solid waste annually. This volume is expected to increase to 2.2 billion tonnes by 2025. The waste generation rate will more than double in the next 20 years in low-income countries. Globally, SWM costs will increase from today's annual \$205.4 billion to approximately \$375.5 billion in 2025. As the world moves towards its urban future, the volume of municipal solid waste (MSW), one of the most important by-products of urban lifestyle, is increasing faster than the rate of urbanization. Ten years ago, there were 2.9 billion urban residents who generated about 0.64 kg of MSW/person/day (0.68 billion tonnes/year). In 2012, MSW increased to about three billion inhabitants, producing 1.2 kg/person/day (1.3 billion tonnes/year). This is expected to be 4.3 billion urban residents by 2025, who will generate approximately 1.42 kg/person/day of MSW (2.2 billion tonnes/year) (Hoornweg & Bhada-Tata, 2012).

Solid waste disposal site suitability analysis using GIS tools is very important to solve all the problems discussed above. Also, in complex decision-making processes

DOI: 10.1201/9781003596080-12

involving multidisciplinary layers and their pairwise comparisons, the analytics hierarchy process (AHP) has proven to be a very efficient decision-making tool (Akbari et al. 2008).

12.2 WHY IS WASTE MANAGEMENT IMPORTANT?

Waste-related problems have both temporal and spatial aspects. The objectives of waste management are to avoid or minimize, as much as possible, harmful or adverse effects on humans, animals, plants, and their environment; to reduce the emission of air pollutants and climate-related gases; to conserve natural resources (raw materials, water, energy, landscapes, land), and to reduce the hazard potential of waste obtained by recycling materials. SWM involves several activities, from its generation to final disposal (Dutta & Goel, 2017). The activities involved can be grouped into six functional elements as shown in Figure 12.1.

To improve the SWM system, various policies and technologies have been implemented, such as waste prevention at the point of generation, reuse, and recycling. Innovation technologies are being used, and certain software, such as RS, GIS, and mathematical optimization methods, can allocate and use resources efficiently (Sudha 2008).

12.3 APPLICATIONS OF RS AND GIS IN SUSTAINABLE SOLID WASTE MANAGEMENT

In recent years, management of the huge amount of solid waste generation has become a major concern, and in this regard, GIS has proved to be a boon for planners. RS and GIS have become essential tools in modern waste management, helping authorities

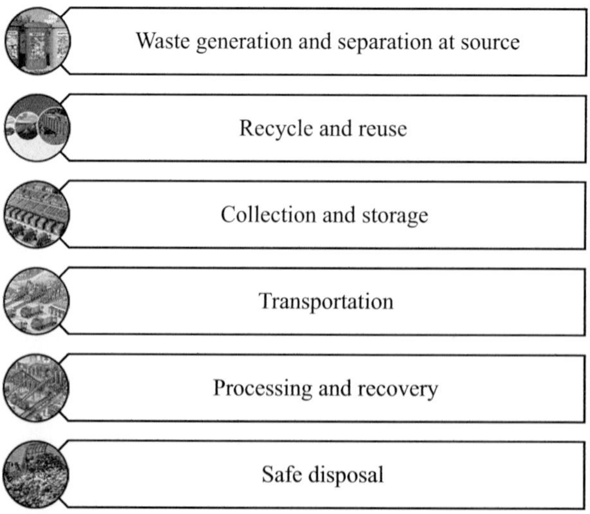

FIGURE 12.1 SWM activities.

efficiently monitor, plan, and implement waste disposal strategies. The integration of these technologies involves data collection, spatial analysis, and decision-making processes in urban and regional waste disposal, addressing critical issues such as landfill site selection, waste bin allocation, environmental impact assessment, and hazardous waste monitoring. As the urban population grows and waste generation increases, RS and GIS offer an innovative approach to managing waste sustainably while minimizing environmental hazards. GIS has proved to be very helpful in reducing time, and it is also very cost-effective. GIS can be used in various fields such as urban planning, transportation, conservation and management of natural disaster management, health sciences, environment modelling, and engineering (Brimicombe 2003). The most common applications of GIS in SWM are the location of waste management and disposal facilities and the optimization of waste collection and transportation (WC&T) (Ghose et al. 2006). Using RS with GIS allows decision-makers to optimize MSWM by improving its efficiency and increasing economic savings. Population distribution data and data on the amount of solid waste generated in a particular area are collected. Different layers of population and solid waste data are created in GIS. All layers are then placed (or superimposed) over the study area for subsequent analysis. The layers can be analyzed, areas with higher solid waste generation can be identified, and classification into different categories and appropriate measures can be implemented. Thus, RS and GIS can be used to identify and classify areas based on solid waste generation rates and to identify sites for the disposal of solid waste.

Mathematical models can also be used to predict solid waste generation rates. A prediction technique based on nonlinear dynamics has been successfully applied to SWM (Navarro-Esbrı et al. 2002). It is compared with the seasonal autoregressive and moving average methods. Both these methods give good quality results, and using these results, MSW managers can make predictions, which is very useful in SWM to optimize the use of resources.

12.3.1 Temporary Storage of Waste

Temporary storage and proper segregation of solid waste after its generation are an important part of an efficient SWM plan. For waste storage, the proper placement of dustbins on the street is very important. Dustbins should be placed to make the collection and transportation of collected waste an easy process. Various GIS techniques have been used to locate dustbins in areas having efficient transportation systems. As per Indian Guidelines (CPHEEO), the distance between two dustbins should not be more than 500 meters, and the distance between the place of waste generation and collection, that is, house and dustbin, should be a maximum of 250 meters (Liebman et al. 1975). The Environmental System Research Institute's (ESRI) Arc/Info program can also be used for such analysis. Detailed information, such as a base map of the study area, the road network present in the area of study, the data providing the road elevations, as well as the information on road class, is needed for planning and managing the system (Vijay et al. 2008). Thematic maps and attributed data are linked to the coverage in GIS, and calculations are done through GIS-based Arc Macro Language (AML) algorithms.

12.3.2 Solid Waste Collection and Routing Efficiency

Waste collection is the responsibility of the local authorities or municipalities. In low-income countries, waste collection and transportation (WC&T) accounts for around 80–90% of the overall SWM budget, while collection efficiency remains below 50%. In contrast, in high-income countries, collection efficiency is over 90% and collection is <10% of the total SWM budget (Hoornweg & Bhada-Tata, 2012). Therefore, optimizing collection routes and bin locations is a vital need in low-income countries. Some of the problems faced during waste collection are changes in waste generation rates over time, large service areas, transportation, labor expenses, and a lack of proper planning.

GIS is used to develop optimal routes for collecting solid waste using analysis. GIS is used to develop optimal routes for the collection of solid waste using network analysis. The Network module of Arc/Info GIS software was then used to provide network-based spatial analysis including routing, travel direction, adjacent facility centers, and service area analysis. ArcGIS Network Analyst helps users dynamically model realistic network conditions, including restrictions on speed, diversions, and limiting transportation facilities, at different times of the day. Network analyzer users can find efficient traffic routes, determine the nearest facility or vehicle, develop travel directions, and find nearby service areas. Using a network analyzer, the optimal route for waste collection was developed based on two criteria: distance and time (Chimote and Bhabhulkar 2012). The optimal routes were delineated using GIS and were less time-consuming and hence cost-effective. The software-based analysis was simple to understand and much faster than manual analysis.

A model for a solid waste monitoring and management system using Radio Frequency Identification (RFID), GPS, and GIS was developed by Purohit and Bothale (2011). In this model, an RFID reader was loaded in the solid waste collection truck. This advanced technology provided information about bin status, collection time, collection truck location, and other parameters in real time, which was useful for managing and monitoring the collection of solid waste from bins. GIS, along with mathematical programs and computerized database management systems, can be used to explore different strategies for the collection of solid wastes in different situations (Chang et al. 1997). Analytical modeling and discrete event simulation were applied to assess various scheduling and routing policies using real-time data. It was found that dynamic scheduling and routing policies result in lower operating expenses, shorter collection and transportation routes, less time consumption, and shorter labor service time, while the static scheduling policy with predetermined routes and pick-up frequency resulted in lower operating expenses, shorter collection and transportation routes, less time consumption, and shorter labor service time (Johansson 2006).

12.3.3 Landfills Location Selection

Irrespective of how MSW has been collected, the final disposal of the residual waste must be in a landfill (Gbanie et al. 2013). Site selection for landfilling is a complex process that requires considering geological, hydrological, land use, sensitive sites, air quality, environmental, geotechnical parameters, and government regulations. The landfill must have enough capacity to hold all the waste that will be deposited within 20–25 years. Various criteria have been proposed for landfill siting (Theisen and Vigil 1993). The selection of the zone segregation method follows the given screening criteria.

- Separation and buffering of regions.
- Conducting 'what-if' data analysis and evaluating various possible scenarios relating to population growth and development of the region, and monitoring the significance of various influencing factors, etc.
- Managing and correlating large amounts of complex geographic data.
- Graphical representation of the results.

12.4 GIS-BASED MULTI-CRITERIA DECISION ANALYSIS

Multi-criteria decision-making (MCDM) techniques have evolved alongside GIS and serve as a key tool for decision-makers. More recently, a combination of MCDM techniques with GIS was used to develop map overlays and advanced site suitability analysis (Jovanovic et. al., 2016; Shahnazari et al., 2020; Garcia-Garcia, 2022). MCDM, also known as multi-criteria decision analysis (MCDA), is a powerful spatial decision-making tool that integrates multiple criteria to identify the best possible solutions to complex geographic problems. This technique is widely used in areas such as environment, site selection, transportation planning, and waste management. By combining GIS with MCDA, decision-makers can systematically evaluate spatial data, analyze competing factors, and prioritize optimal outcomes. The major components of GIS-based MCDA are: decision criteria which depend on multiple factors or parameters such as environmental, social, and economic influence on decision-making; decision options that are possible alternatives or locations evaluated based on the criteria; weighting of criteria which involves assigning relative importance to each criterion on expert opinion or statistical methods; spatial data analysis constituting the processing of geographic data layers in a GIS environment; decision models used for applying mathematical models to integrate and analyze spatial data (Hoang et. al., 2019; Kharat et. al., 2020; ÖZKURT, 2024). The basic functioning of the MCDA is represented by Figure 12.2.

The first stage involves identifying the decision problem, defining the objective, and outlining constraints. For example, in waste management, the objective might be to find the most suitable landfill site while minimizing social impact. In the second stage, the decision criteria are identified based on expert consultation, literature reviews, and stakeholder input. In the selection of a landfill site, relevant criteria may include environmental factors, constituting the distance from water bodies, slope, soil type, and air quality; economic factors taking into account the proximity to the road network, land value; social factor, that is, the distance from residential areas, public acceptance; and regulatory factor compliance with government policies and environmental laws. Then each option (i.e., different locations for landfill sites) is evaluated based on these criteria. Geospatial data in the third stage is collected from a variety of sources, including RS, providing satellite images, aerial photographs, Digital

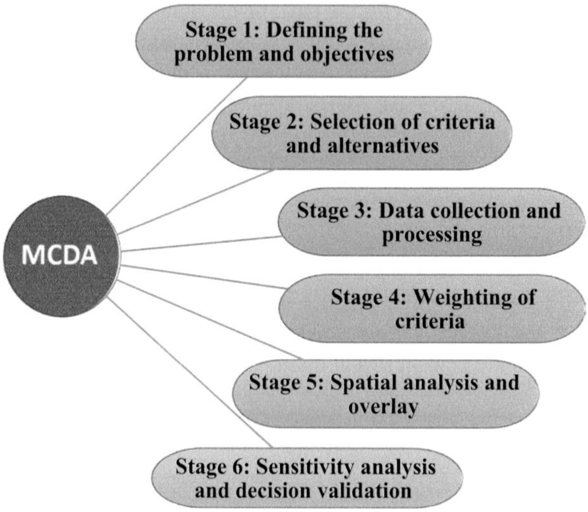

FIGURE 12.2 Methodology adopted by MCDA.

Elevation Models (DEM) used for topographic data for slope analysis; thematic maps indicating land use, hydrology, soil type, and vegetation cover; survey and census data highlighting population density, socio-economic indicators. These datasets are processed in GIS software to generate standardized raster or vector layers. The fourth stage is to ensure an objective decision-making process; different techniques are used to assign importance to each criterion. The most common weighting methods include:

- *AHP:* A structured paired comparison method in which experts assign weights based on the relative importance of criteria.
- *Weighted Linear Combination (WLC):* Criteria weights are normalized and combined into a single decision layer.
- *Fuzzy Logic:* Used to handle uncertainty and ambiguous data by assigning membership values to criteria.
- *Entropy Methods:* These calculate objective weights based on information entropy to minimize bias.

The fifth stage is the use of GIS tools to overlay various weighted criterion layers using mathematical models, for example, Boolean overlay that uses logical (true/false) operations to identify appropriate areas; weighted sum models (WSM) which combine multiple weighted criterion layers to generate to a suitability index; and ordered weighted average (OWA) allowing control over trade-offs between criteria by adjusting the risk level. The resulting map ranks the possible options from the most appropriate. Sensitivity analysis, that is, the sixth stage of the methodology of MCDA, tests the robustness of a decision model by varying the weight values and observing changes in the results. Decision-makers validate the finding using the ground truth verification, expert consultation, obtaining feedback from planners and stakeholders, and comparative analysis.

12.5 CASE STUDY ON WASTE DISPOSAL SITE SELECTION USING SPACE TECHNOLOGY

A study in Lucknow, India, used GIS-based MCDA to identify optimal landfill sites. Factors such as distance from water bodies, urban areas, and road networks were analyzed using the AHP. The results successfully identified a landfill site that minimized environmental and social impact (Kumar et al. 2014). GIS applications have recently been used by municipal authorities as decision support systems to manage waste collection and transportation, reduce fuel consumption, and develop work schedules. GIS applications for workers and vehicles are suggested (Khajuria et al. 2011). GIS applications have been used to analyze already existing maps and data. GIS applications were to generate information about static and dynamic parameters in MSWM and to optimize routing systems (Apaydin and Gonullu 2008).

Decision support systems for SWM with sustainability in mind were adopted to divert residential waste from the landfill site in Sahibzada Ajit Singh Nagar (John 2010). Decision support systems are computer-based information systems (CBIS) designed to identify the best possible solution to a problem. The application of a CBIS makes the decision-making process more effective, especially when large volumes of data have to be handled. Decision Support System (DSS) models integrated with GIS can be used to optimize collection, transportation, processing, and disposal methods for SWM. DSS are a specific class of computerized information systems that provide a structured and logical basis for decision-making. All kinds of information can be compiled in these systems such as raw data, documents, personal knowledge, and business models, to resolve problems and propose suitable decisions. The basic structure of a DSS includes three components: database management, model-based management, and user interface (Bani et al. 2009). In Finland, an MCDM model associated with a centralized/decentralized management model was developed with methods such as incineration, landfilling, composting, and waste-derived fuel production (Hokkanen and Salminen 1997).

An MCDM tool in a study was developed by Hazra et. al. in 2017 to make improvement planning of the SWM system of the Kolkata Municipal Corporation area. Both the direct costs arising from the household waste, as well as indirect costs due to the inconvenience of the system arising with the environmental costs, were evaluated in the study. The direct costs were tried to be minimized by developing a linear model in the study area, as they were the expenditure of municipal authorities in the collection of waste, and the indirect costs, which are generally neglected in the studies, were calculated using state choice data collection from residents of the Kolkata Municipal Corporation area. The environmental costs were calculated by assessing the different solid waste components of the environment. The study concluded that the MCDM tool can be used to evaluate different SWM alternatives and can be recommended for planning of SWM management policies.

12.6 CONCLUSIONS

An increase in the quantity of MSW and poor collection efficiency causes nuisance and health hazards. The application of RS and GIS has proven to be an efficient and effective tool for improving waste collection in urban areas. GIS is capable of handling both spatial and non-spatial data to design successful solid waste collection systems. RS and GIS can provide better or optimal shortest routes that are cheap and efficient. GIS is a dynamic method to reduce operational costs and update information regularly, making the monitoring and management of urban space more transparent, timely, and realistic.

REFERENCES

Aibor, M. S., and Olorunda, J. O. (2006). *A Technical Handbook of Environmental Health in the 21st Century for Professional Students*. Akure: His mercy publisher.

Akbari, V., Rajabi, M. A., Chavoshi, S. H., and Shams, R. (2008). Landfill site selection by combining GIS and fuzzy multi criteria decision analysis, case study: Bandar Abbas, Iran. *World Applied Sciences Journal*, 3(1), 39–47.

Apaydin, O., and Gonullu, M. T. (2008). Emission control with route optimization in solid waste collection process: A case study. *Sadhana*, 33, 71–82.

Bani, M. S., Rashid, Z. A., Hamid, K. H. K., Harbawi, M. E., Alias, A. B., and Aris, M. J. (2009). The development of decision support system for waste management: A review. *International Journal of Chemical and Molecular Engineering*, 3(1), 17–24.

Brimicombe, A. J. (2003, May). A variable resolution approach to cluster discovery in spatial data mining. In *International Conference on Computational Science and Its Applications* (pp. 1–11). Berlin, Heidelberg: Springer Berlin Heidelberg.

Chang, N. B., Lu, H. Y., and Wei, Y. L. (1997). GIS technology for vehicle routing and scheduling in solid waste collection systems. *Journal of Environmental Engineering*, 123(9), 901–910.

Chimote, K., and Bhabhulkar, A. (2012, March). Municipal solid waste (MSW) collection by geographical information system (GIS). In *National Conference on Innovative Paradigms in Engineering & Technology (NCIPET-2012)*. Proceedings published by International Journal of Computer Applications (IJCA).

Dutta, D., and Goel, S. (2017). Applications of remote sensing and GIS in solid waste management–A review. [In *Advances in Solid and Hazardous Waste Management*, 133–151. Springer Nature, Capital Publishing Company, New Delhi.

Garcia-Garcia, G. (2022). Using multi-criteria decision-making to optimise solid waste management. *Current Opinion in Green and Sustainable Chemistry*, 37, 100650.

Gbanie, S. P., Tengbe, P. B., Momoh, J. S., Medo, J., and Kabba, V. T. S. (2013). Modelling landfill location using geographic information systems (GIS) and multi-criteria decision analysis (MCDA): Case study Bo, Southern Sierra Leone. *Applied Geography*, 36, 3–12.

Ghose, M. K., Dikshit, A. K., and Sharma, S. K. (2006). A GIS based transportation model for solid waste disposal–A case study on Asansol municipality. *Waste Management*, 26(11), 1287–1293.

Hazra, T., Maitra, B., and Goel, S. (2017). Development and application of a multi-criteria decision making (MCDM) tool for solid waste management: Kolkata as a case study. In *Advances in Solid and Hazardous Waste Management*, 275–299. Springer Nature, Capital Publishing Company, New Delhi.

Hoang, G. M., Fujiwara, T., Pham Phu, T. S., and Nguyen, L. D. (2019). Sustainable solid waste management system using multi-objective decision-making model: A method for maximizing social acceptance in Hoi An city, Vietnam. *Environmental Science and Pollution Research*, 26, 34137–34147.

Hokkanen, J., and Salminen, P. (1997). Choosing a solid waste management system using multicriteria decision analysis. *European Journal of Operational Research*, 98(1), 19–36.

Hoornweg, D., and Bhada-Tata, P. (2012). What a waste: A global review of solid waste management, World Bank, Washington, DC, 2012-03.

Johansson, O. M. (2006). The effect of dynamic scheduling and routing in a solid waste management system. *Waste Management*, 26(8), 875–885.

John, S. (2010). Sustainability-based decision-support system for solid waste management. *International Journal of Environment and Waste Management*, 6(1-2), 41–50.

Jovanovic, S., Savic, S., Jovicic, N., Boskovic, G., and Djordjevic, Z. (2016). Using multi-criteria decision making for selection of the optimal strategy for municipal solid waste management. *Waste Management & Research*, 34(9), 884–895.

Khajuria, A., Matsui, T., and Machimura, T. (2011). GIS application for estimating the current status of municipal solid waste management system: Case study of Chandigarh city, India. *Our Nature*, 9(1), 26–33.

Kharat, M. G., Murthy, S., Kamble, S. J., and Kharat, M. G. (2020). Selecting sustainable technologies for municipal solid waste treatment and disposal: An expert based MCDM approach. *The Journal of Solid Waste Technology and Management*, 46(1), 44–57.

Kumar, V., Yadav, K., and Rajamani, V. (2014). Selection of suitable site for solid waste management in part of Lucknow city, Uttar Pradesh using remote sensing, GIS and AHP method. *International Journal of Engineering Research*, 3(3), 1461.

Liebman, J. C., Male, J. W., and Wathne, M. (1975). Minimum cost in residential refuse vehicle routes. *Journal of the Environmental Engineering Division*, 101(3), 399–412.

Navarro-Esbrı, J., Diamadopoulos, E., and Ginestar, D. (2002). Time series analysis and forecasting techniques for municipal solid waste management. *Resources, Conservation and Recycling*, 35(3), 201–214.

ÖZKURT, P. (2024). A MCDM approach to the facility location selection problem of sustainable solid waste management. *Present Environment & Sustainable Development*, 18(2), 101–117.

Purohit, S. S., and Bothale, V. M. (2011, September). RFID based solid waste collection process. In *2011 IEEE Recent Advances in Intelligent Computational Systems* (pp. 457–460). IEEE.

Shahnazari, A., Rafiee, M., Rohani, A., Nagar, B. B., Ebrahiminik, M. A., and Aghkhani, M. H. (2020). Identification of effective factors to select energy recovery technologies from municipal solid waste using multi-criteria decision making (MCDM): A review of thermochemical technologies. *Sustainable Energy Technologies and Assessments*, 40, 100737.

Sudha, G. (2008). Municipal solid waste management (MSWM) in India a critical review. *Journal of Environmental Science and Engineering*, 50(4), 319–328.

Theisen, H., and Vigil, S. A. (1993). *Integrated Solid Waste Management: Engineering Principles and Management Issues*. McGraw-Hill.

Vijay, R., Gautam, A., Kalamdhad, A., Gupta, A., and Devotta, S. (2008). GIS-based locational analysis of collection bins in municipal solid waste management systems. *Journal of Environmental Engineering and Science*, 7(1), 39–43.

13 Water Quality Assessment of Ashtamudi Ramsar Wetland System, India, for Achieving Environmental Sustainability

A GIS-Based Approach

A. Krishnakumar, U. Vishnu, Jeenu Jose, and K. Anoop Krishnan

13.1 INTRODUCTION

Wetlands include areas of land that are either covered or saturated with water. They are one of the most threatened ecosystems across the globe (Krishnakumar *et al.*, 2016). Wetlands provide multifaceted ecosystem services that yield substantial benefits for human health, wellbeing, and ecological sustainability, while offering a cost-efficient solution for achieving broader environmental objectives (Krishnakumar *et al*, 2017). They play an important role in recharge of groundwater, controlling water flow, carbon storage and act as an interface between various forms of geography and environment (Krishnakumar *et al*, 2015). They also help in treating of sewage naturally, controlling floods and fish production. The management and conservation of wetlands are in sync with the multiple Sustainable Development Goals (SDGs), which include SDG-6 (clean water and sanitation for all), SDG-11 (sustainable cities and human settlements), SDG-13 (immediate action to counter climate change), and SDG-14 (conservation of marine resources) (SDG, 2016).

In recent decades, the wetlands are used for real estate and construction activities across major cities of India. A report by Animesh Jain says that in between the years of 1970 and 2015, Chennai, Mumbai, Hyderabad, Bengaluru, New Delhi, and Pune lost 90, 71, 55, 55, 38, and 37 percentage of wetlands, respectively. The State Wetland Authority (SWA) acts as the nodal agency for managing wetlands in each

 DOI: 10.1201/9781003596080-13

state. In spite of multiple state and central agencies having administrative control over wetlands, wetland conservation and management are not always a priority within their regulatory frameworks.

The Indian government has put forward various legislations for the conservation and management of wetlands. In 2016, Wetlands (Conservation and Management) Rules (2010), were brought under the Environment Protection Act, 1986, to better safeguard wetlands and improve the way State Wetland Authorities (SWAs) work. In January 2021, for management of wetlands, the National Mission for Clean Ganga made a toolkit that recognizes the role played by wetlands in various functions that include groundwater recharge, helping in controlling erosion and increase in base flow of rivers. The developed toolkit uses Geographic Information System (GIS) and Remote Sensing (RS) technology in making databases of wetlands and associated water bodies. In the year 2023 to conserve and sustainably manage wetlands, the National Plan for Conservation of Aquatic Ecosystems (NPCA) was formulated by the Ministry of Environment, Forest and Climate Change (MoEFCC). Furthermore, the Indian government's new "Amrit Dharohar Capacity Building" scheme launched in the year 2023 promotes unique conservation values of the Ramsar Sites in the country while generating employment opportunities and supporting local livelihood. All these initiatives are in concordance with the Convention on Wetlands, also known as the Ramsar Convention, to which India became a member in the year 1982. India has since designated 85 wetlands covering an area of 13,58,067.757 hectares as Ramsar sites. Presently, India has the maximum number of wetlands all across Asia. Ashtamudi Wetland System (AWS), one of the designated Ramsar wetlands in India situated in Southwestern coast of Kerala, India is the current study area. This study aims to assess the spatiotemporal variation of water quality with special reference to quality indices.

13.2 MATERIALS AND METHODS

13.2.1 Study Area

Ashtamudi Wetland System (AWS), a designated Ramsar site located in the state of Kerala, Kollam district, is a complex of lakes, estuaries, and lagoons covering an area of approximately 61.4 km^2 as shown in Figure 13.1. It is the second-largest wetland ecosystem in Kerala and supports a varied range of flora and fauna, including rare and endangered species. AWS is located between 8°53′14″ N to 9°0′57″ N latitude and 76°31′54″ E to 76°40′2″ E longitude (Figure 13.1). The AWS, along with its various subenvironments, corresponding locations, and the number of samples collected, is shown in Table 13.1. Kallada river system (formed by confluence of three small rivers, namely the Kulathupuzha, the Chendurni and the Kalthuruthy), originating from Kulathupuzha hills of Western Ghats, is the major freshwater flowing to AWS, having an annual discharge of 75×10^9 m^3 of water (Aswathy et al., 2021). This wetland system supports thousands of fisherfolk. The system is experiencing a tropical humid climate and receives both the south-west monsoon (June to September) and the north-east monsoon (October to November). The lithology of AWS consists primarily of Tertiary sedimentary formations (laterite, sandstone, clay of Warkalli) and Quaternary deposits (marine, fluvial, and alluvial source) (Soman, 2002).

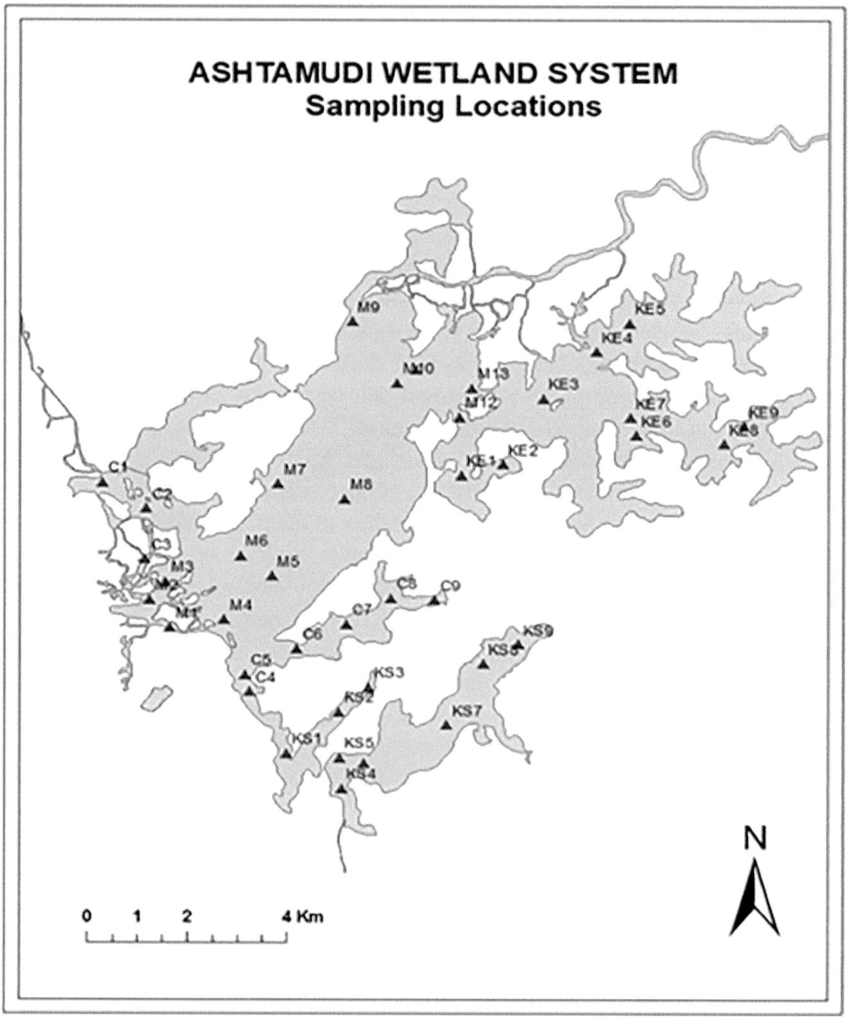

FIGURE 13.1 Study area with sampling locations.

13.2.2 Surface Water Collection and Analysis

In this study, river water samples were collected (n = 80) in 2021 from 13 sampling locations in Ashtamudi Main Lake (AML), nine sampling locations each in Chavara Kayal (CK), Kumbalath-Kanjirakottu Kayal (KE), Kureepuzha-Kandachira Kaayal (KS) in two seasons: monsoon and non-monsoon. Sampling locations were identified and fixed using GPS coordinates. The map shows sampling locations as depicted as Figure 13.1. To ensure accurate data, samples were taken from a depth of 20–30 cm below the water surface to avoid surface scum, debris, and macrophytes. Samples were collected in pre-washed High-Density Polyethylene (HDPE) bottles, which were rinsed three` times with sample water before use. The bottles were properly labeled and sealed at the collection site. Following the protocols set by APHA

TABLE 13.1
Ashtamudi Wetland System with Various Subenvironments and Location Details with Number of Samples Collected

S. No.	Name	Location		Total No. of Samples
		Latitude	Longitude	
1	Ashtamudi Main Lake	8°56′9″–8°59′25″ N	76°32′1″–76°35′46.7″ E	13 × 2
2	Chavara Kayal	8°55′7″–8°57′47″ N	76°32′2″–76°35′59″ E	9 × 2
3	Kureepuzha Kandachira Kayals	8°54′13″–8°55′45″ N	76°34′13″–76°36′57″ E	9 × 2
4	Kumbalathu-Kanjirakottu Kayals	8°57′1″–8°59′58″ N	76°36′1″–76°39′17″ E	9 × 2
				Total n = 80

(1995), the samples were transported to the laboratory for analysis within the required time frame. Physical parameters such as temperature (air and water), pH, Electrical Conductivity (EC), and Total Dissolved Solids (TDS) were measured in situ using an Aquaread multiparameter water quality analyzer (AP-2000-D). Total hardness (TH), total alkalinity, and dissolved oxygen (DO) were estimated by titration method. Biochemical oxygen demand (BOD) was estimated using 5-day BOD method. The predominant anions (NO_3^-, PO_4^{3-}, SO_4^{2-}, Cl^-) were measured using a Continuous Flow Analyzer (CFA, Model: Skalar SAN++), and Ultraviolet-Visible-Near-Infrared (UV-VIS-NIR) spectrophotometer, available at the Central Chemical Laboratory, NCESS, Thiruvananthapuram. Samples for cation analysis were immediately acidified with 2M HNO_3 after collection. ArcGIS version 10.7.1 was utilized to import the sampling site's GPS coordinates within the study area and it was employed to create different maps in the research area. The spatial distribution maps of all water quality parameters were created using the inverse distance weighting (IDW) spatial interpolation model with ArcGIS software.

13.2.3 Evaluating the Water Quality for Different Beneficial Uses

The Water Quality Index (WQI) is a mathematical calculation to integrate physical, chemical and biological parameters into a numerical score that represents the overall status of the water quality under consideration (Saha et al., 2021). Brown et al. (1970) proposed the WQI method based on the National Sanitation Foundation Water Quality Index (NSFWQI).

The important requirements of a WQI are as follows:

1. It should be sensitive to changes in the relevant variables' value(s).
2. It should attain a very low value (perhaps zero), when a relevant single variable reaches a certain value beyond which the water quality for a given use is considered unsuitable.

TABLE 13.2
Relevant Water Quality Variables

Beneficial Use	Relevant Quality Variables
Bathing, swimming	Turbidity, BOD, DO, nitrogen-NH_3, pH
Industry	pH, turbidity, conductivity, hardness
Fish culture and non-contact recreation	Temperature, BOD, DO, chloride

BOD = Biochemical oxygen demand; DO = dissolved oxygen.

3. It should remain unchanged when a relevant variable reaches a value beyond which further change does not affect water use.

In this study of AWS system, the WQI is calculated for the following categories to understand how best the existing water quality is for (1) bathing, swimming (2) industry, and (3) fish culture, wildlife, boating, and other non-contact recreation. The variables that have a significant effect on the use of water are given in Table 13.2. In WQI calculations, the selection of variables and parameters can be personalized to the study's specific goals, local environmental factors, and the data available. It is important to clearly define each variable, specify its unit of measurement, and explain its relevance to ensure the index accurately represents water quality for the intended purpose.

There are various WQIs to compare various variables that include pH, DO, turbidity, temperature, BOD, etc. The arithmetic mean method of WQI does not allow us to sufficiently lower the index if any one critical relevant parameter exceeded the permissible limit (Avvannavar & Shrihari, 2008). Brown et al., (1972) also proposed a multiplicative form of the index. In the multiplicative index, weights assigned to individual parameters are based on a subjective opinion. Those are based on the judgment of the author and a few of his associates. The Bhargava Method (1983) was adopted because of the simplicity involved in handling small to large sets of data for various beneficial uses (Swaroop Bhargava, 1983). Equation 13.1 shows the method

$$WQI = \left[\prod_{i=1}^{n} f_i \left(P_i \right) \right]^{1/n} *100 \tag{13.1}$$

where n = number of variables considered more relevant to the use, $f_i(P_i)$ = sensitivity function of the ith variable. The geometric mean would allow adequate depression in the WQI value due to a low sensitivity function value for any variable relevant to the use.

13.3 RESULTS AND DISCUSSION

13.3.1 Hydrochemistry of the Surface Water Samples

The parameters included for the development of WQIs are pH, DO, BOD, temperature, turbidity, NH_3, conductivity, hardness, and chloride. The value function graphs for the parameters are shown in Figure 13.2. The physicochemical parameters such as pH, EC, TH, DO, BOD, NH_3 were taken into consideration for calculating the water quality index for various beneficial uses. The spatial variability of the different physicochemical parameters during the monsoon and non-monsoon season is shown in Figures 13.3 to 13.5.

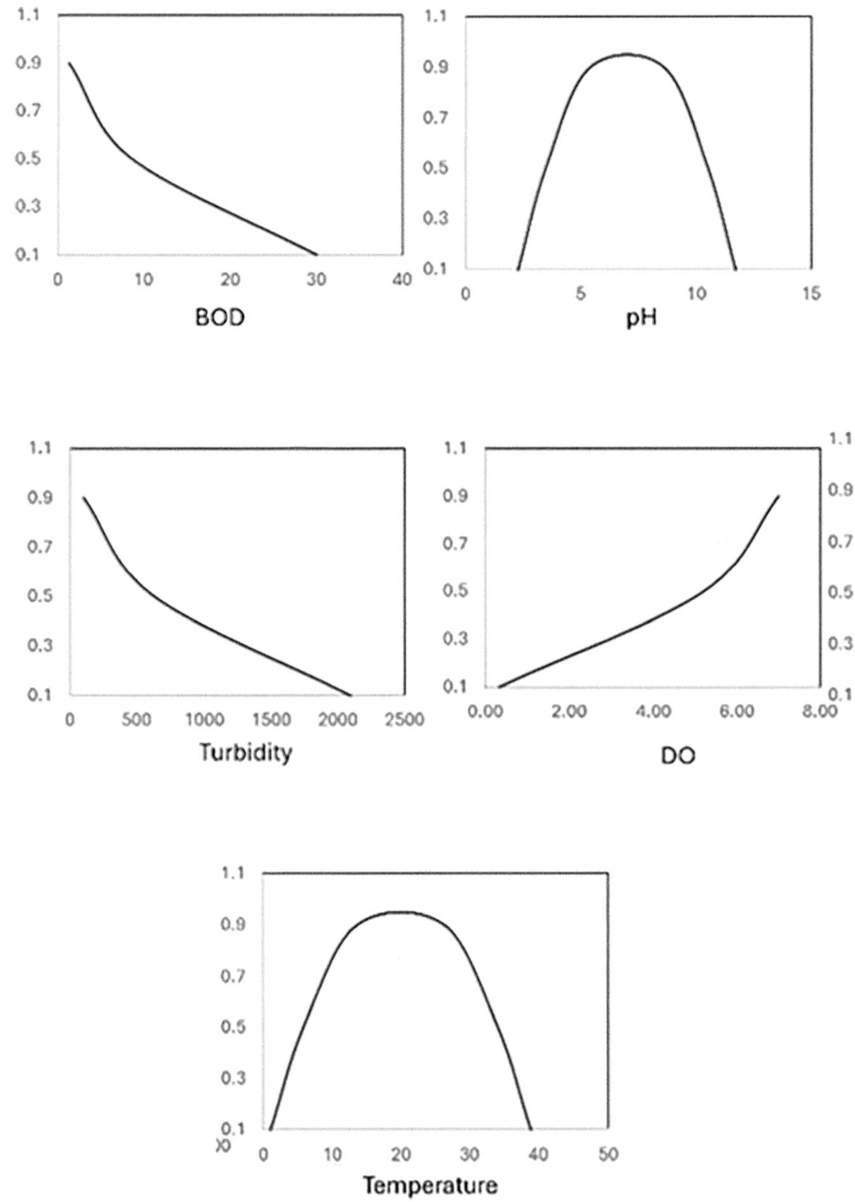

FIGURE 13.2 Sensitive function values for different parameters used for calculating WQI.

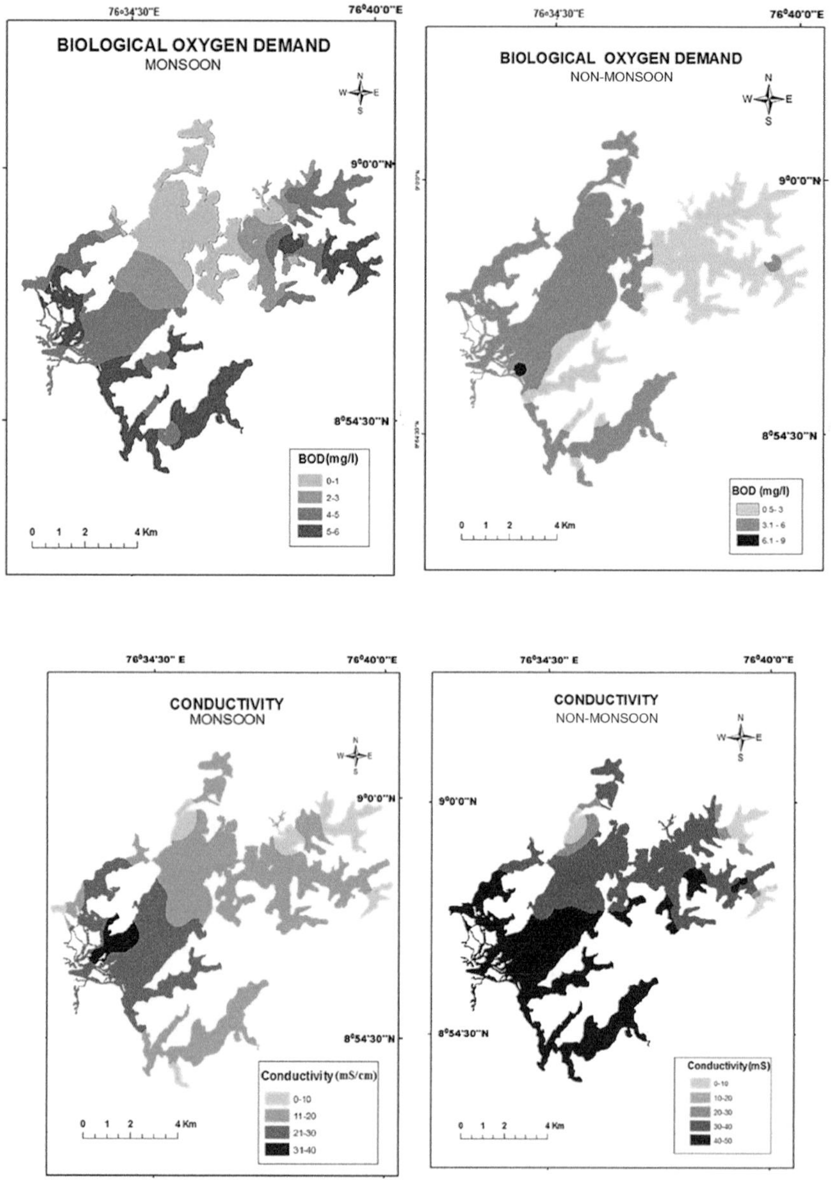

FIGURE 13.3 Spatial variation of BOD and conductivity during monsoon and non-season in AWS.

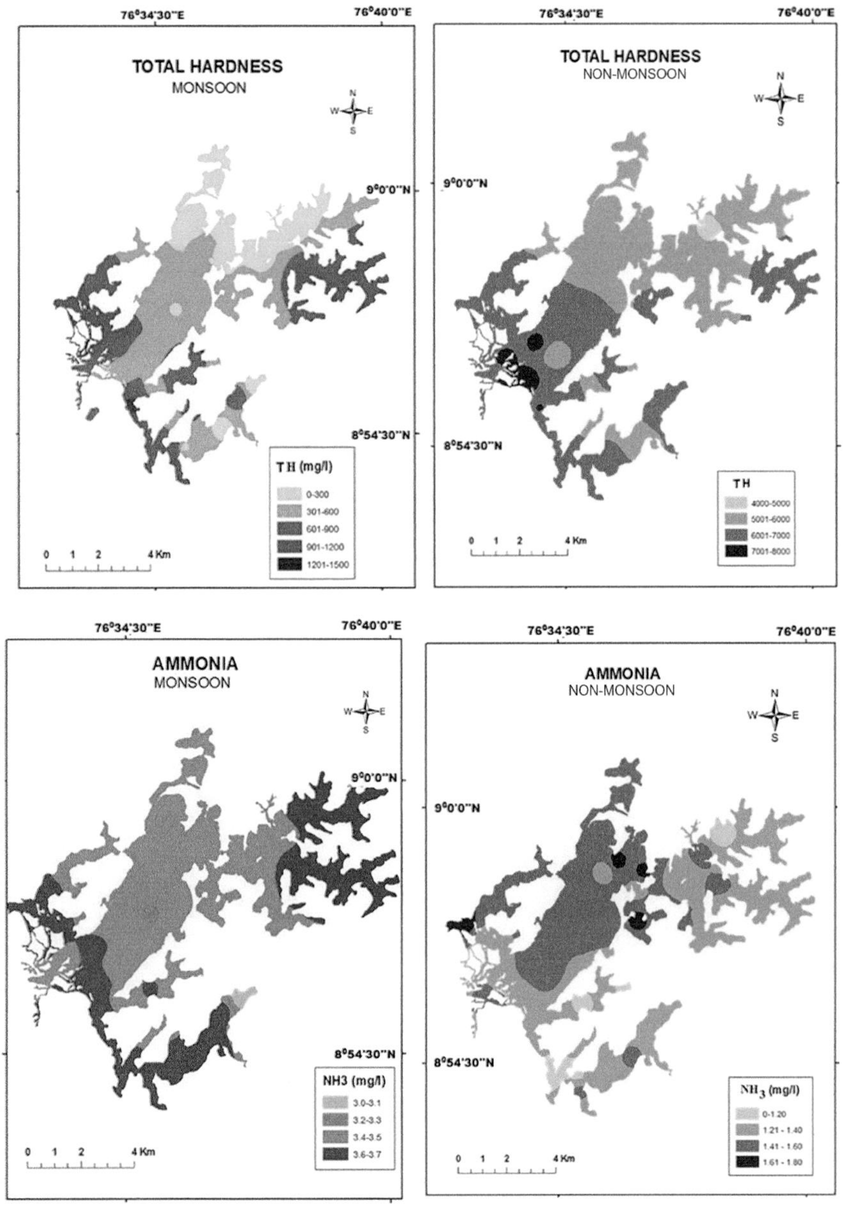

FIGURE 13.4 Spatial variation of total hardness and ammonia during monsoon and non-season in AWS.

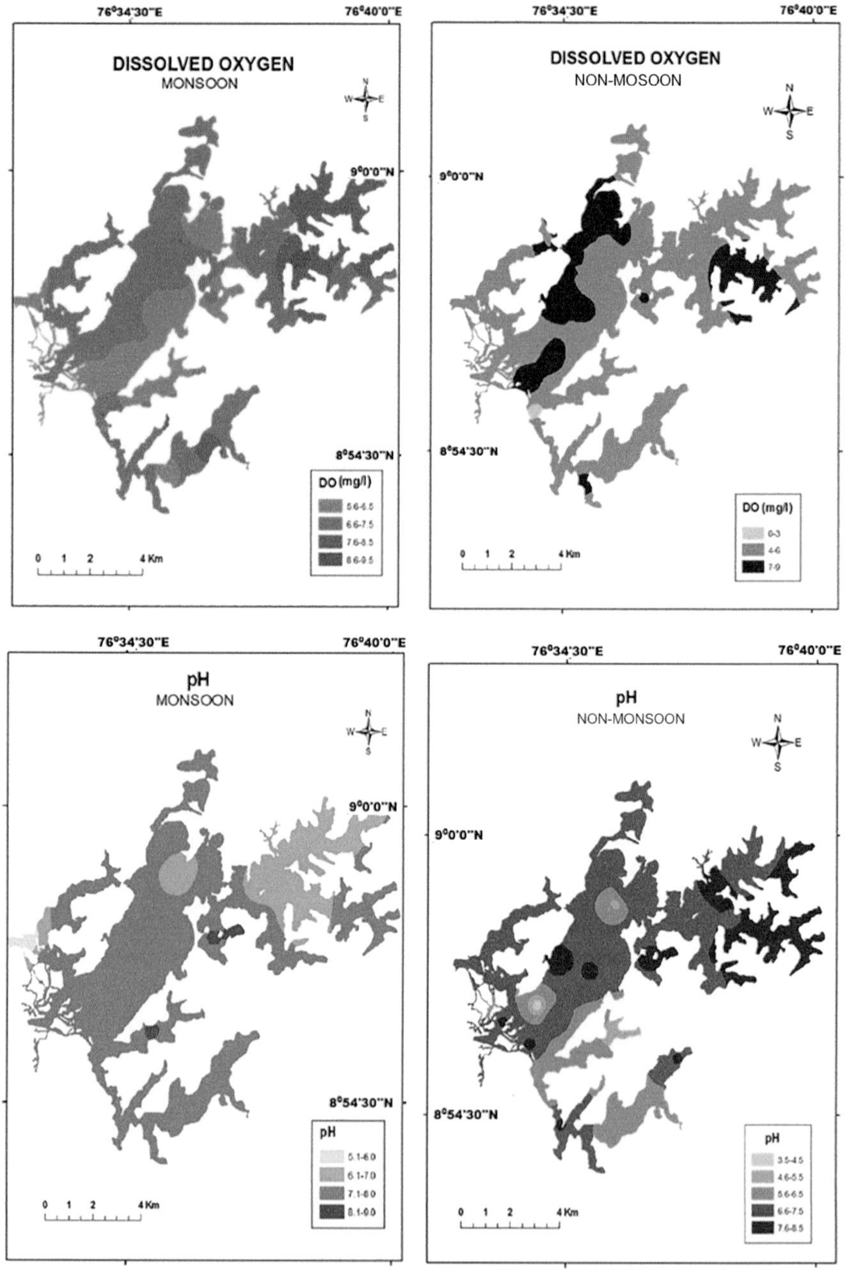

FIGURE 13.5 Spatial variation of DO and pH during monsoon and non-season in AWS.

TABLE 13.3
Annual Statistical Summary of Physicochemical Parameters in Ashtamudi Wetland System

Site	Value	pH	Temperature (°C)	Turbidity NTU	DO (mg/L)	BOD (mg/L)	Chloride (mg/L)	Conductivity (μS/cm)	Hardness (mg/L)	NH_3 mg/L
AML	Range	4.00–8.04	26.00–30.00	2.79–15.12	3.73–7.52	0.38–6.79	6494.00–12368.60	7.65–47.40	430.00–7400.00	1.18–3.60
	Avg ± SD	7.09 ± 0.91	28.19 ± 1.27	6.89 ± 3.15	6.00 ± 1.11	2.91 ± 1.49	9027.30 ± 1738.45	27.33 ± 13.64	3277.31 ± 2774.14	2.44 ± 1.06
CK	Range	4.32–8.12	28.00–30.00	2.30–7.40	1.69–7.54	0.40–5.42	6679.80–9424.60	3.07–45.97	560–7400	1.00–3.61
	Avg ± SD	6.38 ± 1.27	29.11 ± 0.83	4.63 ± 1.84	5.13 ± 1.47	3.11 ± 1.61	8190.07 ± 860.32	29.58 ± 15.48	3739.44 ± 2710.52	2.39 ± 1.18
KE	Range	6.07–8.38	26.00–31.00	3.20–7.20	4.40–9.07	0.08–5.85	6422.30–9240.10	3.10–40.56	300–6800	0.87–3.65
	Avg ± SD	7.25 ± 0.65	28.28 ± 1.41	4.63 ± 1.23	6.27 ± 1.31	2.05 ± 1.59	7951.11 ± 918.06	23.41 ± 15.82	3234.95 ± 2562.74	2.32 ± 1.24
KS	Range	5.24–7.98	26.00–31.00	2.10–7.20	3.04–7.79	0.75–5.53	1424.30–8212.30	5.20–45.11	260–6800	0.98–3.67
	Avg ± SD	6.98 ± 0.82	28.61 ± 1.37	4.00 ± 1.43	5.91 ± 1.42	3.55 ± 1.30	5250.13 ± 2648.55	26.30 ± 19.03	3426.11 ± 2720.10	2.38 ± 1.21

AML = Ashtamudi Main Lake; Avg = average; BOD = biochemical oxygen demand; CK = Chavara Kayal; DO = dissolved oxygen; KE = Kureepuzha Kandachira Kayals; KS = Kumbalathu-Kanjirakottu Kayals; NTU = nephelometric turbidity unit; SD = standard deviation.

13.3.2 Water Quality Assessment

The classification of the WQI category was assessed on the drinking water suitability recommended by Bhargava (1983). The quality criteria are categorized as Class I—Excellent; Class II—Good; Class III—Satisfactory; Class IV—Poor; and Class V—Unacceptable. The physicochemical parameters used for the study included pH, EC, TDS, DO, BOD, N-NH_3, total hardness, turbidity, and chloride. The WQI assessed is on the basis of following requirements:

1. Bathing, swimming
2. Industrial use
3. Fish culture and other recreation

The relevant parameters used for the above categories are different. The parameters considered for: (1) included pH, turbidity, DO, BOD, nitrogen-NH_3; (2) included pH, turbidity, conductivity, and hardness; (3) included water temperature, DO, BOD, and chloride. The annual statistical summary of physicochemical parameters of surface water samples collected from the AWS in monsoon and non-monsoon is shown in Table 13.3. The classification of samples according to WQI values is shown in Table 13.4. The statistical values for WQI across the system during monsoon and non-monsoon are depicted in Table 13.5. The classification based on WQI values across various stations of the Ashtamudi Wetland System is shown in Table 13.6.

The seasonal comparison for monsoon, non-monsoon, and overall WQI generated for the designated uses are listed as shown in Figure 13.6.

The analysis of WQI values reveals significant seasonal and activity-based variations. The assessed activities encompass water for industrial utilization, fish culture, non-contact recreation, and bathing/swimming purposes. The overall WQI during the non-monsoon season falls within the Satisfactory Class (Class III) and poor category (Class IV) during monsoon season. The results indicate that, irrespective of the season, the WQI values consistently fall within the "poor" category (Class IV) for industrial and fish culture activities. Conversely, during the non-monsoon season, WQI values exhibit "satisfactory" (Class III) levels for bathing and swimming in selected areas, primarily those situated inland with minimal estuarine influence. In contrast, the monsoon season is characterized by a decline in WQI values, rendering the water poor (Class IV) for bathing purposes. This deterioration is likely attributed

TABLE 13.4
WQI Ranges for Different Classes

Class	Beneficial Use	Limits of the WQI
I	Excellent	90 and above
II	Good	Between 65 and 89
III	Satisfactory	Between 35 and 64
IV	Poor	Between 11 and 34
V	Unacceptable	10 and below

TABLE 13.5
Computed Statistical Ranges of WQI for Various Beneficial Uses

		WQI Monsoon				WQI Non-Monsoon			
Site	Value	Industry	Fish Culture, Wildlife, Boating and Non-Contact Recreation	Bathing, Swimming	Overall	Industry	Fish Culture, Wildlife, Boating and Non-Contact Recreation	Bathing, Swimming	Overall
AML	Range	30.09–30.37	18.8–28.28	29.11–36.83	26.05–31.82	25.63–30.36	16.82–26.75	58.04–70.15	34.50–40.55
	Avg ± SD	30.3 ± 0.07	23.17 ± 2.7	33.88 ± 2.44	29.12 ± 1.68	29.39 ± 1.38	21.16 ± 3.03	63.73 ± 4.26	38.09 ± 1.97
CK	Range	29.97–30.38	18.8–23.78	28.96–34.88	25.94–29.54	26.35–28.73	16.82–22.36	52.93–71.5	33.43–40.56
	Avg ± SD	30.16 ± 0.18	20.92 ± 1.48	31.63 ± 1.74	27.57 ± 2.55	27.74 ± 0.7	20.03 ± 2.06	64.05 ± 6.21	37.27 ± 2.55
KE	Range	29.76–64.06	22.36–26.75	32.1–35.22	28.38–42.00	28.6–30.37	21.15–25.15	58.15–80.59	36.50–45.37
	Avg ± SD	37.71 ± 14.92	24.45 ± 1.82	34.02 ± 1.34	32.06 ± 2.59	30.07 ± 0.56	22.44 ± 1.63	72.03 ± 6.62	41.52 ± 2.59
KS	Range	30.25–64.17	21.15–23.78	31.84–35.14	27.81–39.13	27.78–28.73	18.8–23.78	62.26–77.1	37.07–43.21
	Avg ± SD	41.52 ± 16.81	21.44 ± 0.88	32.37 ± 1.04	31.78 ± 2.25	28.43 ± 0.39	20.79 ± 1.72	66.55 ± 5.27	38.59 ± 2.25

AML = Ashtamudi Main Lake; BOD = biochemical oxygen demand; CK = Chavara Kayal; DO = dissolved oxygen; KE = Kureepuzha Kandachira Kayals; KS = Kumbalathu-Kanjirakottu Kayals; M = monsoon; NTU = nephelometric turbidity unit.

TABLE 13.6
Water Quality Classification of Ashtamudi Wetland System during Monsoon and Non-Monsoon

	Monsoon		Non-Monsoon	
Subenvironment	**Poor**	**Satisfactory**	**Poor**	**Satisfactory**
AML (n = 13)	13	0	0	13
CK (n = 9)	9	0	2	7
KE (n = 9)	7	2	0	9
KS (n = 9)	9	0	0	9

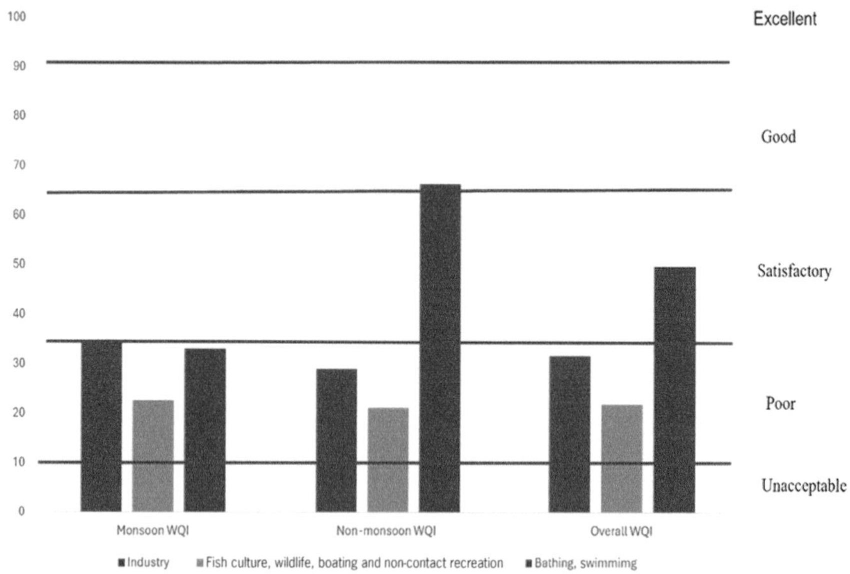

FIGURE 13.6 WQI for monsoon and non-monsoon periods and the designated uses as per the quality ranges.

to increased pollutant loads from runoff during the monsoon season. The mass of "poor" WQI values underscores the imperative need for efficacious water management strategies to safeguard the AWS system. Furthermore, stringent waste disposal practices must be implemented to prevent additional pollution. Only through such measures can the water body be rendered safe for sustaining life for all dependent organisms, including aquatic fauna and human populations.

13.4 CONCLUSION

The WQI analysis reveals significant seasonal and activity-based variations. During the non-monsoon season, WQI values are satisfactory for bathing and swimming in some areas, but decline to poor levels during the monsoon season. Industrial and fish culture activities consistently show poor WQI values. To safeguard the water body, effective water management strategies and strict waste disposal practices are necessary. The preservation and management of wetlands are crucial for achieving multiple SDGs. These ecosystems play a vital role in maintaining ecological balance, supporting biodiversity, and promoting human well-being. By conserving wetlands, we can ensure the provision of clean water and sanitation (SDG-6), foster sustainable urban development (SDG-11), mitigate the impacts of climate change (SDG-13), and safeguard marine resources (SDG-14). The conservation of wetlands also has numerous co-benefits, including the protection of fisheries, support for agriculture, and maintenance of ecosystem services. Furthermore, wetlands have the potential to sequester carbon, reducing greenhouse gas emissions and contributing to global climate change mitigation efforts. Therefore, it is imperative that we prioritize wetland conservation and management, recognizing their intrinsic value and the essential role they play in supporting a sustainable and equitable future.

REFERENCES

APHA (1995) *Standard Methods for the Examination of Water and Wastewater*. 19th Edition, American Public Health Association Inc., New York.

Aswathy, T. S., Achu, A. L., Francis, S., Gopinath, G., Joseph, S., Surendran, U., and Sunil, P. S. (2021). Assessment of water quality in a tropical Ramsar wetland of southern India in the wake of COVID-19. *Remote Sensing Applications: Society and Environment*, 23, 1–7, 100604.

Avvannavar, S. M., and Shrihari, S. J. E. M. (2008). Evaluation of water quality index for drinking purposes for river Netravathi, Mangalore, South India. *Environmental Monitoring and Assessment*, 143(1-3), 279–290.

Bhargava, D. S. (1983). Use of a water quality index for river classification and zoning of Ganga River. *Environmental Pollution, Series B*, 6(1), 51–67.

Brown, R. M., McClelland, N. I., Deininger, R. A., and Tozer, R. G. (1970). A water quality index—Do we dare? *Water and Sewage Works*, 117(10), 339–343.

Brown, R. M., McClelland, N. I., Deininger, R. A., and Tozer, R. G. (1972). A water quality index-crashing the psychological barrier. In *Proceedings of the 6th International Conference on Water Pollution Research* (pp. 787–797).

Krishnakumar, A., Saranya, P., and Das, R. (2017). Hydrogeochemistry and environmental issues of the wetlands of Kerala, Southwestern India. *Wetland Science: Perspectives from South Asia* (pp. 145–161).

Krishnakumar, A., Saranya, P., Prasad, P. G., and Rakhi, C. (2015). Holocene records of human driven geological impacts in a Ramsar Wetland of India. *Aquatic Procedia*, 4, 373–380.

Saha, A., Ramya, V. L., Jesna, P. K., Mol, S. S., Panikkar, P., Vijaykumar, M. E., Sarkar, U. K., and Das, B. K. (2021). Evaluation of spatio-temporal changes in surface water quality and their suitability for designated uses, Mettur Reservoir, India. *Natural Resources Research*, 30(2), 1367–1394.

Soman, K. (2002). *Geology of Kerala*. Bangalore: Geological Society of India. 1–336.

United Nations. (2016). Sustainable Development Goals. https://share.google/UKurAspGvtXcJfpf1.

14 GIS-Based Modeling of Flash Flood Susceptibility and Water Resource Management

A Case Study of the Gulf of Aqaba Drainage Basins, Egypt

Kamal Srogy Darwish and Alaa Hassan

14.1 INTRODUCTION

Natural hazards are worldwide phenomena that have the power to transform Earth's landscapes and profoundly affect human environments. While water is crucial for sustaining life and maintaining ecosystems, its impact can also bring significant challenges (Sohail et al. 2022). In recent years, the impact of climate change and extreme weather events has increased, causing a series of hazards, particularly in mountainous regions. Climate hazards and disasters have clearly increased, posing significant threats to large sectors of human settlements and urbanization. Flash floods, which constitute 85% of all flood events and leading to over 5,000 fatalities each year, are among the deadliest natural disasters in the world (WMO 2016). Several countries worldwide have been threatened by flash floods, resulting in extensive damage to infrastructure. Around 2,466 individuals died as a result of flash floods in Europe alone during the last 40 years (Petrucci et al. 2019).

In several arid areas, flash floods are a common natural hazard. They frequently damage livelihoods and interfere with economic activity, posing serious hazards to human life and infrastructure (Deoli et al. 2023). Floods that occurred in Syria destroyed properties and caused severe damage to municipalities (Abdo et al. 2024). Egypt is among the leading countries that have experienced numerous devastating flash flood events in recent years. They also threaten historic monuments, which are essential to Egypt's tourism sector, thereby generating a significant portion of the nation's economy. Multiple flash floods have occurred, causing significant fatalities and extensive damage to vital infrastructure and buildings, particularly in areas such as the Sinai Peninsula, the northern coast, and the Red Sea region.

 DOI: 10.1201/9781003596080-14

Notable events include the 1979 flash flood in El-Quseir and Marsa Alam, which claimed 19 lives and destroyed a coastal road along the Red Sea; the 1991 flash flood in Marsa Alam; the 1993 flood in Alexandria, which resulted in 21 fatalities; and the November 1994 flash flood in Assiut, which caused further loss of life and damage to infrastructure (El-Rawy et al. 2022). In January 2010, one significant event occurred in the Sinai Peninsula, resulting in numerous fatalities and missing persons (Youssef et al. 2011; Mohamed and El-Raey 2020; Helmi and Zohny 2020). On October 26 and 27, 2016, excessive rainfall at Ras Gharib on the Red Sea coast caused catastrophic flooding that killed numerous individuals and severely damaged property as well as infrastructure (Elnazer et al. 2017). Additionally, on November 14, 2019, Wadi El-Sukkari experienced flooding due to severe rainfall that also reached the Idfu-Marsa Alam Road. Thankfully, no fatalities were reported (Kamel and Arfa 2020). In November 2020, a devastating flood storm struck the Minya area in central Egypt, causing extensive damage, especially to the eastern El-Geish Highway (Darwish 2023).

Recently, the integration of remote sensing and Geographic Information Systems (GIS) has become essential for mapping and assessing flash flood hazards. GIS is a powerful tool for analyzing, modeling, and mapping flash flood susceptibility while also supporting sustainable water resource management (Deoli and Kumar, 2019). El-Rayes et al. (2023) applied these methods to assess flood hazards in the Safaga region, Egypt. GIS-based modeling approaches are crucial for identifying areas vulnerable to flash floods and have been widely applied in numerous studies worldwide. These approaches utilize analytical methods, including morphometric parameters and equations, to define hazard zones. For instance, Bashir (2023) used morphometric parameters and geospatial analysis for flash flood susceptibility assessment in Jeddah City, Saudi Arabia. Additionally, advancements in multicriteria decision analysis (MCDA) techniques, such as the Analytical Hierarchy Process (AHP), have been employed to enhance flood hazard mapping. Riaz and Mohiuddin (2025) applied GIS-based multi-criteria decision analysis of hydro-geomorphological factors for assessing flash flood susceptibility mapping in Bangladesh. Similarly, in Pakistan, Ur Rahman et al. (2023) utilized a GIS-based flood susceptibility mapping approach using a bivariate statistical model in the Swat River Basin. Geospatial artificial intelligence models, such as machine learning and deep learning, have significantly enhanced GIS's ability to accurately map flood hazard zones. Demissie et al. (2024) integrated machine learning and GIS for flood susceptibility mapping and improved risk assessment. In this study, a geospatial modeling approach based on morphometric parameters was applied across the study area to assess and map vulnerable flash flood zones. El Afandi and Morsy (2020) developed an Early Warning System for Flash Flood hazard in the Sinai Peninsula.

14.2 STUDY AREA

The study area stretches along the Gulf of Aqaba in the southeastern part of the Sinai Peninsula, located between latitudes 27°51′01″ N and 29°35′57″ N, and longitudes 34°16′12″ E and 34°52′02″ E (Figure 14.1). It spans from Wadi Taba in the north to

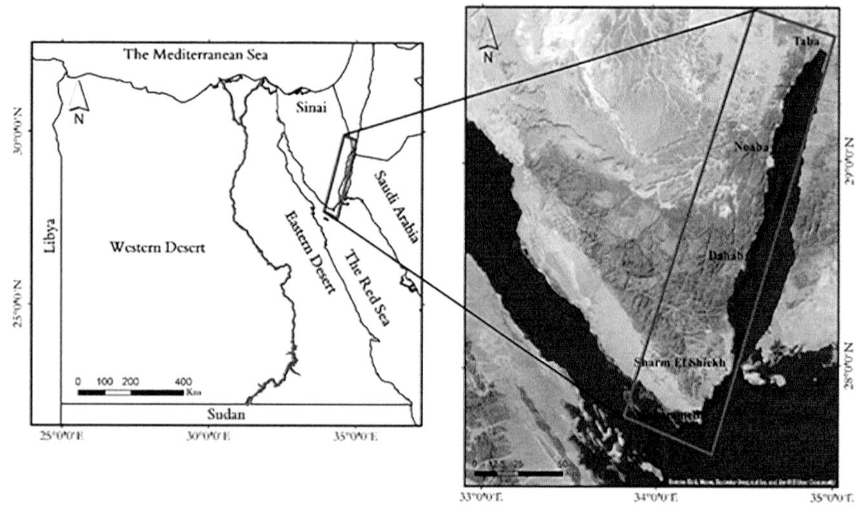

FIGURE 14.1 Location of study area.

Wadi Ras Om Sad in the south. The region is distinguished by its rugged topography, with elevations rising to approximately 2,500 m above sea level, including Egypt's highest peak, Saint Catherine Summit. It is traversed by three major watersheds—Kid, Dahab, and Wateir from south to north. The coastal plain, though relatively narrow, is home to four significant cities: Sharm El-Sheikh, Dahab, Nuweiba, and Taba. The area also boasts unique coral reefs and marine natural protected zones, underscoring its environmental importance (Figure 14.2a).

The Sinai Peninsula's climate is similar to that of desert areas around the world, with lengthy hot and dry summers, mild winters, and intense aridity. Some regions receive heavy, relatively quick rainfall in the winter, causing Wadi basins to overflow and catastrophic flash floods that can sometimes cause fatalities and damage to infrastructure (Omran 2020). According to TRMM 3B42.v7A 3-hourly data collected from 1998 to 2013 (Mohamed, 2015), the mountainous regions in the mid-southern zones of the Sinai Peninsula receive annual precipitation ranging from 50 to 150 mm, while the southern highlands record rainfall between 50 and 75 mm/year. Central Sinai is the driest area, with annual precipitation seldom exceeding 30 mm. On average, the Sinai Peninsula receives between 80 and 100 mm of rainfall annually (Omran 2020) (Figure 14.2b).

The study area is geologically linked to the basement rocks that form part of the Nubian-Arabian Shield. As shown in Figure 14.3, Precambrian rocks dominate the southern Sinai region, comprising the majority of the study area. The area has been influenced by the Red Sea Rift, which introduced a fault system oriented in a NE-SW direction, rendering it tectonically active.

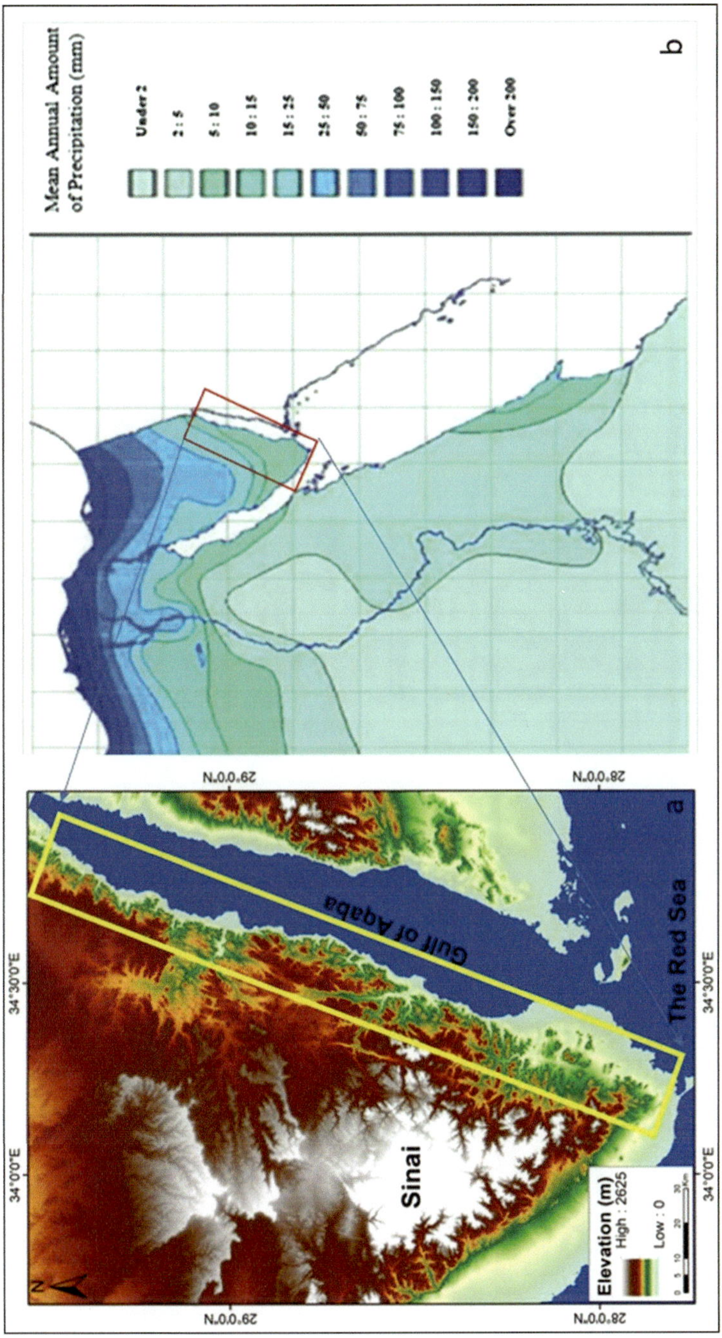

FIGURE 14.2 (a)Topography of study area, (b) average rainfall (mm/year). (Source: (a) SRTM-1 arc second, (b) Afandi and Morsy 2020.)

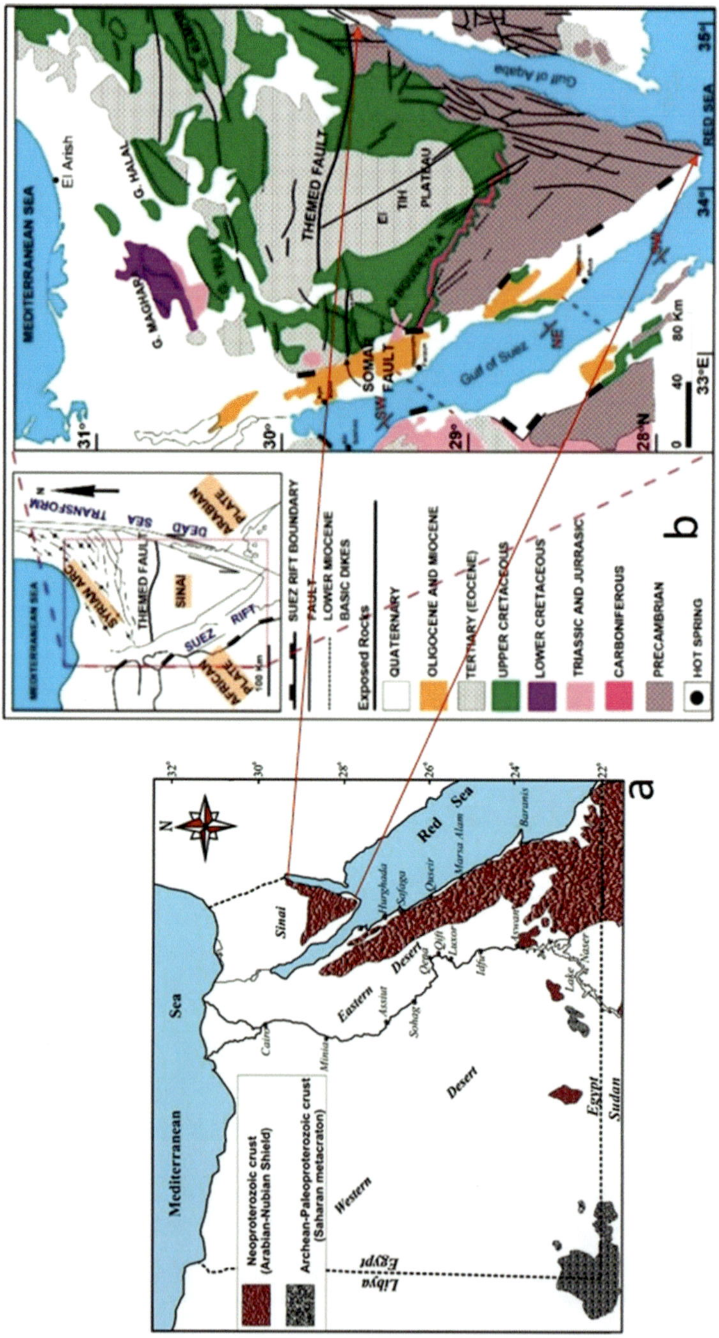

FIGURE 14.3 Geological settings of study area. (Source: (a) El-Bialy 2019; (b) Fahil et al. 2020.)

14.3 DATA USED AND METHODOLOGY

14.3.1 Data

An integration of multisource spatial raster and vector data was utilized for flash flood susceptibility mapping in the study region. This dataset included Shuttle Radar Topography Mission (SRTM) 1-arc second data, Multitemporal Google Earth Pro Data, and topographical maps at a scale of 1:50,000. These resources were collected, examined, and analyzed to derive various morphometric and topographic parameters.

14.3.2 Methods

The use of GIS for creating flood hazard maps and supporting disaster decision-making has seen continuous improvement and widespread adoption since the early 2000s, driven by the growing availability of spatial databases and GIS software (Zerger and Smith 2003). This study employs a methodological approach based on mathematical modeling using the Watershed Modeling System, integrated with GIS tools and a relative ranking technique for comprehensive analysis. Figure 14.4 illustrates the flowchart used in this study to calculate hazard levels within the study area. The first stage involves extracting watersheds and streams from digital elevation models (SRTM) with a 30-m spatial resolution. This extraction process includes several steps, such as hydrological preprocessing, SRTM calculations, gap filling, flow direction analysis, flow accumulation, and stream order determination, as shown in Figures 14.5 and 14.6. The second stage focuses on topographical analysis, which involves generating slope maps, aspect maps, hillshade models, and conducting hypsometric analysis. The third stage applies morphometric parameters using a field calculator, including aerial, form, and linear parameters, which are summarized in Table 14.1. Finally, all assigned parameters were used to calculate the flash flood susceptibility map using the equation outlined by Davis (1975) and Bajabaa et al. (2014) (Equations 14.1 and 14.2):

$$Y = \frac{4\left(X - X_{max}\right)}{\left(X_{min} - X_{max}\right)} + 1 \tag{14.1}$$

In this context, Y represents the relative hazard degree, with Y_{max} and Y_{min} denoting the upper and lower bounds of the projected scale. Similarly, X_{max} and X_{min} correspond to the maximum and minimum values of a given parameter, while X represents a specific parameter value falling within this range. The flash flood hazard score for each basin varied between 1 and 5, depending on the parameter's contribution to flood susceptibility.

$$Y = \frac{\left(Y_{max} - Y_{min}\right)\left(X - X_{max}\right)}{\left(Y_{min} - X_{max}\right)} + Y_{min} \tag{14.2}$$

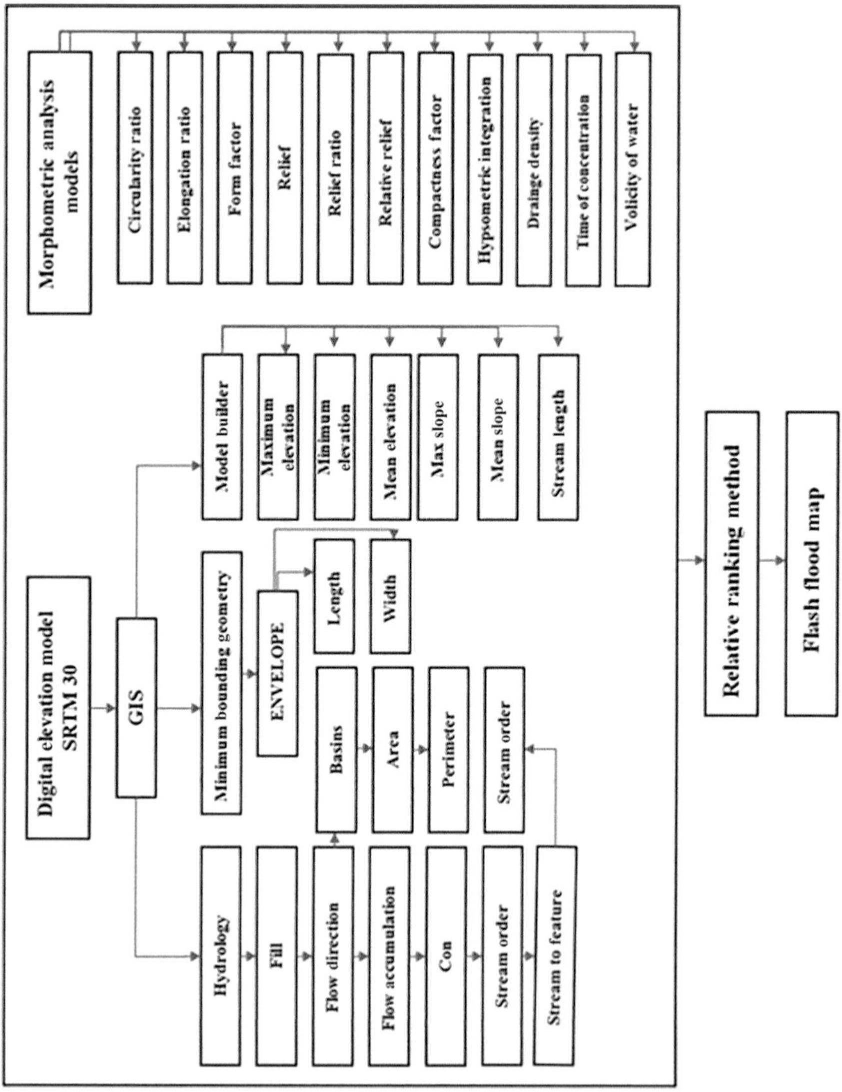

FIGURE 14.4 Flow diagram illustrating the methodology of the study.

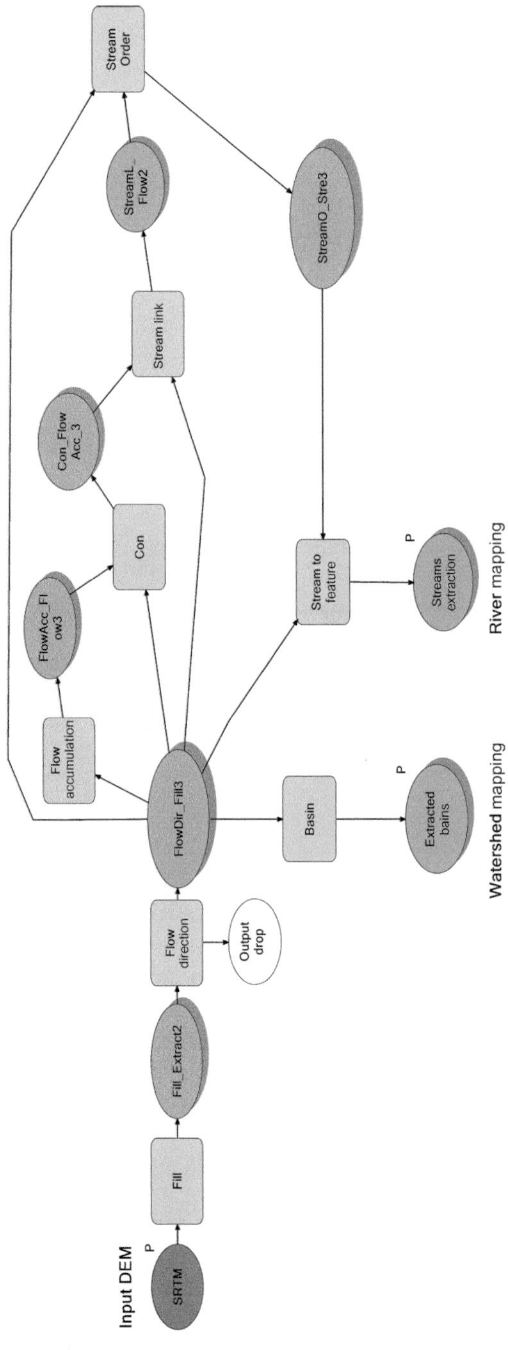

FIGURE 14.5 Spatial model illustrating the digital extraction of watersheds and stream networks using ArcGIS ModelBuilder.

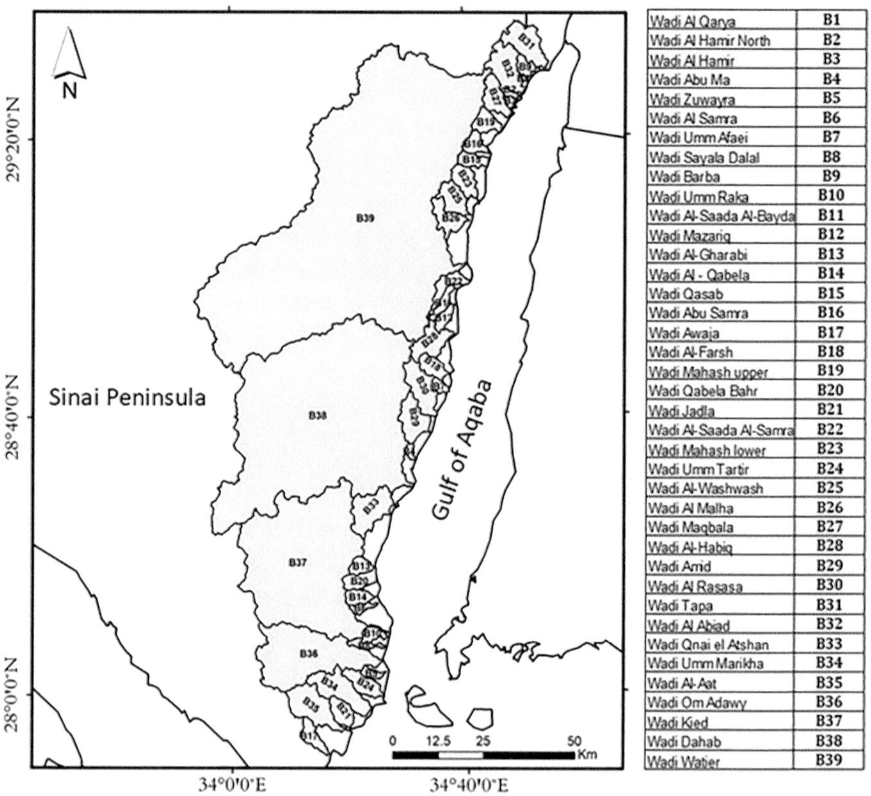

FIGURE 14.6 Drainage basins of the study area identified using SRTM 1-arc second.

The morphometric analysis of watersheds is essential for geomorphological analysis to understanding the complex interactions between the relief, linear, and areal characteristics of a watershed. By quantifying various drainage and land surface parameters, a comprehensive database is created to support this analysis (Shekar and Mathew, 2024). By calculating morphometric parameters such as relief, basin shape, drainage density, and stream lengths, researchers can identify quantitative relationships and patterns that provide valuable insights into the behavior and characteristics of a watershed. This analysis enhances the understanding of hydrological processes and plays a vital role in flood risk assessment and effective watershed management. The morphometric parameters can be classified into two main categories:

14.3.3 Linear Factors

The morphometric linear factors refer to parameters that characterize the linear dimensions and hierarchical structure of a watershed's drainage system. These factors

TABLE 14.1
Morphometric Parameters Applied in This Study

	Morphometrical Parameters	Calculation Method	Reference
1	Area (A; km^2)	GIS Analysis (Calculate Geometry—Area)	DEM (SRTM 30)
2	Perimeter (P; km)	GIS Analysis (Calculate Geometry—Perimeter)	DEM (SRTM 30)
3	Length (L; km)	GIS Analysis (Minimum Bounding Geometry—ENVELOPE)	DEM (SRTM 30)
4	Width (W; km)	GIS Analysis (Minimum Bounding Geometry—ENVELOPE)	DEM (SRTM 30)
5	Circularity ratio	$4\pi A / P^2$	(Miller 1953)
6	Elongation ratio	$2\sqrt{A/\pi}\ /L$	(Schumm 1956)
7	Form factor	A/L^2	(Horton 1945)
8	Maximum elevation (+M)	GIS Analysis (Model Builder)	DEM (SRTM 30)
9	Minimum elevation (–M)	GIS Analysis (Model Builder)	DEM (SRTM 30)
10	Relief (H)	Maximum Elevation- Minimum Elevation	(Strahler 1957)
11	Compactness factor (C)	$P/\sqrt{\pi A}$	(Horton 1945)
12	Mean elevation (mM)	GIS Analysis (Model Builder)	DEM (SRTM 30)
13	Hypsometric integration	$\text{Hyptho} = \frac{\text{Mean Elev} - \text{Min Elevation}}{\text{Max Elevation} - \text{Min Elevation}}$	(Patton 1988)
14	Max slope	GIS Analysis (Model Builder)	DEM (SRTM 30)
15	Mean slope	GIS Analysis (Model Builder)	DEM (SRTM 30)
16	Stream order (U)	GIS Analysis	DEM (SRTM 30)
17	Stream length (LU; km)	GIS Analysis (Model Builder)	DEM (SRTM 30)
18	Drainage density (Dd)	LU/A	Horton (1932)
19	Time of concentration (TC)	$TC = (L)^{1.15} \div 7700\ H^{0.38}$	Wong (2005)
20	Velocity of water	$VW = L \div TC$	Wong (2005)

include Stream Order (U), Stream Number (Nu), Stream Length (Lu), Mean Stream Length (Lsm), Stream Length Ratio (RL), and Bifurcation Ratio (Rb) as shown in Table 14.2.

14.3.4 Areal Factors

The morphometric areal factors refer to parameters that evaluate the spatial arrangement and surface attributes of a watershed. These include Drainage density (DD), Stream

TABLE 14.2
Morphometric Analysis of Drainage Basins

	Area (A; km²)	Perimeter (P; km)	Length (L; km)	Width (W; km)	Circularity Ratio	Elongation Ratio	Form Factor	Compactness Factor
B1	2.20	8.22	2.28	2.28	0.41	0.73	0.42	3.13
B2	7.13	15.54	4.80	3.17	0.37	0.63	0.31	3.29
B3	7.43	13.79	3.80	3.73	0.49	0.81	0.52	2.85
B4	7.85	14.03	4.47	3.32	0.50	0.71	0.39	2.83
B5	13.04	20.88	7.21	3.76	0.38	0.57	0.25	3.26
B6	13.20	21.49	8.13	3.46	0.36	0.50	0.20	3.34
B7	13.72	17.45	5.09	4.48	0.57	0.82	0.53	2.66
B8	15.69	31.80	11.23	3.56	0.19	0.40	0.12	4.53
B9	19.79	23.59	6.50	5.36	0.45	0.77	0.47	2.99
B10	20.41	24.05	7.17	4.61	0.44	0.71	0.40	3.00
B11	21.26	31.95	9.71	8.18	0.26	0.54	0.23	3.91
B12	21.51	27.39	8.36	6.18	0.36	0.63	0.31	3.33
B13	23.39	21.61	7.56	5.28	0.63	0.72	0.41	2.52
B14	26.17	23.79	8.59	4.58	0.58	0.67	0.35	2.62
B15	26.26	28.03	8.66	5.16	0.42	0.67	0.35	3.09
B16	27.45	27.30	7.52	6.49	0.46	0.79	0.49	2.94
B17	33.47	33.41	10.37	8.09	0.38	0.63	0.31	3.26
B18	34.48	28.63	8.45	7.52	0.53	0.78	0.48	2.75
B19	34.64	29.99	8.52	7.20	0.48	0.78	0.48	2.88
B20	34.73	29.23	9.61	6.16	0.51	0.69	0.38	2.80
B21	37.69	29.95	8.61	7.02	0.53	0.8	0.51	2.75
B22	38.39	48.77	12.71	12.13	0.20	0.55	0.24	4.44
B23	43.99	37.37	10.71	7.92	0.40	0.70	0.38	3.18
B24	44.65	38.36	12.21	8.22	0.38	0.62	0.30	3.24
B25	50.86	39.84	9.85	9.79	0.40	0.82	0.52	3.15
B26	51.14	47.85	12.12	10.74	0.28	0.67	0.35	3.78
B27	51.37	41.69	11.91	8.03	0.37	0.68	0.36	3.28
B28	62.75	51.66	12.43	11.73	0.30	0.72	0.41	3.68
B29	69.19	45.98	13.95	9.15	0.41	0.67	0.36	3.12
B30	80.86	55.17	16.13	9.56	0.33	0.63	0.31	3.46
B31	83.46	60.65	13.44	12.94	0.28	0.77	0.46	3.75
B32	89.05	62.71	14.12	11.82	0.28	0.75	0.45	3.75
B33	96.58	55.8	14.31	12.64	0.39	0.77	0.47	3.20
B34	99.34	65.13	18.57	13.92	0.29	0.61	0.29	3.69
B35	105.88	61.98	17.57	13.04	0.35	0.66	0.34	3.4
B36	365.51	124.49	35.55	18.88	0.30	0.61	0.29	3.67
B37	1038.45	208.69	43.99	42.17	0.30	0.83	0.54	3.65
B38	2071.53	293.14	58.85	56.61	0.30	0.87	0.60	3.63
B39	3515.46	411.75	88.20	82.40	0.26	0.76	0.45	3.92
Min	2.20	8.22	2.28	2.28	0.19	0.40	0.12	2.52
Max	3515.46	411.75	88.20	82.40	0.63	0.87	0.60	4.53

TABLE 14.3
Topographic Characteristics of Drainage Basins

	Maximum Elevation (m)	Minimum Elevation (m)	Relief (m) (H)	Mean Elevation (M)	Hypsometric Integration	Max Slope	Mean Slope
B1	550	51	499	317.6	0.53	49.06	19.32
B2	740	25	715	332.6	0.43	51.66	17.86
B3	556	9	547	219.5	0.38	59.29	15.35
B4	984	12	972	437.6	0.44	68.03	23.91
B5	376	11	365	111.5	0.28	47.02	5.22
B6	512	10	502	168.5	0.32	45.03	9.07
B7	930	11	919	423.8	0.45	61.86	23.49
B8	583	0	583	204.8	0.35	45.52	8.93
B9	778	16	762	321.4	0.40	63.35	17.73
B10	481	9	472	112.8	0.22	41.05	7.07
B11	995	40	955	637.4	0.63	61.95	21.90
B12	986	10	976	652.4	0.66	67.34	24.07
B13	1066	16	1050	479.2	0.44	58.78	25.10
B14	767	12	755	267.1	0.34	46.72	14.14
B15	1033	8	1025	461.8	0.44	65.17	17.16
B16	1039	19	1020	608.2	0.58	63.24	18.82
B17	884	8	876	349.6	0.39	60.67	12.69
B18	930	28	902	560.3	0.59	61.18	19.39
B19	1064	17	1047	617.2	0.57	70.08	21.62
B20	898	0	898	358.7	0.4	55.14	19.37
B21	532	10	522	133.4	0.24	66.98	4.96
B22	955	7	948	399.2	0.41	64.99	14.23
B23	1126	0	1126	615.4	0.55	59.87	20.53
B24	606	8	598	152.4	0.24	59.23	6.76
B25	1167	22	1145	670.6	0.57	67.67	22.09
B26	1051	20	1031	531.8	0.50	64.13	17.93
B27	986	17	969	606	0.61	70.09	18.98
B28	976	15	961	678.9	0.69	66.19	18.68
B29	944	16	928	564.4	0.59	67.08	19.68
B30	991	0	991	620.9	0.63	66.19	19.29
B31	826	16	810	528.1	0.63	63.77	11.03
B32	1024	15	1009	585.5	0.57	70.65	14.11
B33	1299	26	1273	593.3	0.45	70.87	19.37
B34	822	11	811	245.4	0.29	54.54	6.90
B35	1370	0	1370	472.1	0.34	70.70	14.91
B36	2048	10	2038	663.2	0.32	69.66	16.92
B37	2421	9	2412	933.9	0.38	72.09	22.02
B38	2507	11	2496	1061.8	0.42	72.51	15.18
B39	1630	12	1618	911.3	0.56	73.35	10.57
Min	376	0	365	111.5	0.22	41.05	4.96
Max	2507	51	2496	1061.8	0.69	73.35	25.1

frequency (Fs), Drainage texture (Rt), Circularity ratio (Rc), Elongation ratio (Re), Length of overland flow (Lg), Compactness Factor (C), Form Factor. These factors offer valuable information about the hydrological processes and geomorphological characteristics of the watershed as shown in Table 14.3.

14.4 RESULTS AND DISCUSSION

The morphometric analysis of the drainage basins in the study area revealed significant geomorphological characteristics of the watersheds and their hydrological response to extreme rainfall events. A total of 39 drainage basins were delineated using Digital Elevation Models and topographic maps, as illustrated in Figure 14.6. These basins exhibit variations in size, with Wadi Watier, Wadi Dahab, and Wadi Kid identified as the largest watersheds, encompassing areas of 3,515.46 km², 2,071.53 km², and 1,038.45 km², respectively. All other basins in the study area cover less than 1,000 km², as summarized in Table 14.2 and presented in Figure 14.7. The size of each drainage basin significantly influences its perimeter and valley dimensions, including length and width. Notably, Wadi Watier features the longest valley, stretching 88 km, while Wadi Al Qarya has the shortest valley, measuring only 2.3 km. These morphometric attributes play a critical role in understanding the hydrological dynamics and assessing the flood susceptibility of the study area.

The form of a drainage basin is evaluated by examining its elongation, which is particularly useful for comparing different basins. For basins of equal area, those with a more elongated shape typically experience lower flood peaks but longer flood durations. In this study, three morphometric areal parameters were analyzed to assess the form of the basins. The circularity ratio ranged from a maximum of 0.63 in Wadi Al-Gharabi to a minimum of 0.19 in Wadi Sayala Dalal. These results suggest that the drainage basins in the study area are predominantly elongated rather than circular. The shape of the drainage pattern plays a crucial role in calculating runoff time and

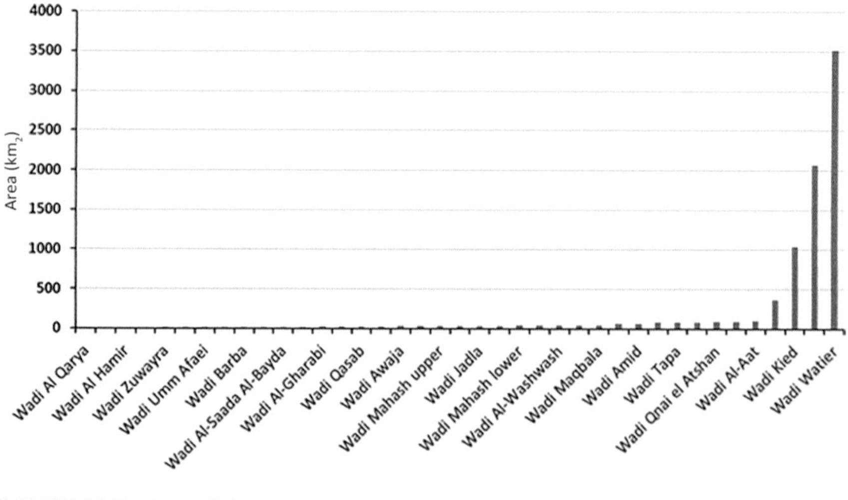

FIGURE 14.7 Area of drainage basins in square kilometers.

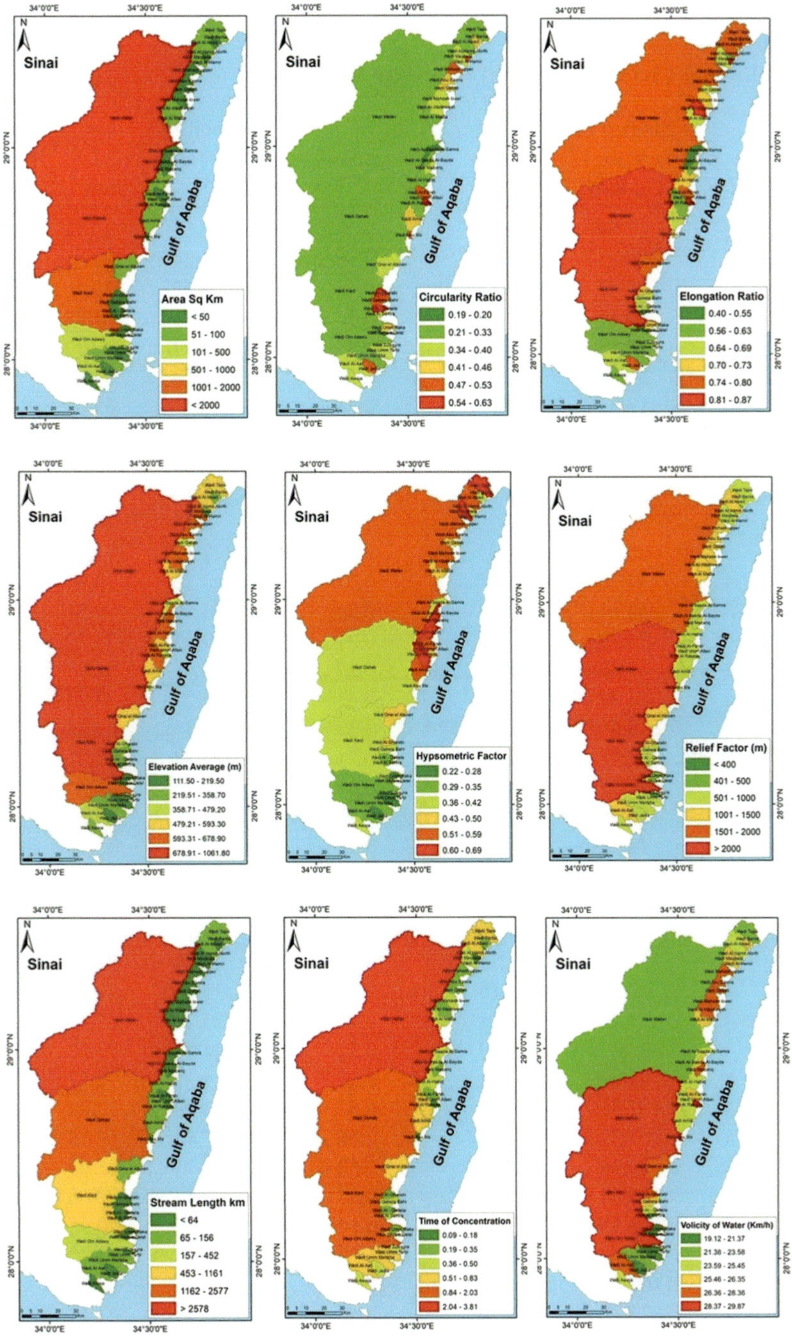

FIGURE 14.8 Geographical distribution of aerial parameters of the study area.

TABLE 14.4
Hydrological Characteristics of Drainage Basins

	Stream Order	Stream Length (Lu; km)	Drainage Density (DD)	Time of Concentration (TC)	Velocity of Water (km/h)	Sum	Hazard Degree
B1	3	2.59	1.18	0.09	25.59	76.19	2
B2	3	8.75	1.23	0.18	26.24	79.15	2
B3	3	8.47	1.14	0.15	24.55	79.26	2
B4	4	9.47	1.21	0.15	29.81	71.37	2
B5	4	20.13	1.54	0.38	19.12	86.90	1
B6	4	20.92	1.58	0.38	21.19	85.36	1
B7	4	17.8	1.3	0.18	28.61	69.81	3
B8	4	23.07	1.47	0.53	21.37	86.28	1
B9	4	20.1	1.02	0.25	25.68	75.71	2
B10	6	31.13	1.53	0.34	21.10	81.70	1
B11	3	26.24	1.23	0.37	26.35	72.29	2
B12	3	28.96	1.35	0.31	27.17	71.24	2
B13	4	27.26	1.17	0.27	28.36	70.97	2
B14	4	33.61	1.28	0.35	24.55	79.51	2
B15	4	35.99	1.37	0.31	27.54	73.22	2
B16	4	33.86	1.23	0.27	28.08	68.87	3
B17	4	43.47	1.30	0.41	25.25	77.73	2
B18	4	39.4	1.14	0.32	26.33	69.82	3
B19	4	37.48	1.08	0.31	27.83	68.51	3
B20	4	43.27	1.25	0.37	25.78	76.44	2
B21	5	63.53	1.69	0.40	21.33	76.33	2
B22	5	46.18	1.20	0.50	25.24	75.57	2
B23	4	46.37	1.05	0.39	27.64	72.81	2
B24	4	72.56	1.63	0.57	21.31	81.55	1
B25	4	60.12	1.18	0.35	28.17	66.14	3
B26	4	58.74	1.15	0.46	26.24	72.25	2
B27	4	60.74	1.18	0.46	25.70	70.77	2
B28	4	73.95	1.18	0.49	25.45	69.32	3
B29	4	82.89	1.20	0.57	24.69	71.52	2
B30	4	97.64	1.21	0.65	24.77	72.57	2
B31	4	102.31	1.23	0.57	23.58	72.04	2
B32	5	107.52	1.21	0.56	25.44	68.53	3
B33	5	109.96	1.14	0.52	27.73	65.95	3
B34	5	155.84	1.57	0.83	22.47	77.66	2
B35	5	136.15	1.29	0.64	27.65	70.75	2
B36	6	451.6	1.24	1.23	28.93	62.82	3
B37	6	1160.83	1.12	1.47	29.87	50.94	4
B38	6	2577.12	1.24	2.03	28.97	44.43	5
B39	6	4516.13	1.28	3.81	23.13	42.70	5
Min	3	2.59	1.02	0.09	19.12		
Max	6	4516.13	1.69	3.81	29.87		

TABLE 14.5
Flash Flood Susceptibility Degrees

M	Parameter	Degree of Hazard	Sig. (1-tailed)	
1	Degree of hazard	1	0	
2	Mean elevation (M)	0.84	0.014	
3	Relief (m; H)	0.80	0	
4	Perimeter (P; km)	0.78	0	
5	Length (L_b; km)	0.75	0	
6	Area (A; km^2)	0.74	0	
7	Stream length (LU; km)	0.73	0.002	
8	Time of concentration (TC)	0.69	0	
9	Max slope	0.67	0.023	Statistically significant
10	Elongation ratio	0.65	0	
11	Velocity of water (km/h)	0.54	0	
12	Stream order	0.49	0	
13	Hypsometric integration	0.35	0	
14	Mean slope	0.32	0.001	
15	Minimum elevation (m)	0.13	0	
16	Compactness factor	0.10	0	
17	Drainage density (DD)	-.44	0	
18	Circularity ratio	-.11	0	
19	Form factor	0.66	0.271	Non-statistically significant
20	Width (W; km)	0.79	0.249	
21	Maximum elevation (m)	0.80	0.215	

determining the time of concentration to the destination, as it influences the speed at which water flows through the watershed, thereby affecting the timing and duration of flood peaks. Table 14.2 and Figure 14.8 illustrate the geographic distribution of areal parameters across the drainage basins in the study area.

Topographic factors play a crucial role in evaluating hazard levels for flash flood zonation. The topographic analysis of the watersheds reveals that the relief varies significantly, ranging from 2,507 m in Wadi Dahab to 376 m in Wadi Zuwayra. This variation highlights the predominantly mountainous terrain of the study area. The topographic characteristics of the drainage basins in the study area are detailed in Table 14.3 and illustrated in Figure 14.8.

Table 14.4 outlines the hydrological parameters of the drainage basins, emphasizing the significant influence of a drainage basin's extent on both stream order and stream length. The analysis shows that Wadi Watier has the longest total stream length at 4,516.13 km, while Wadi Al Qarya has the shortest at 2.59 km.

The drainage density varies from 1.02 in Wadi Barba to 1.69 in Wadi Jadla. This parameter reflects the texture and roughness of the valley surfaces, serving as a crucial factor in water resource management, dam site selection, and aquifer design. The

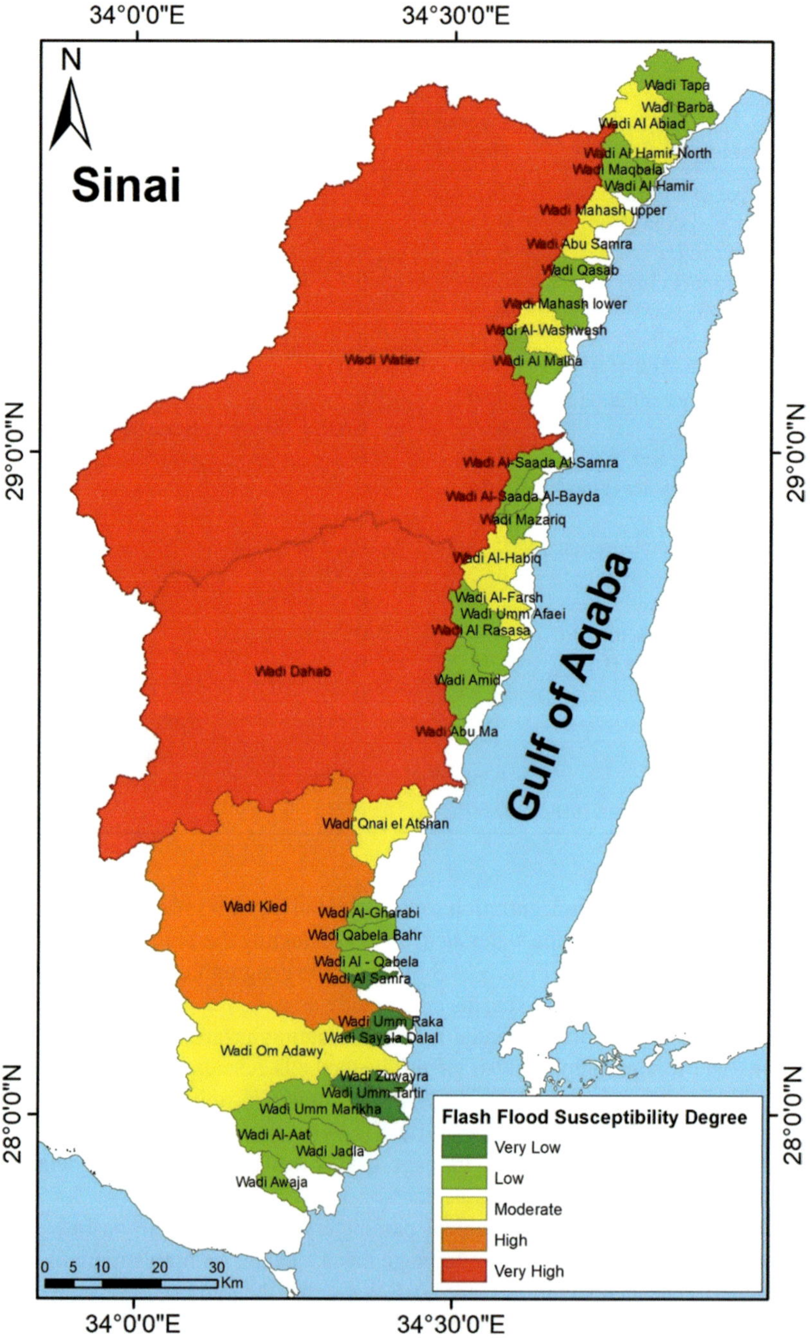

FIGURE 14.9 The flash flood susceptibility map of study area.

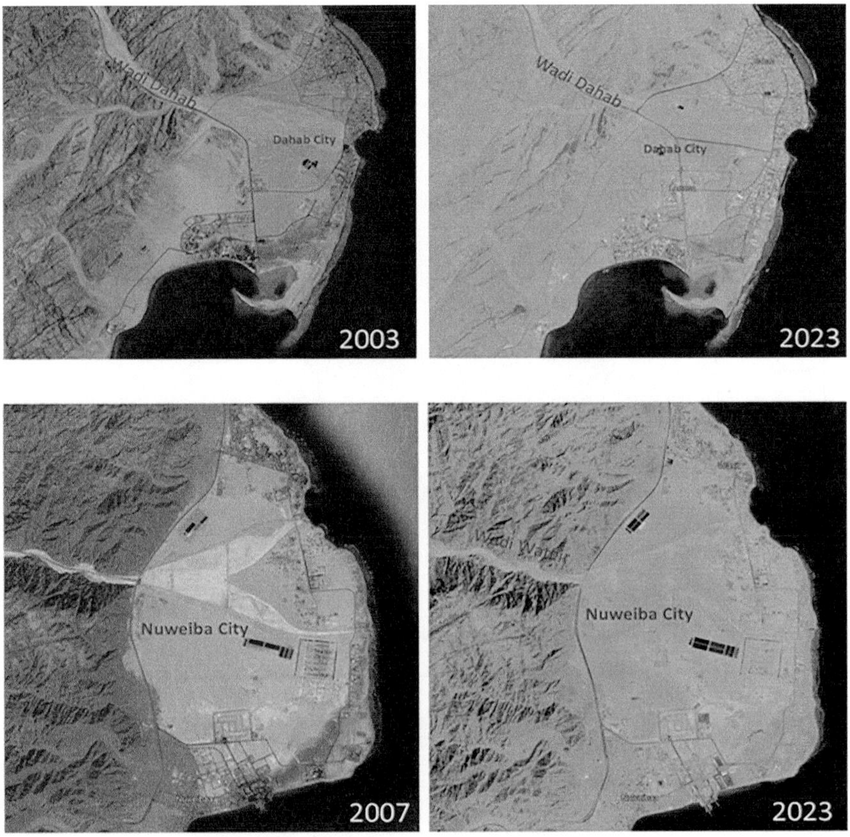

FIGURE 14.10 Urban growth of Dahab and Nuweiba Cities in the study area during the last 20 years. (Source: Google Earth Pro Images.)

time of concentration and water velocity are critical parameters for hazard management and watershed control. In the study area, water velocity varies significantly, with a maximum of 29.87 km/h observed in Wadi Kid and a minimum of 19.12 km/h in Wadi Zuwayra, as illustrated in Figure 14.8 and detailed in Table 14.5. These findings are crucial for understanding flood dynamics and developing effective mitigation strategies.

The flash flood susceptibility map was generated by integrating all available parameters. The 21 parameters used to calculate and model the hazard degrees in this study are summarized in Table 14.5. The results indicate that two drainage basins, Wadi Wateir and Wadi Dahab, are categorized as very high risk. One drainage basin, Wadi Kid, is ranked as high-hazard risk. Eight drainage basins are classified as moderate risk, while the remaining basins are identified as low and very low risk zones, as illustrated in Figure 14.9 and detailed in Table 14.4. Areas located in downstream directions and along runoff paths are considered the most hazardous, particularly the alluvial fans of valleys. Figure 14.10 highlights the urban expansion of Dahab and

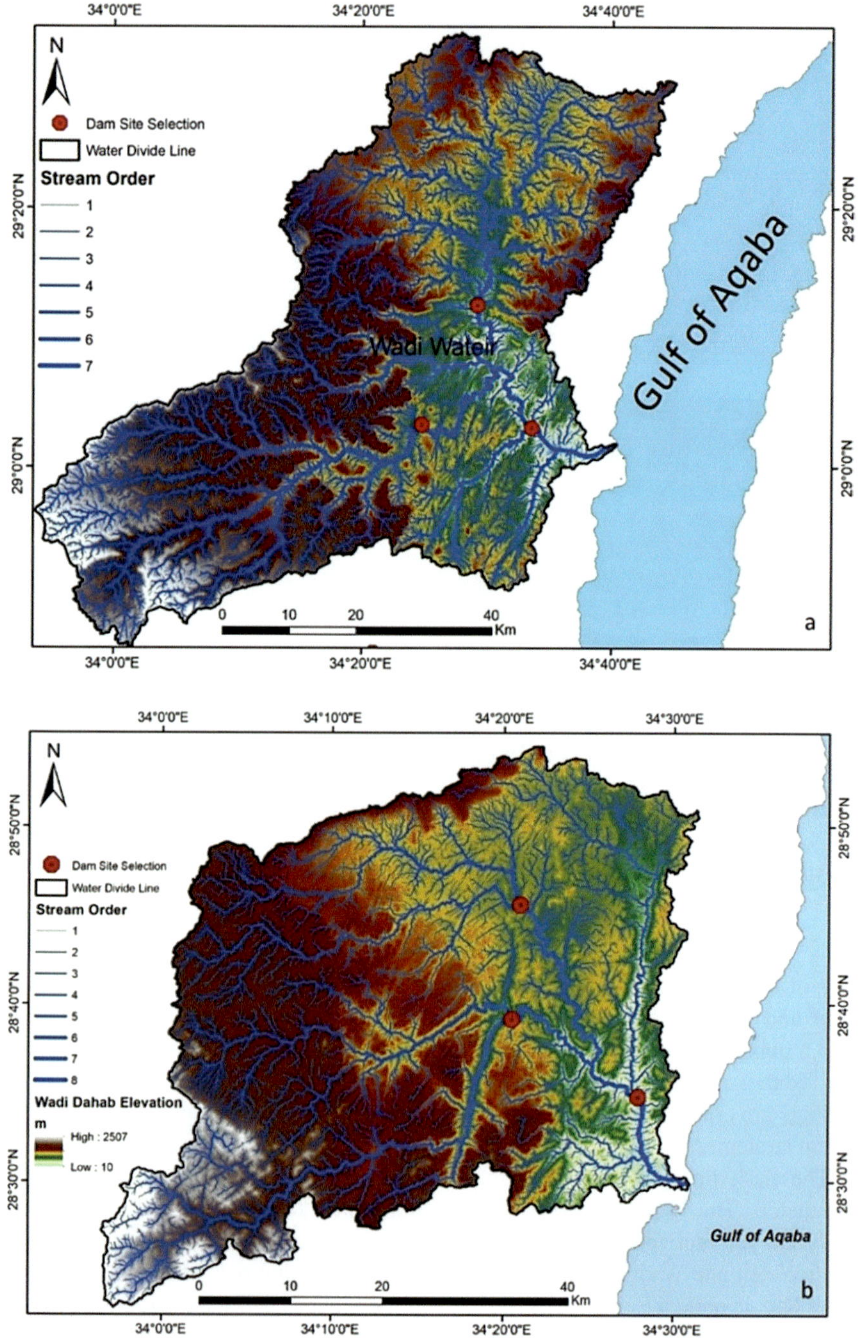

FIGURE 14.11 The watershed systems of the Dahab and Wateir drainage basins, along with the proposed dam sites.

Nuweiba cities, which are built on the alluvial fans of Wadi Dahab and Wadi Wateir, making them highly vulnerable to flash floods.

For water resource management in the study area, proposed dam site selection was conducted for the largest drainage basins, Wadi Wateir and Wadi Dahab. The selection of dam locations was guided by multiple criteria, including geology, topography, soil characteristics, slope gradients, streamflow direction, and drainage density.

Wadi Wateir and Wadi Dahab, the largest and most vulnerable drainage basins in the study area, were chosen for water management and harvesting initiatives. Damming and water storage are critical methods for water conservation and management, as well as for ensuring the safety of human activities, such as land reclamation, road development, and settlements (Figure 14.11).

14.5 CONCLUSIONS

The use of GIS for quantitative analysis and modeling of watersheds and flash flood hazards has significantly increased in recent years, owing to the higher accuracy of results and the integrated approach provided by multisource and multiscale geospatial data. In this study, a quantitative analysis and geo-computational approach incorporating multiple morphometric parameters was employed to assess and map vulnerable areas along the Sinai Gulf of Aqaba region, with a focus on flash flood hazards and water resource harvesting and management. More than 20 morphometric factors, including aerial, topographical, and hydrological parameters, were analyzed to delineate flash flood susceptibility zones in the study area. The results were classified into five categories, ranging from very low to very high vulnerability. A positive relationship was observed between drainage basin area and vulnerability. Wadi Wateir, Wadi Dahab, and Wadi Kid were identified as the most highly vulnerable basins, while smaller basins exhibited lower risk levels. Most of human activities in the downstream areas and runoff paths are considered at significant risk, particularly in alluvial fan regions, including urban areas, roads, reclaimed lands, and tourist sites. To address these risks, this study proposed dam sites in two large drainage basins as a strategic approach for water resource management and mitigating the impacts of flash flood hazards and risks.

REFERENCES

Abdo, H. G., Darwish, K. S., Bindajam, A. A., Niknam, A., Youssef, Y. M., Ahmed, M. F. M., and Mallick, J. (2024). Detection of flood vulnerable areas in urban basins using multi-criteria analysis and geospatial tools: A case study from eastern Mediterranean. *Environmental Earth Sciences*, 83(17), 524.

Bajabaa, S., Masoud, M., and Al-Amri, N. (2014). Flash flood hazard mapping based on quantitative hydrology, geomorphology, and GIS techniques (case study of Wadi Al Lith, Saudi Arabia). *Arabian Journal of Geosciences*, 7, 2469–2481.

Bashir, B. (2023). Morphometric parameters and geospatial analysis for flash flood susceptibility assessment: A case study of Jeddah City along the Red Sea Coast, Saudi Arabia. *Water*, 15(5), 870.

Darwish, K. (2023). GIS-based multi-criteria decision analysis for flash flood hazard and risk assessment: A case study of the Eastern Minya Watershed, Egypt. *Environmental Science Proceedings*, 25, 87.

Davis, J. C. (1975). *Statistics and Data Analysis in Geology*. New York: Wiley.

Demissie, Z., Rimal, P., Seyoum, W. M., Dutta, A., and Rimmington, G. (2024). Flood susceptibility mapping: Integrating machine learning and GIS for enhanced risk assessment. *Applied Computing and Geosciences*, 23, 100183.

Deoli, V., and Kumar, D. (2019). Remote sensing and GIS approach for spatiotemporal mapping of Ramganga reservoir. *International Journal of Current Microbiology and Applied Sciences*, 8(5), 775–783.

Deoli, V., Kumar, D., Shikha, D., Saw, S., and Patel, R. (2023). Mapping and trend analysis in water spread area of upper and lower lakes of Bhopal, using remote sensing technique. In *Surface and Groundwater Resources Development and Management in Semi-arid Region: Strategies and Solutions for Sustainable Water Management* (pp. 177–189). Cham: Springer International Publishing.

El Afandi, G., and Morsy, M. (2020). Developing an early warning system for flash flood in Egypt: Case study Sinai Peninsula. In *Flash Floods in Egypt* (pp. 45–60). Cham: Springer International Publishing.

El-Bialy, M. Z. (2019). Precambrian basement complex of Egypt. *In The Geology of Egypt*, 37–79. Springer Nature Switzerland.

Elnazer, A. A., Salman, S. A., and Asmoay, A. S. (2017). Flash flood hazard affected Ras Gharib City, Red Sea, Egypt: A proposed flash flood channel. *Natural Hazards*, 89, 1389–1400.

El-Rawy, M., Elsadek, W. M., & De Smedt, F. (2022). Flash flood susceptibility mapping in Sinai, Egypt using hydromorphic data, principal component analysis and logistic regression. *Water*, 14(15), 2434.

El-Rayes, A. E., Arnous, M. O., and Helmy, A. M. (2023). GIS-based flash flooding susceptibility analysis and water management in arid mountain ranges: Safaga Region, Red Sea Mountains, Egypt. *Journal of Mountain Science*, 20, 3665–3686.

Fahil, A. S., Ghoneim, E., Noweir, M. A., and Masoud, A. (2020). Integration of well logging and remote sensing data for detecting potential geothermal sites along the Gulf of Suez, Egypt. *Resources*, 9(9), 109.

Helmi, A. M., and Zohny, O. (2020). Flash flood risk assessment in Egypt. In A. Negm (Ed.), *Flash Floods in Egypt* (pp. 253–312). *Advances in Science, Technology & Innovation*. Cham: Springer.

Horton, R. E. (1932). Drainage basin characteristics. *Transactions of the American Geophysical Union*, 13, 350–361.

Horton, R. E. (1945). Erosional development of streams and their drainage basins: Hydrophysical approach to quantitative morphology. *Geological Society of America Bulletin*, 56, 275–370.

Kamel, M., and Arfa, M. (2020). Integration of remotely sensed and seismicity data for geo-natural hazard assessment along the Red Sea Coast, Egypt. *Arabian Journal of Geosciences*, 13(1195) , 1–23.

Miller, V. G. (1953). A quantitative geomorphic study of the drainage basin characteristics in the Clinch Mountain area, Virginia and Tennessee. Columbia University Geology Department Project NR 389-042, Technical Report No. 3, 30.

Mohamed, L. (2015). Structural controls on the distribution of groundwater in Southern Sinai, Egypt: Constraints from geophysical and remote sensing observations. Dissertations Paper, 593.

Mohamed, S. A., and El-Raey, M. E. (2020). Vulnerability assessment for flash floods using GIS spatial modeling and remotely sensed data in El-Arish City, North Sinai, Egypt. *Natural Hazards*, 102, 707–728.

Omran, E. S. E. (2020). Egypt's Sinai Desert cries: Flash flood hazard, vulnerability, and mitigation. In A. Negm (Ed.), *Flash Floods in Egypt. Advances in Science, Technology & Innovation*. Cham: Springer.

Patton, P. C. (1988). Drainage basin morphometry and floods. *Flood Geomorphology*. New York: Wiley. 51–64.

Petrucci, O., Aceto, L., Bianchi, C., Bigot, V., Brázdil, R., Pereira, S., and Zêzere, J. L. (2019). Flood fatalities in Europe, 1980–2018: Variability, features, and lessons to learn. *Water*, 11(8), 1682.

Rahman, Z. U., Ullah, W., Bai, S., Ullah, S., Jan, M. A., Khan, M., and Tayyab, M. (2023). GIS-based flood susceptibility mapping using a bivariate statistical model in Swat River Basin, Eastern Hindukush region, Pakistan. *Frontiers in Environmental Science*, 11, 1178540, 1–17.

Riaz, R., and Mohiuddin, M. (2025). Application of GIS-based multi-criteria decision analysis of hydro-geomorphological factors for flash flood susceptibility mapping in Bangladesh. *Water Cycle*, 6, 13–27.

Schumm, S. A. (1956). Evolution of drainage systems and slopes in badlands at Perth Amboy, New Jersey. *Bulletin of the Geological Society of America*, 67, 597–646.

Shekar, P. R., and Mathew, A. (2024). Morphometric analysis of watersheds: A comprehensive review of data sources, quality, and geospatial techniques. *Watershed Ecology and the Environment*, 6, 13–25.

Sohail, M. T., Hussan, A., Ehsan, M., Al-Ansari, N., Akhter, M. M., Manzoor, Z., and Elbeltagi, A. (2022). Groundwater budgeting of Nari and Gaj formations and groundwater mapping of Karachi, Pakistan. *Applied Water Science*, 12, 1–24.

Strahler, A. N. (1957). Quantitative analysis of watershed geomorphology. *Transactions of the American Geophysical Union*, 38, 913–920.

WMO (World Meteorological Organization). (2016). Flash flood guidance system (FFGS) with global coverage. Retrieved December 20, 2024.

Wong, T. S. W. (2005). Assessment of time of concentration formulas for overland flow. *Journal of Irrigation and Drainage Engineering*, 131(4), 383–387.

Youssef, A. M., Pradhan, B., and Hassan, A. M. (2011). Flash flood risk estimation along the St. Katherine Road, southern Sinai, Egypt using GIS-based morphometry and satellite imagery. *Environmental Earth Sciences*, 62, 611–623.

Zerger, A., and Smith, D. I. (2003). Impediments to using GIS for real-time disaster decision support. *Computers, Environment and Urban Systems*, 27, 123–141.

15 Evaluating Detrending Methods of Crop Yield Time Series for Drought Studies

A Machine Learning-Based Approach

Aekesh Kumar, Anil Kumar, Vishnu Prasad, and Mahesh Chand Singh

15.1 INTRODUCTION

Drought is a shortage of water below a certain level for a prolonged period. It is a natural disaster characterized by a deficiency in water availability that significantly affects ecosystems, agriculture, economies, and human life. Droughts can vary in intensity, duration, and geographic scope, ranging from localized shortages to regional or even global crises. Among the sectors most affected, agriculture is particularly vulnerable, leading to crop failures, livestock losses and food insecurity (Sharafi et al. 2021; Deb et al. 2022; Prodhan et al. 2022). Additionally, droughts can impact water resources for drinking, sanitation, industry and energy generation, posing serious challenges to communities and ecosystems.

In the past, droughts exerted a significant economic impact on the agricultural sector in India. Droughts in 2002, 1965, 1972, 1918, and 2009 were designated as the five most severe drought years witnessed in the country (Chuphal et al. 2024). India's dependence on monsoon-driven agriculture makes it particularly vulnerable, especially with staple crops such as rice, which play a pivotal role in national food security and economic stability. For instance, the drought experienced in the year 2000 led to a decrease of 22% in rice production, whereas the yield of wheat and oilseeds decreased by 9.5% and 23.5%, respectively (Shah et al. 2009). The drought that occurred in 2002 had a significant impact on food grain production, resulting in a steep decline of 38.7 million tonnes (Gautam and Bana 2014). This drought not only affected the agricultural sector but also had severe consequences for the population, with 300 million humans and 150 million cattle suffering from a lack of food, fodder, and water. Consequently, the country's agricultural gross domestic product (GDP) experienced a decrease of 3.1%. Similarly, in 2009, another drought led to a 7% reduction in food grain production compared to the previous year (2008–09) (Gautam and

DOI: 10.1201/9781003596080-15

Bana 2014; Chuphal et al. 2024). Prodhan et al. (2022) have projected a heightened risk of yield reduction in the presence of exceedingly dry weather conditions. In the future, rice, wheat, and maize crops are expected to experience respective losses of 54.15%, 29.30%, and 50.66%. Previous research has documented the overall agricultural losses that occur during specific events. However, only a limited number of studies differentiate these losses from different events due to challenges posed by evolving technology and other non-climatic factors that impact crop yield.

The concept of trend is widely utilized in the analysis of time series data. It refers to the temporal fluctuations observed in the mean value of a particular variable. Zhao et al. (2019) focused on discerning whether these trends are a result of natural or human-induced factors. Conversely, others make a prediction model by decomposing the time series data using trend modeling (Ben-Ari and Makowski 2014; Chang et al. 2021). Selection of a statistical test is important because it may influence the trend magnitude (Adamowski et al. 2010). Previous studies suggested that trend model can significantly impact the outcome of decomposition, therefore it necessitates careful selection. Typically, a trend model is applied to historical yield time series and subtract the predicted values obtained from the trend model from the actual observed yield values to obtain the residual. Conversely, the influence of trend models has been largely overlooked, as the reasons behind selecting a particular trend model for their study and why it was thought to be better than other models are rarely mentioned.

Detrending plays a crucial role in the validation of drought by accounting for the unpredictable losses that are observed in historical time series data. The term technological yields is commonly used to refer to trend yields, with weather conditions and disasters being attributed as the factors responsible for the residuals (Lobell and Field 2007). Detrending not only separates the trend and residuals but also aids in attribution. While statistical methods can be powerful tools for analyzing data and identifying trends, they have limitations in distinguishing between trends influenced by natural and other factors. Does the selection of a trend model have a considerable influence on the outcomes? And if it does, which trend model or models can accurately identify the genuine trend is still a question.

Spatiotemporal computation and differentiating drought losses pose significant challenges due to the multitude of factors that influence crop yields and production (Xu et al. 2021). These factors include scientific and technological advancements in areas such as genetic variation, pesticides, fertilizers, and irrigation resources, with climate conditions. The overarching trend indicates an increase in yield, primarily attributed to technological progress, while a greater number of fluctuations are primarily driven by extreme weather. The complex interplay of these factors results in non-stationary and non-linear long-term crop yield data, characterized by varying means and standard deviations. Consequently, effectively comparing and visually representing the influence of drought on crop yield becomes arduous. The effects of droughts in agriculture cannot be quantitatively extracted and compared by simply looking at the original crop yield, as technical developments and geographical distributions of crop production affect yield in different ways. To accurately model and visually represent the influence of drought on crop yield, it is crucial to differentiate the technological advancement and climate variability in time series data.

This study aims to explore and propose a methodology for determining long-term trends and detrending in crop yield data and isolating the significant climate effect on crop yield.

Before undertaking other fundamental applications, such as calculating the correlation function, it is crucial to remove the trend (Wu et al. 2007). In the previous study, the detrending of crop yield was accomplished by utilizing a predetermined function, for example, simple linear regression (SLR) model or a polynomial regression model. Finger (2010) used a robust regression technique for crop yield data detrending. Tao et al. (2015) conducted experiments on simulated yield trends, and it was found that detrending methods that rely on smoothing techniques demonstrated an overall advantage over linear, loglinear, and autoregressive integrated moving average (ARIMA) models. The most outstanding results were produced by the moving average and robust locally weighted regression models. Choudhury et al. (2015) applied piecewise regression for detrending the crop yield for agriculture insurance and found it better than SLR. Lu et al. (2017) used a six-trend simulation model and two decomposition models for detrending crop yield and identified spatial and temporal drought impact on the United States. Irawan et al. (2023) applied Locally Weighted Scatterplot Smoothing (LOWESS) for detrending crop yield for drought visualization in west Java province of Indonesia.

The topic of detrending crop models using machine learning (ML) is rarely discussed. Previous studies have predominantly utilized traditional regression techniques for detrending purposes. The strengths and weaknesses of these methods have been well comprehended, as evidenced by their algorithmic foundations and practical applications (Sidhu et al. 2023). SLR, which is primarily an adopted traditional technique, has the advantage of interpretability but is limited in its predictive capabilities. It is unable to accommodate the nonlinearity observed in crop yield time series. Conventional regression techniques face challenges in addressing these issues, but ML models can provide potential solutions. These models are capable of fitting intricate functions and exhibiting superior predictive accuracies (Hu et al. 2023).

This study compares conventional regression methods and ML techniques for detrending and assesses their respective merits and drawbacks in terms of their suitability across space and time, reliability, and efficiency. This study examines a suitable method for data self-adaptive detrending, which has the capability to automatically simulate long-term non-stationary and nonlinear yield trends primarily influenced by technological advancements, thereby eliminating trends due to extreme weather.

15.2 MATERIALS AND METHODS

In this section, a brief overview of the study area has been presented, followed by information on the source of data used in this study. Further, a detailed description of trend simulation methods has been presented.

15.2.1 Study Area and Data Collection

The time series data of crop yield was obtained from the Directorate of Economics and Statistics (DES), Ministry of Agriculture and Farmers Welfare, Government of India (www.desagri.gov.in), which maintains an extensive database of crop area, production, land use and yield. Rice yield was evaluated in this study because it is one of the most extensively cultivated crops in India. This study examined detrending techniques on rice crop yield from 1966 to 2020 at the state level for five major rice-growing states (Andhra Pradesh (AP), Bihar (BH), Chhattisgarh (CG), Maharashtra (MH), and Uttar Pradesh (UP)) for India (Figure 15.1).

15.2.2 Trend Simulation Methods

Six distinct detrending approaches were evaluated to eliminate the rising trend in rice yield. The initial step in detrending involves simulating the inherent trend in crop yield. The trend simulation approaches encompassed an SLR, polynomial regression model

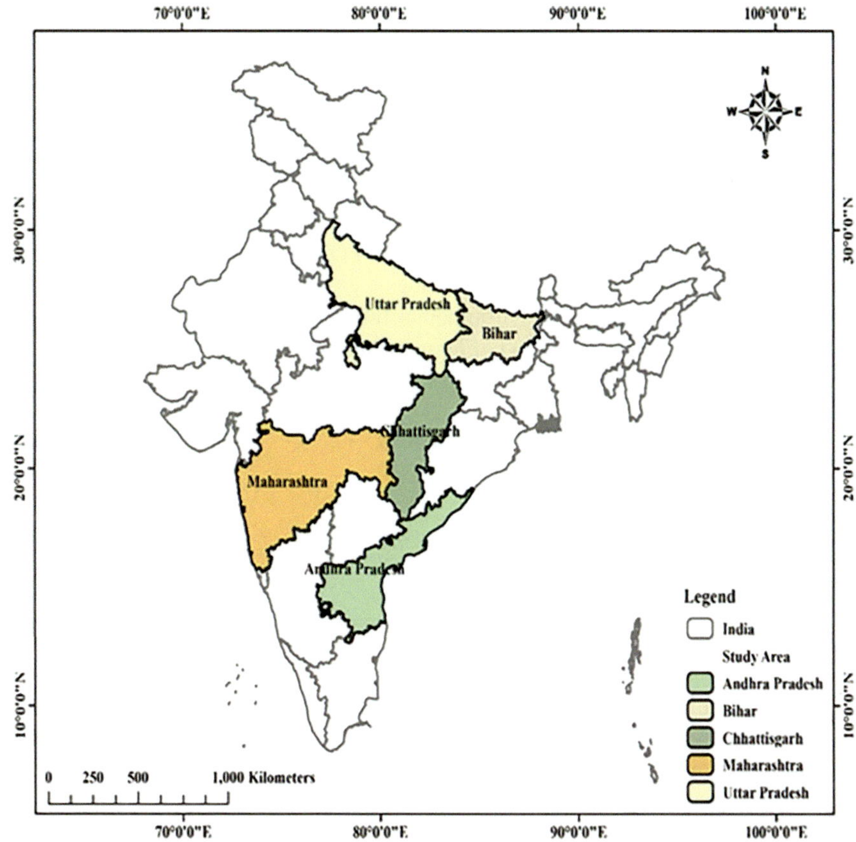

FIGURE 15.1 Location of India and selected states.

of the second order (POLYR), central moving average (CMA) model, LOWESS, and two ML models, namely Random Forest (RF) and Support Vector Regression (SVR). Subsequent to trend simulation, to detrend the data two decomposition models were used and compared. These techniques were tested individually for selected states. All data analyses were conducted using the R programming language with necessary packages.

15.2.3 Simple Linear Regression Model

The SLR is a statistical approach that enables you to model and simulate a linear trend if the relation between two variables is linear. The goal is to model the connection between the dependent (yield) and the independent variable (time) using a linear equation. The SLR model is represented by Equation 15.1:

$$Y_t = \beta_0 + \beta_1 t + \varepsilon \tag{15.1}$$

where Y_t is rice crop yield at time t; β_0 is y-intercept; β_1 is the slope of the line; ε is the error term. SLR analysis estimates the values of β_0 and β_1 that minimizes the sum of square differences between observed values of Y_t and predicted values by the model.

15.2.4 Second-Order Polynomial Regression Model

The POLYR is generally used to model trends when the association between the dependent and independent variable is more complex than a straight line (Bahrami and Mahmoudi 2022; Huang et al. 2017; Liu et al. 2016). The quadratic trend observed in time series data would necessitate the use of a POLYR model. Polynomial regression, as opposed to linear regression, provides a greater degree of versatility examination of yield dependent on environmental humidity and temperature conditions (Morozova et al. 2019). The method of least squares can be employed to fit a polynomial regression model of second order against time (Equation 15.2):

$$Y_t = \beta_0 + \beta_1 t + \beta_2 t^2 + \varepsilon \tag{15.2}$$

where β_0, β_1, and β_2 are the coefficients.

15.2.5 Central Moving Average Model

The CMA model can be utilized for a time-series analysis technique that smoothens data by calculating the average of a certain number of consecutive data points, with the central point being the focus. Moving averages can adapt to changing patterns in the data over time. As the window moves along the time series, the model can capture shifts in trends or identify emerging patterns. As the duration of the moving average extends, the trend exhibits a higher level of smoothness but may lose some responsiveness to short-term fluctuations. CMAs are computed here to determine the trend, with calculations conducted for time spans of 3 years. The formulas employed for each respective time span is provided in Equation 15.3:

$$3CMA:\widehat{Y}_t = \frac{1}{3}\sum_{j=-1}^{-1} y_{t+j} \tag{15.3}$$

where $\widehat{Y}_t$ is CMA model simulates crop yield.

15.2.6 Locally Weighted Scatter Plot Smoothing Model

The LOWESS is a nonparametric smoothing approach that extracts the underlying nonlinear behavior in a scatter plot (Cleveland, 1979). This method characterizes the application of local regression employing linear recursive least squares. However, the modeling accuracy is compromised when outliers are present in the data. By utilizing local regression, the algorithm mitigates the impact of outliers, leading to enhanced modelling performance. Notably, this approach involves assigning significantly reduced weights to the outliers. LOWESS necessitates both a weight function and a parameter denoting a fraction of points in the neighborhood (*f*).

In this study, a locally weighted quadratic fitting method was employed. In this approach, $0 < f \leq 1$ condition was established. Moreover, d_{max}, as distance between t_i and *r*th closest neighbor was defined, where *r* denotes the number of points employed in estimating the point of interest t_i. Subsequently, for every t_i weight function w was established, which is defined for all t_x (where *x* ranges from 1 to *n*) in Equations 15.4 and 15.5:

$$w_{t_x}(t_i) = \left(1 - \left|\frac{t_x - t_i}{d_{max}}\right|^3\right)^3 \tag{15.4}$$

$$w_{t_x}(t_i) = 0, for\ |t_x - t_i| \geq d_{max} \tag{15.5}$$

For every t_i, the estimates $\hat{\beta}_0(t_i)$,$\hat{\beta}_1(t_i)$, and $\hat{\beta}_2(t_i)$, of $\hat{\beta}_0(t_i)$, $\hat{\beta}_1(t_i)$ and $\hat{\beta}_2(t_i)$, are determined using the technique of weighted least squares where a weight function *W* is employed:

$$\sum_{x=1}^{n} w_{t_x}(t_i)\left(y_{t_x} - \beta_0(t_i) - \beta_1(t_i)t_x - \beta_2(t_i)t_x^{\,2}\right)^2 \tag{15.6}$$

Thus, the fitted value $\hat{y}_{t_i}$ is given in Equation 15.7:

$$\hat{y}_{t_i} = \hat{\beta}_0(t_i) - \hat{\beta}_1(t_i)t_i - \hat{\beta}_2(t_i)t_i^2 \tag{15.7}$$

As the proportion of nearby points (*f*) increases, the regression analysis of the point of interest becomes more extensive, encompassing a larger number of points. The selection of the parameter *f* holds a crucial aspect in LOWESS. To determine the optimal parameter *f*, cross-validation offers an appropriate approach. In this study, the parameter *f* was established using the *k*-fold cross-validation technique. The procedure entails dividing the original dataset into *k* equal-sized folds or subsets. The model is then trained and assessed *k* times, each time with a different fold serving as the validation set and the remaining folds as the training set. During each iteration, the model is trained on the training set and verified on a certain fold. The mean absolute error across these *k* iterations serves as a measure of cross-validation. The optimal parameter *f* is chosen based on the minimum mean absolute error derived from this process.

15.2.7 Random Forest

The RF algorithm is a well-known supervised ML technique that utilizes ensemble learning principles to address both categorical and regression problems (Dang et al. 2021). Here, this algorithm was employed as a tool for regression analysis. The name "decision tree" is derived from the visual representation of the algorithm's flow, resembling the structure of trees. Initially, a predetermined number of training sets are randomly sampled, and distinct trees are generated. At each node, a random selection of independent variables is utilized to partition the data into more units within each tree. These trees are grown to their full extent, and the predicted value of a constant response is determined by averaging the fitted reactions from all the different trees. The effectiveness of the RF algorithm in mitigating the issue of overfitting trees contributes to its success as an ML technique (Everingham et al. 2016).

Observed rice crop yield was trained and applied RF for trend simulation. The effectiveness of the RF algorithm primarily relied on a number of decision trees utilized and collection of potential attributes included in a subset (Ramosaj and Pauly 2019). The correlation between a number of trees and out of bag (OOB) error was employed to ascertain the optimal count of decision trees, which yielded the most favorable outcomes.

15.2.8 Support Vector Regression

This study employs SVR which has emerged as a robust and versatile regression technique. It provides a principled approach to modeling nonlinear relationships and high-dimensional data. The fundamental concept underlying this algorithm is to maximize the data fit while adhering to the margin constraint (Shafiee et al. 2021). By finding the appropriate hyperplane from the given data points, SVR determines the closest relationship between the support vectors and the hyperplane's location, as well as the function used to describe them (Khosla et al., 2020). In certain scenarios, SVR aims to achieve the best possible line by restricting the number of violation constraints using hyperparameter tuning parameters like epsilon (ε), gamma (γ), and the cost function (C), which serves as the regularization parameter in kernel transformation. This study explores various types of kernels, including linear, polynomial, and radial basis functions and selects the best kernel based on minimum root mean square error. Grid search is a hyperparameter optimization technique that is employed to systematically search a predefined set of hyperparameters and identify the combination that yields the optimal performance for the ML model (Paidipati et al., 2021).

15.2.9 Decomposition Model

After employing appropriate statistical models to simulate the trend, decomposition model is utilized to eliminate simulated trend and acquire detrended data. The significance of decomposition models lies in their ability to extract meaningful information from data by revealing underlying patterns and trends. These models serve as invaluable tools in the realm of data analysis, offering a systematic approach to unraveling intricate time series or observational data into distinct components. Detrending represents one of the primary applications of decomposition models; it separates the observed data from the underlying patterns or trends (Kartikasari and Hikmah 2022). Two methods are available for accomplishing this task. They are the additive and multiplicative models.

Additive model generates detrended data by deducting trend line values from the observed data. This particular model is deemed appropriate in scenarios where the variations in the data remain relatively consistent across various levels. It is worth noting that the unit of measurement for the residuals corresponds to that of the original data (Lu et al. 2017).

A different approach involves utilizing multiplicative decomposition, in which detrended data is obtained by calculating the ratio between the observed data and trend line values. These detrended data points are devoid of dimension and illustrate the percentage discrepancies when compared to the trend line values. This particular model is better suited for situations where the magnitude of the fluctuations in the data varies across different levels.

15.2.10 Performance of Trend Simulation Models

Performance of trend simulation models was checked by Root Mean Square Error (RMSE), Nash–Sutcliffe efficiency (NSE), Mean Absolute Error (MAE), and Index of Agreement (IOA). RMSE and MAE are extensively used metrics for assessing the effectiveness of prediction models, notably in regression analyses. NSE is a statistical metric often used to evaluate the performance of hydrological or environmental models. It indicates how accurately the model predictions fit the observed data. NSE values range from $-\infty$ to 1. A greater NSE indicates better model performance, with 1 representing the optimal value (Nash and Sutcliffe 1970). IOA provides a comprehensive evaluation of model performance, taking into account both systematic errors and random errors in predictions (Willmott, 1981). IOA and NSE are applicable to a broader range of models and not limited to linear relationships which provide advantages over the widely utilized coefficient of determination (R^2). On the other hand, both IOA and NSE have the capacity to identify disparities in the means and variances of observed and model simulated data. The equations for these four quantitative measures can be found in Table 15.1.

15.3 RESULTS AND DISCUSSION

The recorded rice yield exhibited a rising pattern attributed to technological advancements, management practices, and the cumulative impact of prolonged

TABLE 15.1
Performance Parameters Used for Models Accuracy

S. No.	Parameter	Equation
1	RMSE	$RMSE = \sqrt{\frac{1}{n}\sum_{i=1}^{n}\left(y_i - \hat{y}_i\right)^2}$
2	NSE	$E = 1 - \frac{\sum_{i=1}^{n}\left(y_i - \hat{y}_i\right)^2}{\sum_{i=1}^{n}\left(y_i - \bar{y}_i\right)^2}$
3	MAE	$MAE = \frac{1}{n}\sum_{i=1}^{n}\left\lvert y_i - \hat{y}_i\right\rvert$
4	IOA	$IOA = 1 - \frac{\sum_{i=1}^{n}\left(y_i - \hat{y}_i\right)^2}{\sum_{i=1}^{n}(\lvert \hat{y}_i - \bar{y}_i\rvert + \lvert y_i - \bar{y}_i\rvert)^2}$

climate variations. The detrending of yield serves as a method to manage the temporal fluctuations within statistical models.

15.3.1 Trend Simulation Comparison

Figure 15.2 illustrates the temporal sequence of rice yield in selected states spanning from 1966 to 2020, alongside the outcomes of six models employed for trend simulation. In each state, the rice yield time series exhibited a nonlinear upward trend that governs long-term time series of crop yield patterns. This trend was predominantly attributed to advancements in technology and the augmentation of inputs, which continued to increase. Furthermore, the series displayed substantial high-frequency fluctuations, primarily driven by weather-related factors that intensified over time. To grasp the year-to-year fluctuations, it became essential to remove the technological trend embedded in the time series, thereby standardizing the crop yield.

Due to nonlinearity of the technology trend, the efficacy of an SLR model in elucidating fluctuations in rice yield within the study regions (Figure 15.2) was found notably deficient and was bereft of any logic or reason. While a quadratic trend did improve the association but it remained inadequate in capturing the gradual increase evident in the trend. These predetermined models lacked the requisite flexibility to eliminate nonlinear and non-stationary trends across every state.

Inspecting the visual representation of rice yield revealed the need for implementing a CMA model to address the inconsistencies observed in the crop yield data. A specific time period was required to perform the CMA model. Nevertheless, the process of determination and selection of this time span remained subjective. Moreover, the utilization of a CMA model presented a boundary issue. To implement a 3-year CMA model, it was essential to have 3-year data prior to and following year of interest. Consequently, when data points shift to initial or latest years, absence of adequate data for estimation leads to the identification of the first three and last three data

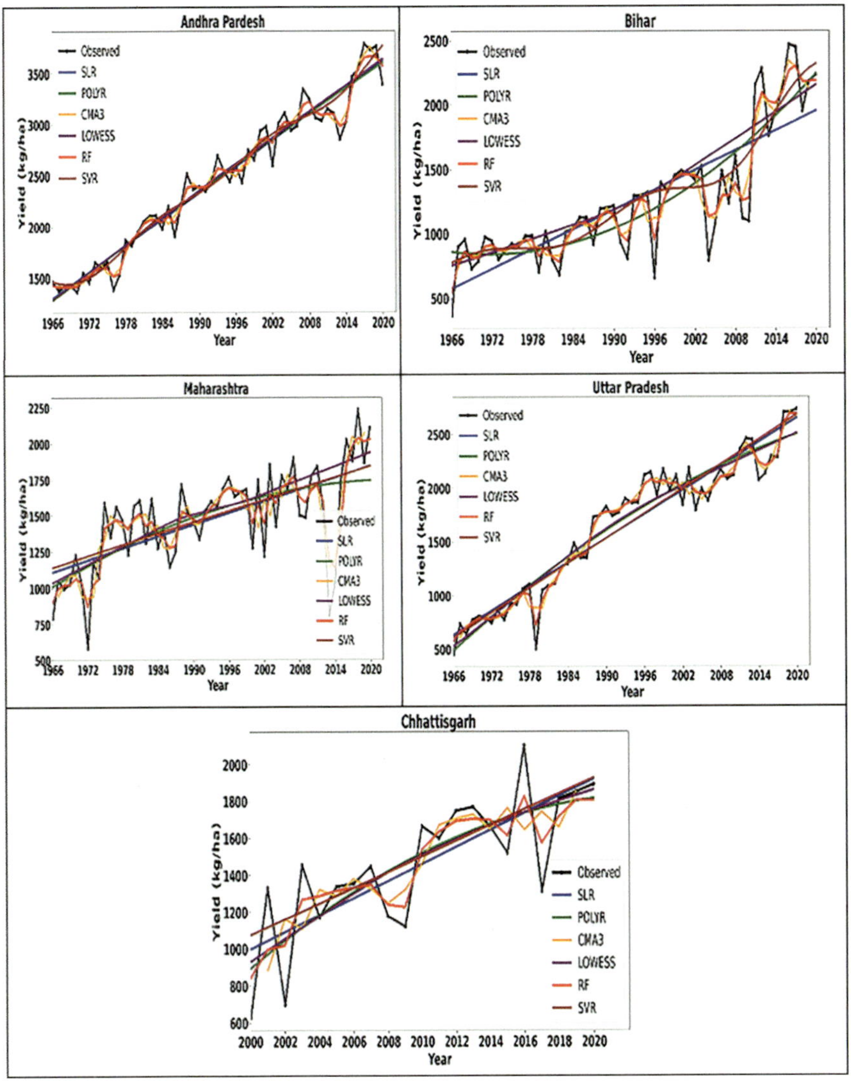

FIGURE 15.2 Trend simulation model (SLR, POLYR, CMA3, LOWESS, RF, and SVR) comparison of rice crop yield (1966–2020) for Andhra Pradesh, Bihar, Maharashtra, Uttar Pradesh, and Chhattisgarh.

points as missing data. Furthermore, the presence of a single missing data point in crop yield series could result in an additional three data points being designated as missing data for CMA model. Even in the absence of missing data in time series, CMA model for three years remained sacrificed two data points at the starting and ending positions in time series. It is important to highlight that the effectiveness of the CMA model was compromised or biased near the border of time series data.

TABLE 15.2
Optimized Hyperparameters Used in SVR

States	Kernel	Cost Function (*C*)	Epsilon (ε)
Andhra Pradesh	Radial basis function	10	0.1
Bihar	Radial basis function	10	0.2
Chhattisgarh	Linear	1	0.2
Maharashtra	Linear	10	0.5
Uttar Pradesh	Linear	1	0.1

On the contrary, LOWESS models enabled the incorporation of adjacent data points at the time series boundary, leading to the estimation of boundary points instead of encountering missing values. This model might exhibit either a linear or polynomial character. When there is significant curvature in the regression surface, such as in the case of rice yield over time, locally weighted quadratic fitting outperforms other methods.

In ML model, RF performed best among all other models. Trend curve predicted by RF model closely mirrors the time series of rice yield and demonstrates strong performance for all states (Figure 15.2). The support vector regression model performance was lower than RF model. This study incorporated the most suitable kernel and user-defined hyperparameters C and ε (Table 15.2) for better accuracy of this model. The SVR model employed different kernel functions, namely linear, radial basis, and polynomial functions. Grid search optimization and k-fold cross-validation techniques were utilized to optimize these hyperparameters, aiming to minimize error estimates while reducing bias and variance in the dataset.

15.3.2 Evaluation Metrics for Model Performance

Figure 15.3 presents the performance metrics used to evaluate various trend simulation models. Among these, the RF model demonstrates the best performance, exhibiting the lowest RMSE value for Uttar Pradesh (76.52), followed closely by the CMA model in Andhra Pradesh (102.90). Conversely, SLR model recorded the highest RMSE (285.04) for Bihar, indicating poor model performance. A similar trend was observed for the Mean Absolute Error (MAE). The IOA heatmap highlighted that the RF model achieved the highest IOA (99.65%), followed by CMA3 (99.43%) in Andhra Pradesh. A comparable pattern was evident in the NSE values. These evaluation metrics collectively confirm that SLR was the least accurate model for trend simulation. In comparison, POLYR models performed moderately better than SLR. The central moving average CMA method consistently outperformed SLR, POLYR, and LOWESS models by fitting the rice yield trends more accurately. Notably, ML-based models, particularly RF, outperformed conventional regression techniques across all states. CMA3 typically emerged as the second-best performing model, especially in states such as Uttar Pradesh and Andhra Pradesh. The relatively lower performance of all models in Chhattisgarh was attributed to data limitations in that state.

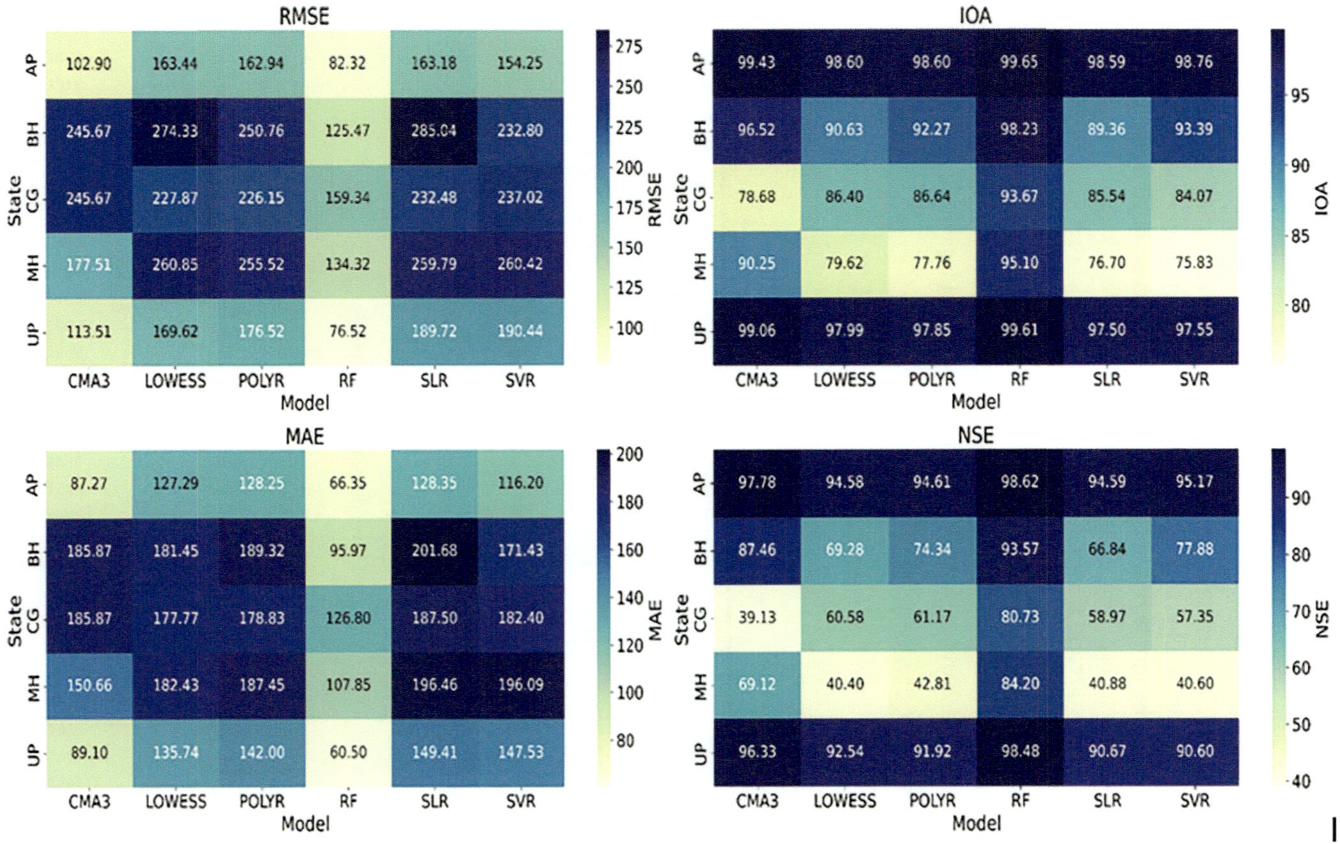

FIGURE 15.3 Model comparison based on RMSE, MAE, IOA, and NSE.

15.3.3 Decomposition Models Comparison

To depict the deviation of crops from the normal condition, Guttormsen (1999) and Zhang and Qi (2005) considered detrended data obtained by subtracting them from the regression line. These studies made the assumption that fluctuations and trends were additively composed.

Lu et al. (2017) suggested multiplicative model for detrending long-term crop yield. With the application of a multiplicative decomposition approach to eliminate the trend from the time series, the detrended rice production in all states exhibited an increasing variance over time (Figure 15.4). As the detrended data's variance was standardized to the scale of crop output, it underwent a transition toward increased stationarity over time, an additive decomposition model make more sense (Figure 15.5). It was observed that the rice yield anomalies exhibited a correlation with drought years, as evidenced by the significant occurrence of major drought years (1979, 2002, and 2009) as shown in Figure 15.5. Rice yield anomalies could capture

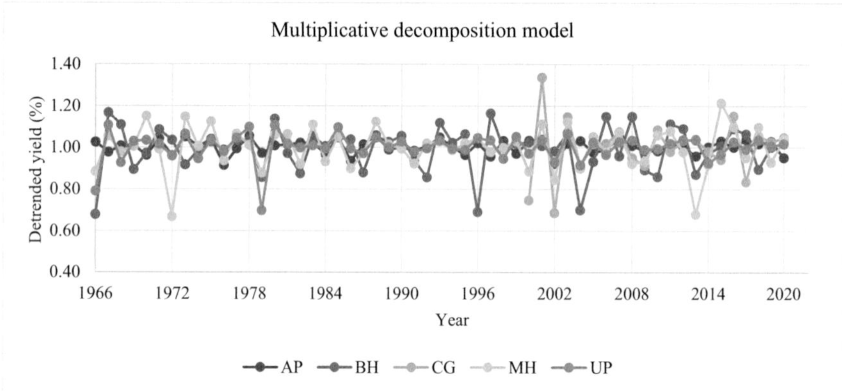

FIGURE 15.4 Multiplicative decomposition model based on RF model output.

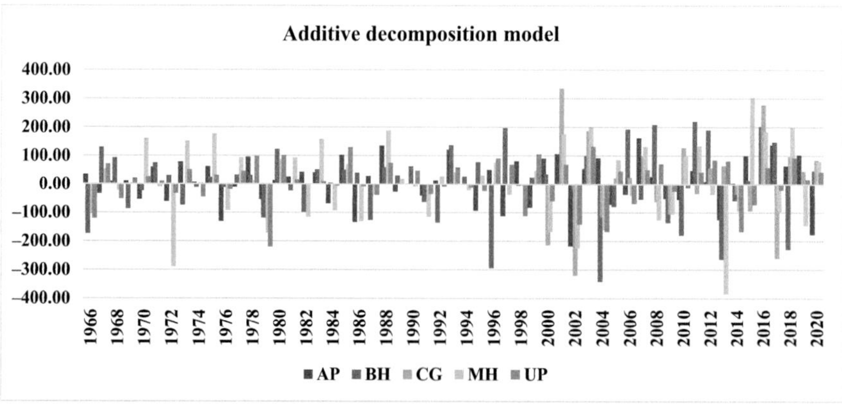

FIGURE 15.5 Additive decomposition model based on RF model output.

the variations in rice yield that were lower than the normal levels during drought years. Therefore, an additive decomposition model was used to detrend rice yield after putting an acceptable trend simulation method.

In this study, the selection of detrending model was guided by considerations of performance, reliability, and efficiency. The analysis presented above highlights the underwhelming performance of SLR, POLYR, and LOWESS models. Though the CMA model proved superior among the conventional models, it has certain limitations and is significantly affected by missing data. Conversely, two ML-based models, namely RF and SVR, surpass conventional detrending techniques. These ML-based models excel in automatically capturing underlying patterns of non-stationary and nonlinear rice yield time series, thereby offering effective trend fitting of crop yields. These detrended crop yields were considered as crop yield anomaly, which showed higher and lower than the normal crop yield. Thus, the combination of the RF model with an additive decomposition approach stands out as the preferred technique for detrending rice yield for all states.

15.4 CONCLUSIONS

The current study aimed to determine the suitable detrending technique in ML for standardizing and detrending rice yield. The analysis compares different conventional and ML trend simulation models using four quantitative metrics. The results indicated that SLR, POLYR, and LOWESS models exhibit poorest fit. Comparatively, CMA outperforms other three conventional regression techniques with constrained by boundary issues. However, when considering ML models, the RF model demonstrated greater accuracy than the others. Additionally, two decomposition models were compared and the results revealed that the additive decomposition model was more suitable for detrending crop yield. This was because the variance of the detrended crop yield was modified in accordance with the scale of the crop yield magnitude, resulting in increased stationarity over time. Therefore, the most appropriate approach for detrending rice crop yield was to employ the RF model in conjunction with the additive decomposition model.

The methodology employed in this study involved standardizing the rice yield, thereby enabling a quantitative assessment of the association with drought and rice yield. Additionally, this approach facilitated the visualization of the impacts of drought on rice yield. Further correlation studies comparing several drought indicators and rice yield anomalies could be done for validation of drought. This detrending methodology was not constrained to the study of rice and drought scenarios but was equally applicable to different crops and various natural disasters. It was important to note that deviations in crop yield could arise due to factors beyond drought, including flooding, abrupt and extreme weather events, pest invasions, and diseases. This study effectively distinguished the influences of weather and environmental related elements from those attributed to technological factors other than drought.

This study provides a way to evaluate how drought affects crop production. The findings could help policymakers and stakeholders to develop robust strategies for risk adaptation and comprehensive management plans to reduce the impact of extreme weather on agriculture. It can be further enhanced using more ML and deep learning algorithms and also be employed to make short-term predictions regarding the impact of technological advancements on crop yields. Lastly, the anomalies in crop yield obtained through this method can also be employed in agricultural studies focusing on the impacts of climate change.

REFERENCES

Adamowski, J., Adamowski, K., and Bougadis, J. (2010) Influence of trend on short duration design storms. *Water Resources Management*, 24, 401–413.

Bahrami, M., and Mahmoudi, M. R. (2022). Long-term temporal trend analysis of climatic parameters using polynomial regression analysis over the Fasa Plain, southern Iran. *Meteorology and Atmospheric Physics*, 134(2), 42.

Ben-Ari, T., and Makowski, D. (2014). Decomposing global crop yield variability. *Environmental Research Letters*, 9(11), 114011.

Chang, M., Thellufsen, J. Z., Zakeri, B., Pickering, B., Pfenninger, S., Lund, H., and Østergaard, P. A. (2021). Trends in tools and approaches for modelling the energy transition. *Applied Energy*, 290, 116731.

Choudhury, A. H., Jones, J. R., Choudhury, R. L., and Spaulding, A. D. (2015). Association of rainfall and detrended crop yield based on piecewise regression for agricultural insurance. *Journal of Economics and Economic Education Research*, 16(2), 31–44.

Chuphal, D. S., Kushwaha, A. P., Aadhar, S., & Mishra, V. (2024). Drought atlas of India, 1901–2020. *Scientific Data*, 11(1), 7.

Cleveland, W. S. (1979). Robust locally weighted regression and smoothing scatterplots. *Journal of the American Statistical Association*, 74(368), 829–836..

Dang, C., Liu, Y., Yue, H., Qian, J., and Zhu, R. (2021). Autumn crop yield prediction using data-driven approaches:-support vector machines, random forest, and deep neural network methods. *Canadian Journal of Remote Sensing*, 47(2), 162–181.

Deb, P., Moradkhani, H., Han, X., Abbaszadeh, P., & Xu, L. (2022). Assessing irrigation mitigating drought impacts on crop yields with an integrated modeling framework. *Journal of Hydrology*, 609, 127760.

Everingham, Y., Sexton, J., Skocaj, D., & Inman-Bamber, G. (2016). Accurate prediction of sugarcane yield using a random forest algorithm. *Agronomy for Sustainable Development*, 36, 1–9.

Finger, R. (2010). Revisiting the evaluation of robust regression techniques for crop yield data detrending. *American Journal of Agricultural Economics*, 92(1), 205–211.

Gautam, R. C., and Bana, R. S. (2014). Drought in India: Its impact and mitigation strategies A review. *Indian Journal of Agronomy*, 59(2), 179–190.

Guttormsen, A. G. (1999). Forecasting weekly salmon prices: Risk management in fish farming. *Aquaculture Economics & Management*, 3(2), 159–166.

Hu, T., Zhang, X., Bohrer, G., Liu, Y., Zhou, Y., Martin, J., ... and Zhao, K. (2023). Crop yield prediction via explainable AI and interpretable machine learning: Dangers of black box models for evaluating climate change impacts on crop yield. *Agricultural and Forest Meteorology*, 336, 109458.

Huang, H., Wang, Z., Xia, F., Shang, X., Liu, Y., Zhang, M., and Mei, K. (2017). Water quality trend and change-point analyses using integration of locally weighted polynomial regression and segmented regression. *Environmental Science and Pollution Research*, 24, 15827–15837.

Irawan, A. N. R., Komori, D., and Hendrawan, V. S. A. (2023). Correlation analysis of agricultural drought risk on wet farming crop and meteorological drought index in the tropical-humid region. *Theoretical and Applied Climatology*, 153(1), 227–240.

Kartikasari, M. D., and Hikmah, N. (2022). Decomposition method with application of Grey Model GM (1, 1) for forecasting seasonal time series. *Pakistan Journal of Statistics and Operation Research*, 18(2), 411–416.

Khosla, E., Dharavath, R., and Priya, R. (2020). Crop yield prediction using aggregated rainfall-based modular artificial neural networks and support vector regression. *Environment, Development and Sustainability*, 22(6), 5687–5708.

Liu, Y., Wang, Y., Du, Y., Zhao, M., and Peng, J. (2016). The application of polynomial analyses to detect global vegetation dynamics during 1982–2012. *International Journal of Remote Sensing*, 37(7), 1568–1584.

Lobell, D. B., and Field, C. B. (2007). Global scale climate–crop yield relationships and the impacts of recent warming. *Environmental Research Letters*, 2(1), 014002.

Lu, J., Carbone, G. J., and Gao, P. (2017). Detrending crop yield data for spatial visualization of drought impacts in the United States, 1895–2014. *Agricultural and Forest Meteorology*, 237, 196–208.

Morozova, S. V., Polyanskaya, E. A., Kononova, N. K., Denisov, K. E., and Poletaev, I. S. (2019). The study of the dependence of spring crops yield on the abiotic environmental factors using nonlinear interpolation. In *IOP Conference Series: Earth and Environmental Science*, 381(1), 012065.

Nash, J. E., and Sutcliffe, J. V. (1970). River flow forecasting through conceptual models part I—A discussion of principles. *Journal of Hydrology*, 10(3), 282–290.

Paidipati, K. K., Chesneau, C., Nayana, B. M., Kumar, K. R., Polisetty, K., & Kurangi, C. (2021). Prediction of rice cultivation in India—Support vector regression approach with various kernels for non-linear patterns. *AgriEngineering*, 3(2), 182–198.

Prodhan, F. A., Zhang, J., Sharma, T. P. P., Nanzad, L., Zhang, D., Seka, A. M., and Mohana, H. P. (2022). Projection of future drought and its impact on simulated crop yield over South Asia using ensemble machine learning approach. *Science of The Total Environment*, 807, 151029.

Ramosaj, B., and Pauly, M. (2019). Consistent estimation of residual variance with random forest Out-of-Bag errors. *Statistics & Probability Letters*, 151, 49–57.

Shafiee, S., Lied, L. M., Burud, I., Dieseth, J. A., Alsheikh, M., and Lillemo, M. (2021). Sequential forward selection and support vector regression in comparison to LASSO regression for spring wheat yield prediction based on UAV imagery. *Computers and Electronics in Agriculture*, 183, 106036.

Shah, T., Kishore, A., and P, H. (2009). Will the impact of the 2009 drought be different from 2002?. *Economic and Political Weekly*, 11–14.

Sharafi, L., Zarafshani, K., Keshavarz, M., Azadi, H., and Van Passel, S. (2021). Farmers' decision to use drought early warning system in developing countries. *Science of the Total Environment*, 758, 142761.

Sidhu, B. S., Mehrabi, Z., Ramankutty, N., & Kandlikar, M. (2023). How can machine learning help in understanding the impact of climate change on crop yields?. *Environmental Research Letters*, 18(2), 024008.

Tao, F., Zhang, S., Zhang, Z., & Rötter, R. P. (2015). Temporal and spatial changes of maize yield potentials and yield gaps in the past three decades in China. *Agriculture, Ecosystems & Environment*, 208, 12–20.

Willmott, C. J. (1981). On the validation of models. *Physical Geography*, 2(2), 184–194.

Wu, Z., Huang, N. E., Long, S. R., and Peng, C. K. (2007). On the trend, detrending, and variability of nonlinear and nonstationary time series. *Proceedings of the National Academy of Sciences*, 104(38), 14889–14894.

Xu, R., Li, Y., Guan, K., Zhao, L., Peng, B., Miao, C., and Fu, B. (2021). Divergent responses of maize yield to precipitation in the United States. *Environmental Research Letters*, 17(1), 014016.

Zhang, G. P., and Qi, M. (2005). Neural network forecasting for seasonal and trend time series. *European Journal of Operational Research*, 160(2), 501–514.

Zhao, Y., Sun, R., and Ni, Z. (2019). Identification of natural and anthropogenic drivers of vegetation change in the Beijing-Tianjin-Hebei megacity region. *Remote Sensing*, 11(10), 1224.

16 Hydrological Modeling and Its Limitations for Sustainable Land and Water Resource Management

Shubhi Khare, Deepak Kumar, and Benson Turyasingura

16.1 INTRODUCTION

Degradation of land and water resources is one of the most pressing problems faced by the world in the 21st century and demands attention to achieve sustainability. Land, being one of the prime resources for fulfilling human food requirements, is under the menace of degradation as approximately 20 million ha of land is degrading each year in which salinity and waterlogging being the crucial contributors. This land degradation not only impacts the sustainability of resources but also threatens global food security significantly. According to Abdel Rahman (2023), the global food shortage may increase by 12% every resulting in a hike of food prices to 30% in the next 25 years, which is a concerning figure. The annual cost of global degraded land is estimated to be approximately 3.4 trillion Euros. The overexploitation of water resources is another prominent problem globally, especially in India. Though rich in surface as well as groundwater resources, India is still emerging as one of the major developing countries facing water scarcity. The increased thrust of groundwater extraction can be assumed by the figures that in 2020, the groundwater in the country was 244.92 billion cubic meters, which dropped to 239.16 billion cubic meters in 2022. The annual extraction rate of groundwater in the country is assumed to be 241.34 billion cubic meters in 2023, and the severity of the problem is supposed to increase in the upcoming years.

The major reason for this exaggerating exploitation of resource is lack of proper planning and policies. A high magnitude of arable land is getting converted into degraded land, soil erosion problem is advancing annually and groundwater is depleting at a much faster pace. This indicates even more need to adapt sustainable management practices to conserve the depleting resources exorbitantly. Hydrological modeling, the method of representing and hydrological phenomenon in empirical form, plays an important role in sustainable planning and management of land and

DOI: 10.1201/9781003596080-16

water resources. The various hydrological models must integrate with the sustainable development goals to create proper planning for the management of land and water resources sustainably. The hydrological models used by various researchers may be classified as groundwater models, surface runoff models, soil erosion models, evapotranspiration (ET) models, and watershed models. This chapter provide a brief overlay of these hydrological models along with its limitations.

16.2 GROUNDWATER FLOW MODELING

Groundwater modeling is widely utilized within environmental science. These models are employed to analyze various hydro geological settings. In recent times, they have become essential tools in forecasting contaminant movement for the purpose of risk analysis. Models serve as theoretical representations or simplified versions of real-world systems, expressed through mathematical formulas. They do not replicate the actual system precisely, but offer a structured way to simulate and analyze hydrogeological conditions. By formulating a simplified mathematical model of a groundwater system, different potential scenarios can be projected, evaluated, and compared (Kinzelbach, 1986). The relevance and reliability of any model depend significantly on how closely its mathematical framework mirrors the actual physical system. To properly evaluate a model's validity, one must have a deep understanding of both the real-world system and the assumptions made during the model's development. Groundwater models simulate subsurface water flow and the movement of pollutants by employing mathematical formulations built upon certain assumptions. These simplifications often relate to flow direction, aquifer geometry, geological variations like anisotropy or heterogeneity, as well as contaminant behavior and chemical interactions. Due to these inherent assumptions and uncertainties in input data, models should be regarded as approximations rather than exact reflections of field conditions (Baalousha, 2009). Practical uses of groundwater models include water budgeting, understanding the unsaturated zone's characteristics, simulating fluid and solute movement in the saturated zone (including river–aquifer exchanges), evaluating the ecological consequences of groundwater fluctuations, designing and optimizing monitoring systems, and delineating protected groundwater zones. Besides having advanced models to stimulate groundwater, modeling efforts have largely remained within the domain of academic and research bodies. Practicing engineers and planners often still rely on simplified, lumped models for groundwater management and recharge planning. These models typically overlook the spatial complexity of groundwater systems, adhering to outdated notions such as safe yield, and failing to incorporate factors like stream–aquifer interactions and lateral recharge influenced by groundwater levels. As a result, they cannot accurately determine permissible over-extraction or sustainable long-term yields. The key focuses of groundwater modeling include (1) evaluating recharge potential, (2) tracking water table dynamics, (3) understanding interactions between streams and aquifers, and (4) investigating sea-water intrusion, etc. (Singh, 2013).

The process of groundwater modeling starts with developing a conceptual understanding of the physical system in question. After this foundation is established, the next phase involves converting the physical conditions into a mathematical framework. Typically, this results in the formulation of standard groundwater flow and solute transport equations (Rushton, 2004). For three-dimensional (3D) flow in fully saturated porous media, the main governing equation is the groundwater flow equation (Equation 16.1).

$$\frac{\partial}{\partial x}\left(K_{XX}\frac{\partial h}{\partial x}\right)+\frac{\partial}{\partial y}\left(K_{YY}\frac{\partial h}{\partial y}\right)+\frac{\partial}{\partial z}\left(K_{ZZ}\frac{\partial h}{\partial z}\right)-Q=S_s\frac{\partial h}{\partial t} \tag{16.1}$$

where K_{XX}, K_{YY}, K_{ZZ} represent the hydraulic conductivity along the x,y,z axes which are assumed to be parallel to the major axes of the hydraulic conductivity; h is the piezometric head; Q is the volumetric flux per unit volume representing source/sink terms; S_s is the specific storage coefficient.

The transport of solutes in the saturated zone is governed by the advection-dispersion equation for a porous medium with uniform porosity distribution, as formulated in following Equation 16.2 (Benson et al., 2000):

$$\frac{\partial c}{\partial t}=-\frac{\partial}{\partial x_i}\left(cv_i\right)+\frac{\partial}{\partial x_i}\left(D_{ij}\frac{\partial c}{\partial x_i}\right)+R_c \tag{16.2}$$

where c is concentration of the solute; R_c is sources or sinks; D_{ij} is dispersion coefficient tensor; and v_i velocity tensor.

To develop a groundwater flow or transport model using Equations 16.1 and 16.2, boundary and initial conditions are needed. The core processes include groundwater movement, solute migration, and thermal transport. Groundwater models are generally constructed using either deterministic approaches, which are based on specific cause-and-effect or input–output relationships, or stochastic models, which address the inherent uncertainty and randomness within groundwater systems. The equations that govern groundwater systems are usually solved through either analytical or numerical methods. Analytical models offer exact solutions to the governing equations, maintaining continuous representation in both space and time. Conversely, numerical models provide approximate solutions by discretizing space and time and applying numerical techniques to solve the partial differential equations that govern the system. Several numerical methods are employed in groundwater modeling, with three of the most commonly used being the Finite Difference Method (FDM), Finite Element Method (FEM), and Analytical Element Method (AEM) (Mohtashami et al., 2025). Each method has its own set of advantages and disadvantages regarding cost, accessibility, ease of use, application, and the level of expertise required from the user (Arnold et al., 1993).

16.2.1 FEFLOW (Finite Element Subsurface Flow System)

FEFLOW represents an engaging finite element modeling platform for simulating groundwater systems in three dimensions (3D) or two dimensions (2D), encompassing scenarios that are horizontally averaged (aquifer-focused), vertically cross-sectional, or radially symmetric. It can handle time-varying (transient) or stable (steady-state) conditions, as well as flow coupled with fluid density or operating linearly, and the transport of mass, heat, or the fully interconnected processes of thermohaline transport within subsurface water resources. This software suite boasts a completely visual and interactive environment, integrating the stages of preparing input data, running simulations, and analyzing results. It provides a data link to Geographic Information System (GIS) platforms and an interface for programming extensions (Diersch, 2013). The incorporated numerical techniques enable the resolution of substantial and complex problems. FEFLOW provide captivating 3D perspectives, in-depth 2D slices, and accurate 1D line graphs. This capability equips professionals to attain exceptional understanding and clarity. FEFLOW's ability to work together with prominent geological software such as GOCAD, Leapfrog, and GeoModeller 3D guarantees a fluid transition in work processes.

The extensive array of features and settings within FEFLOW makes it a sophisticated piece of software that can be challenging for new users to learn quickly, particularly in comparison to simpler alternatives like Three-Dimensional Finite-Difference Ground-Water Flow (MODFLOW) model. Configuring models and comprehending the various parameters may require more time and expert knowledge. Utilizing finite element methods, FEFLOW often demands more processing power than MODFLOW, which relies on finite difference methods, particularly for simulations involving a large number of cells or complex shapes. Consequently, simulation times can be longer, especially for transient and 3D models (Koskinen et al., 2025).

16.2.2 MODFLOW (Three-Dimensional Finite-Difference Groundwater Flow Model)

MODFLOW refers to the Modular Three-Dimensional Groundwater Flow developed by the U.S. Geological Survey (USGS). It has gained widespread recognition as the global standard for simulating groundwater systems, owing to its versatility in handling diverse hydrogeological conditions and comprehensive open-access documentation. This model is widely employed in applications such as water resource planning, environmental cleanup, and dewatering operations in mining. When correctly applied, MODFLOW is considered the benchmark tool in its field application (Harbaugh et al., 2000). The key goals behind MODFLOW's development included creating software that is easy to modify, user-friendly, compatible across various computer platforms with minimal adjustments, and capable of efficiently managing the extensive datasets required for large-scale simulations. The input and output mechanisms in MODFLOW are designed to offer maximum adaptability. MODFLOW simulates groundwater movement through an aquifer using a block-centered finite-difference technique. It can handle confined, unconfined, or mixed aquifer layers. Additionally, it supports simulation of various external influences such as pumping from wells, surface recharge, ET, flow into drainage systems, and interactions with river channels (Sisay et al., 2023).

MODFLOW's reliance on a structured, block-centered finite-difference grid can be inefficient and present difficulties in accurately representing intricate geological boundaries, irregular aquifer configurations, and localized details. Traditional versions of MODFLOW primarily focus on simulating saturated groundwater flow. Although supplementary packages and more recent iterations (such as MODFLOW-NWT and MODFLOW-OWHM) have integrated the ability to simulate unsaturated flow and surface water interactions, these were not inherent to the original core functionality (Niswonger et al., 2011). Furthermore, the fundamental equation solved by MODFLOW assumes uniform fluid density and viscosity. While extensions like SEAWAT address variable-density flow scenarios (e.g., saltwater intrusion), this is an added capability, and the core model operates under this assumption. Modeling solute transport and geochemical reactions necessitates the integration of separate programs such as MT3DMS, RT3D, or PHREEQC, which can introduce additional complexity to the overall modeling workflow. Similarly, while MODPATH can perform particle tracking based on the flow field calculated by MODFLOW, it is a distinct program, and intricate scenarios may require further processing steps (Brunner et al., 2010; Kumar, 2012). MODFLOW 6 is the latest version of this software series.

16.2.3 MT3D (A Modular 3D Solute Transport Model)

MT3D is an advanced 3D modeling tool developed to simulate the movement and transformation of dissolved substances in complex groundwater systems. Its modular architecture allows users to simulate different transport mechanisms either separately or in combination. The model is equipped to handle a wide range of processes, including advection in both steady and unsteady flow conditions, anisotropic dispersion, first-order kinetic reactions (both decay and generation), and sorption behaviors that can be either linear or nonlinear (Guo et al., 2019). In addition, MT3D supports simulations involving more complex chemical interactions such as those found in bioplume scenarios, monad-type reactions, and the formation of secondary (daughter) products. This allows for modeling of multi-species chemical dynamics and facilitates analysis of natural attenuation processes occurring within contaminant plumes. MT3D works in conjunction with MODFLOW, the USGS groundwater flow model, and is specifically optimized for transport scenarios where advection is the dominant process (Morway et al., 2023).

MT3D's accuracy hinges on the assumption of a consistent or gradually changing groundwater flow pattern, typically supplied by models like MODFLOW, throughout the duration of the transport simulation. Rapid shifts in this flow may impact the reliability of the transport predictions. Furthermore, because MT3D uses a grid-based approach, it averages subsurface characteristics (such as permeability, void space, and spreading tendencies) within each grid cell. This simplification can introduce errors, particularly in complex aquifer systems where property variations occur within distances smaller than the grid cell size. Although MT3D can simulate several basic chemical processes (like adsorption, degradation, and some simple reaction kinetics), it has limited capacity for modeling intricate geochemical interactions, such as mineral precipitation/dissolution or microbial activity involving multiple chemical species, unlike more specialized reactive transport models (Ben Simon et al., 2015).

16.2.4 SEAWAT (Three-Dimensional Variable-Density Groundwater Flow)

SEAWAT is a simulation tool designed to model 3D, time-dependent groundwater flow in porous materials where fluid density varies. It was created by merging the capabilities of MODFLOW and MT3DMS into one integrated code that simultaneously solves equations for both fluid movement and solute transport. The structure of SEAWAT is modular, allowing for the easy integration of new features with minimal adjustments to the core program. It is compatible with standard MODFLOW and MT3DMS input and output files, though certain simulations may require additional data. This compatibility enables the use of many existing tools for preparing input files and interpreting simulation results. For users already experienced with MODFLOW and MT3DMS, adapting to SEAWAT for modeling density-dependent groundwater flow should be relatively straightforward (Lyu et al., 2024).

SEAWAT, similar to MT3D, often assumes a steady or quasi-steady flow field from MODFLOW within a given time step. While SEAWAT offers implicit coupling to better handle systems with rapidly changing density-driven flow, a purely sequential approach might introduce errors. The averaging of subsurface properties within grid cells remains a simplification in heterogeneous aquifers. The assumption of instantaneous and complete mixing at sources and sinks might not always be realistic. Although often a reasonable simplification, this linear approximation may not accurately capture density variations in highly concentrated solutions or under extreme temperature/pressure conditions where non-linear effects become important. While more recent versions of SEAWAT (V4 and later) offer the capability to account for viscosity changes with solute concentration and temperature, earlier versions assumed a constant viscosity, which could be a limiting factor in scenarios with significant temperature gradients or very high solute concentrations (Kumar, 2012).

16.2.5 SUTRA (2D Saturated/Unsaturated Transport Model)

SUTRA represents a 2D model for groundwater flow in both saturated and unsaturated zones, encompassing comprehensive capabilities for simulating saltwater intrusion and energy transfer. This model facilitates the simulation of fluid movement and the transport of either thermal energy or dissolved constituents within subsurface settings. SUTRA utilizes a 2D hybrid approach, combining the finite-element and integrated finite-difference methods, to approximate the fundamental equations that depict the two interconnected phenomena being modeled. A 3D iteration of SUTRA has been made available recently (Narayanamurthi and Ramasamy, 2023).

Setting up a SUTRA model necessitates detailed hydrogeological and transport parameters for both the saturated and unsaturated zones, which can be a significant undertaking due to data availability and acquisition challenges. While SUTRA can model unsaturated flow, its numerical methods are not specifically tailored for the highly non-linear behavior characteristic of vadose zone processes. SUTRA typically employs a linear equation of state to relate fluid density to solute concentration or temperature, an approximation. Standard SUTRA generally handles the air phase in an implicit manner and does not explicitly simulate air flow or its potential influence on water movement and solute transport. This simplification can be significant in certain unsaturated zone processes where air dynamics play a crucial role (Provost and Voss, 2019).

16.2.6 SWIM (Soil Water Infiltration and Movement Model—Simulate Soil Water Balances)

SWIM model is a software collection designed for the simulation of water entry into and movement within soils. Mirroring real-world scenarios, SWIM incorporates water input into the system as rainfall and its removal via surface water flow, subsurface drainage, vaporization from the soil surface, and water uptake by plants. SWIM assists researchers in comprehending the soil water budget, enabling them to evaluate potential impacts of activities such as deforestation, strip mining, and irrigation control. SWIM holds significance for scientists and consultants engaged in land use planning and land resource management (Chavira et al., 2022). For instance, if a construction project involving tree removal is under consideration, SWIM can be utilized to predict salinity or surface water runoff issues that might arise from alterations in the soil water budget linked to the absence of trees. SWIM (version 2) is a model grounded in mechanistic principles, formulated to tackle soil water and dissolved substance budget concerns associated with both agricultural output and its ecological ramifications. SWIMv2 utilizes rapid, computationally efficient methods for resolving Richards' equation for water flow and the convection-dispersion equation for dissolved substance transport, making it suitable for personal computer applications. It can be employed to simulate surface runoff, water entry, water content redistribution, dissolved substance transport and redistribution, plant absorption and transpiration, evaporation, deep percolation, and the washing away of solutes (Krysanova et al., 2000).

These models operate under the assumption that the soil profile is consistent horizontally, disregarding any lateral changes in soil properties, topography, and water movement. This simplification restricts their usefulness in landscapes where these factors vary spatially. Their 1D nature limits the ability to simulate intricate flow pathways, such as macropore flow or bypass flow, which can substantially influence water infiltration and solute transport in actual soils. Solute concentrations are computed as uniform across the entire soil layer, preventing the simulation of concentration gradients within that layer, especially near plant roots where they can affect water and nutrient absorption (Hattermann et al., 2005).

16.3 SURFACE RUNOFF AND WATERSHED MODELS

Surface runoff is the flow of water over the surface of the land, usually happening when the soil is saturated and will no longer accept further rainfall, or water falling on impervious surfaces such as pavement or rock. Rather than seeping into the ground, the excess water overflows the land and ultimately finds its way into streams, rivers, and other water bodies. This is an important part of the water cycle and has a strong impact on natural and human habitant systems. Surface runoff has the potential to erode soil, move pollutants, and cause flooding, particularly in urban habitats where natural land surfaces are being covered with asphalt and concrete. Surface runoff management by having adequate drainage systems and green infrastructure is necessary to minimize its adverse effects. There are different types of models for watershed modeling some of which are mentioned below.

16.3.1 Hydrological Engineering Center, Hydrologic Modeling System (HMS)

This model was firstly designed by US Army Corps of Engineers and was released by Hydrological Engineering Center (HEC) in 1992. It was initially designed to simulate the rainfall-runoff process in dendritic watersheds. The open-source software HEC HMS 1.0 was the initial release of this software which was a replacement of FORTAN based HEC 1. This initial version was capable of performing the event-based simulations and gridded runoff representation (Sahu et al., 2023). It is an updated version, that is, HEC HMS 2.0 and incorporates continuous event simulations along with the integration of soil moisture accounting methods. In HEC HMS version 3.0, a new graphical user interface was introduced along with the new snowmelt and ET models for a smoother simulation process. The next version, HEC HMS 4.0 focused mainly on computation and included various erosion and sediment transport models for channels and reservoirs (Chu and Steinman, 2009). Integration of nutrient water quality simulation was also a major highlight of this version. There have been various enhancements in this version and the refinements were released as 4.1, 4.2, 4.3, 4.4, 4.5, 4.6, 4.7, 4.8, 4.9, 4.10, 4.11, and the latest version 4.12. Also, HEC GeoHMS is another version of HEC HMS with the inclusion of GIS with it (Gebre, 2015).

Even though the model is widely used, there exist several limitations to the model. This model fails to simulate rainfall-runoff for backwater effects. Also, this model was designed for dendritic watershed and hence it struggles with the branched or looped stream networks. The supplemental tool, HEC-GeoHMS, requires ArcGIS with the Spatial Analyst Extension, which can be a barrier for some users. Another core limitation of this model is that it is not publicly available and its code is limited to the US Army Corps of Engineers. The models also fail to assess the cross-channel or off-channel variation which is required in the assessment of floodplains.

16.3.2 TOPKAPI (Topographic Kinematic Approximation and Integration) Model

TOPKAPI is a totally distributed-lumped system. The distributed TOPKAPI model is employed to determine the medium regulating the dynamics of the impregnated area to the face runoff as a function of the total water storehouse, thereby carrying a law supporting a derivative of the lumped model. The model is innovated on the conception of incorporating the kinematic system with the receptacle geomorphology; the ultimate is defined by a Digital Elevation Model (DEM) whose grid size generally grows with the overall size. Every grid cell of the DEM is given a value for every one of the physical parcels modelled. The inflow paths and pitches are reckoned from the DEM, grounded on a neighborhood relation innovated on the minimal energy cost principle (Todini and Ciaraphica, 2001; Liu and Todini, 2002). A schematic diagram representing the detailed working of TOPKAPI model is illustrated in Figure 16.1.

The precipitation's spatial uniformity assumption is not considering rainfall variability at small scales, potentially with a marked influence on runoff, particularly within large or heterogeneous catchments. Allying all rain to infiltration until the soil gets saturated eliminates the overland flow process (runoff resulting from excessive infiltration capacity) typical in dry or urban basins. Because infiltration is considered to be infinite except when saturation is reached, the model does not take into consideration more than infiltration rates of rainfall intensity, one of the primary causes of surface runoff in most environments. Employing the kinematic wave method simplifies the Saint Venant equations by omitting acceleration and pressure gradient terms, potentially resulting in error in areas where backwater effect or dynamic flow occurs. The model structure is limited for the study of macropores, root channels, and cracks that can quickly pass water vertically or laterally and circumvent the assumed uniform process of infiltration (Abbenante et al., 2025).

A watershed is defined as a geographical area of land where all of the water that falls in it drains into a form of runoff to a common or single outlet. Basically, it is a natural drainage basin. Watersheds range in size from small local watersheds draining into a stream, to regional or even continental watersheds that receive water over vast distances. Watershed modeling involves modelling of all the hydrologic processes at the watershed scale and coupling them in such a manner as to calculate the watershed response. The watershed scale in hydrology may be as small as a few square meters in area or as large in the order of hundreds of thousands of square kilometers. Due to the large variations between various size watersheds, watershed models are sometimes referred to as small watershed, medium watershed, and large watershed models. The hydrologic response complexity is greatly dependent on the size of the watershed. More so, the interactions among land surface, pedologic, lithospheric, hydrosphere, and atmospheric processes have a direct correspondence with the size of the watershed (Singh and Frevert, 2003). There are different types of models for watershed modeling some of which are mentioned below.

16.3.3 SCS CN (Soil Conservation Service Curve Number)

This model or method has a chief parameter, which is CN, and depends on cumulative rainfall, soil cover, land use, and initial water content. The SCS method is suitable for the estimation of surface runoff for small watersheds and can determine values of CN under different antecedent moisture levels with generally high practicability. Consequently, other altered models, although with intricate structures, cannot be directly transferred to actualities because of the issue of model closure, so the SCS method ranks as one of the most widely used models for several decades. The SCS model possesses the advantages of simplicity, stability, reliable predictability, and precision with regard to being able to express the properties of watersheds. It is then sensible to test the hypotheses by choosing an infiltration model that is able to represent the real scenarios in a theoretical manner. A correction factor is added to the

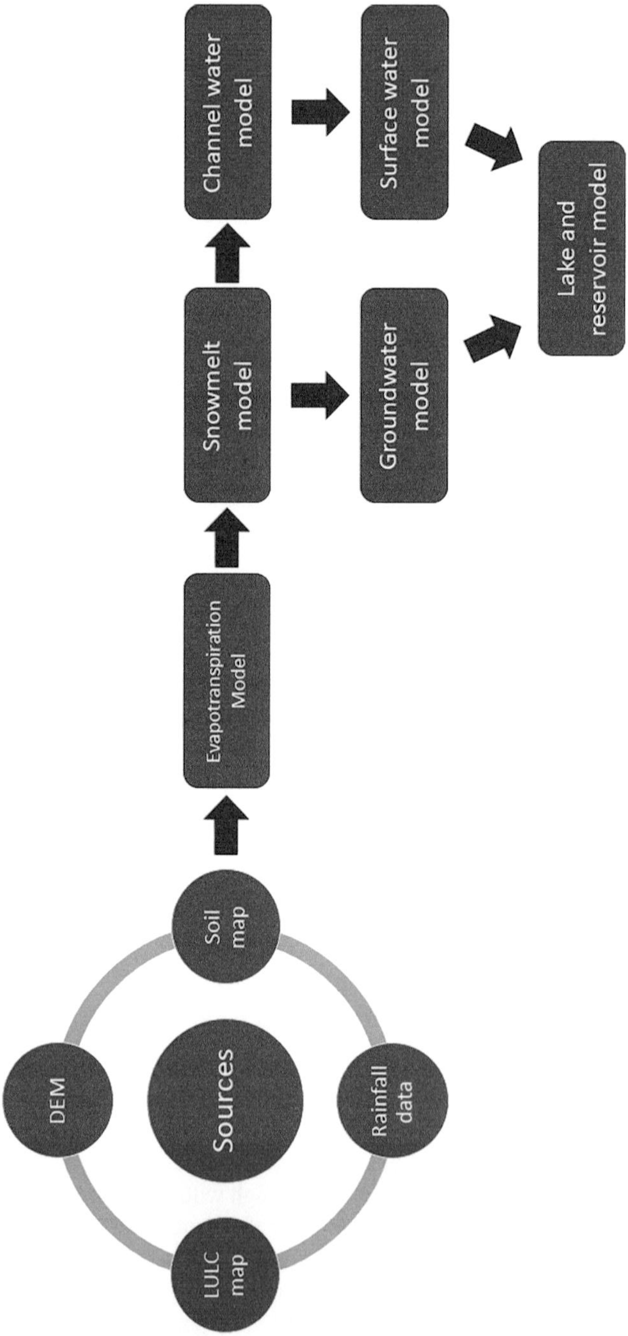

FIGURE 16.1 Functioning of TOKAPI model.

SCS model to minimize uncertainty brought about by experiential hypothesis (Geetha et al., 2008).

The results are better if SCS method is performed in heavy rainfall conditions. It was originally developed and calibrated with high-intensity storm data, so it does not necessarily reflect average or light rain conditions well. Under low rainfall conditions, it can over- or underpredict runoff by a considerable amount. This renders it less accurate in areas where light to moderate rain prevails. The procedure takes for granted that rainfall lasts long enough to fill the ground and produce runoff efficiently. During long rainfall events or heavy storms, the model resolves that soil moisture conditions prior to the storm do not play an important role. This may not be the case in most actual situations where soil moisture can affect the quantity of rain that enters or flows over the ground. Disregarding these effects could result in runoff predictions being either too high or too low based on the actual field conditions. The model presumes that increasing rainfall intensity will proportionally increase infiltration, but this is not always the case. Actually, intense rainfall can rapidly overflow infiltration capacity and result in surface runoff irrespective of the soil (Soulis and Valiantzas, 2012).

16.3.4 Rational Method

The rational method in empirical form is expressed in Equation 16.3:

$$Q = CIA \tag{16.3}$$

where Q is the peak discharge for recurrence interval T; C is rational runoff coefficient; I is the rainfall intensity; and A is the watershed area. The drainage area, A , can be read from a topographic map and is usually the most accurate of the three inputs to the rational formula. The rainfall intensity, I, is chosen from intensity-duration-frequency tables with a recurrence interval identical to that for the target peak discharge and with a duration identical to an effective rainfall averaging time. The watershed time of concentration, T_C, is typically utilized as the rainfall averaging time. There are number of empirical equations for T_C which take into account the length (L), the slope (S), and sometimes the roughness of the path of flow; the estimation method utilized for T_C is given in Equation 16.4 (Young et al., 2009):

$$T_c = 0.01947\ L^{0.77} S^{-0.385} \tag{16.4}$$

The rational method is generally most accurate for small watersheds (typically less than 200 acres). For larger watersheds, the method becomes less reliable because it does not account for the complexities of larger drainage areas, such as varying land uses, elevation changes, or longer runoff travel times. The method assumes that the entire watershed reacts uniformly to rainfall, meaning it does not consider spatial variations in land use, soil type, or slope. In reality, watersheds can have heterogeneous characteristics, and these variations can significantly affect the runoff response (Dhakal et al., 2011). The rational method simplifies the runoff process by assuming that runoff occurs only after the rainfall intensity exceeds a threshold. It does not account for intermediate or gradual runoff processes, such as baseflow or infiltration before the runoff threshold is reached. The method heavily relies on accurate data for rainfall intensity (usually for a specific duration), the runoff coefficient (*C*), and the drainage area (*A*). Errors in these data can lead to inaccurate runoff estimates, and in many regions, rainfall data might not be precise or available for the exact conditions needed. The rational method assumes that runoff happens instantaneously once the rainfall starts, which is unrealistic in many cases. The runoff coefficient (*C*) used in the rational method is a simplification of the actual runoff process. It is an average value based on land use, soil type, and slope. However, in reality, runoff is influenced by many more factors, such as antecedent moisture conditions, vegetation type, or specific storm characteristics. The method assumes a uniform distribution of rainfall across the entire watershed, which may not be the case in reality.

16.4 SOIL EROSION MODELING

Soil erosion modeling is a method used to estimate the rate, location, and extent of soil detachment, transportation, and deposition caused by both natural processes and human activities. These models can be software-based, conceptual, empirical, or field-scale, and are selected based on their complexity, input requirements, and intended use. Key factors influencing these models include rainfall, soil type, wind speed, topographic features, land cover, and management practices. For researchers, erosion models act as structured tools that help translate their conceptual understanding of soil detachment, movement, and deposition into measurable and predictive frameworks. By enhancing our understanding of global soil erosion patterns, these models assist in identifying erosion hotspots and play a vital role in shaping, implementing, and evaluating effective soil conservation policies (Borrelli et al., 2021; Batista et al., 2019; Merritt et al., 2003).

16.4.1 Universal Soil Loss Equation

The Universal Soil Loss Equation (USLE) is developed by Wischmeier and Smith in 1978, to predict soil losses due to water erosion, was based on an extensive analysis of over 10,000 plot-years of runoff and soil loss data collected from 49 locations across the United States. To compare the equation's structure, a reference condition called the "unit plot" was used. This unit plot measured 22.1 m in length and 1.83 m in width, had a 9% slope, and was maintained in a continuously tilled fallow state with up-and-down slope tillage. It represented the worst-case scenario for soil erosion and served as the baseline for comparing soil loss under different conditions (Majhi et al., 2021). From Equation 16.5, the USLE can be defined as:

$$A = R.K.L.S.C.P \tag{16.5}$$

where A is the annual average soil erosion, Mg ha^{-1} yr^{-1}; R is the rainfall-runoff erosivity factor, MJ mm h^{-1} ha^{-1} yr^{-1}; K is the soil erodibility factor, Mg h MJ^{-1} mm^{-1};

L is the slope length factor, dimensionless; S is the slope steepness factor, dimensionless; C is the land cover and management factor, dimensionless; P is the soil conservation or prevention practices factor, dimensionless.

The USLE, although widely used, has several limitations that affect its accuracy and scope. It was originally developed to estimate long-term average soil loss, such as over a crop rotation or growing season, rather than for short-term or event-based predictions. Its limitations became more apparent with increased global use, especially in rangeland areas where it fails to estimate erosion accurately (Wischmeier and Smith, 1978; Devatha et al., 2015). USLE is mainly effective for evaluating on-site soil erosion on individual agricultural fields and is not designed for larger catchment-scale analysis. As a simplified empirical model, it does not individually address the natural processes involved in erosion, such as plant growth, infiltration, runoff, or sediment transport. Moreover, the model was developed based on unit plot conditions where the land is ploughed, left without plant residue, and kept bare for extended periods which do not reflect typical farming practices. Despite these shortcomings, it remains a widely used tool for assessing sheet and rill erosion under specific field conditions (Nearing et al., 2005; Majhi et al., 2021).

In the USLE model, runoff is not directly considered, even though erosion is strongly influenced by the flow of water carrying sediment. As sediment is transported with runoff, and its concentration varies with the amount of runoff, this omission limits USLE's ability to estimate sediment yield accurately. Runoff is actually one of the best indicators for predicting sediment yield (Arekhi et al., 2012; Gwapedza et al., 2018, 2021). To address this, the Modified Universal Soil Loss Equation (MUSLE) was developed. MUSLE replaces the rainfall factor with a runoff factor, removing the need for a delivery ratio, and allowing for better estimation of sediment yield from individual storm events. Despite this improvement, MUSLE also has limitations. When used without proper calibration, it can produce large errors and results that are inconsistent with the principles of dimensional analysis (Luo et al., 2016). RUSLE is the revised version of MUSLE and is widely considered as the best soil erosion model.

16.4.2 Soil and Water Assessment Tool

The Soil and Water Assessment Tool (SWAT) model is most effective when applied at a finer spatial scale to accurately capture local hydrological processes such as infiltration rates, preferential flow paths, early saturation zones, and other site-specific responses. It is based on the use of homogeneous land use, soil types, and slope characteristics, which makes it suitable for relatively uniform landscapes but limits its applicability in heterogeneous regions (Nyeko Martine, 2015). This assumption of internal uniformity within Hydrologic Response Units (HRUs) reduces the model's ability to reflect real-world variability. To address these structural limitations and improve flexibility, a more advanced version, SWAT+, was developed after decades of evolution. While SWAT has demonstrated good performance in simulating streamflow in temporary rivers, it tends to be less reliable under extremely low-flow conditions (Aloui et al., 2023). In addition, calibrating the model using only water flow data is insufficient; sediment transport data must also be incorporated to enhance accuracy (Kateb et al., 2020). SWAT's performance is also limited during extreme weather events, especially in cases of short-duration, high-intensity rainfall (Ramos et al., 2015).

16.4.3 Water Erosion Prediction Project

The Water Erosion Prediction Project (WEPP) is a process-based model designed to represent soil erosion processes, including inter-rill and rill erosion, as well as sediment deposition. It uses variables that continuously change according to a set of differential equations. The model tracks erosion caused by surface runoff on hillslopes and concentrated flow in small channels (rills), which may include features like grassed waterways or road ditches, and also accounts for sediment deposition in both natural and man-made structures. Along with erosion modeling, WEPP includes a weather component that relies on a stochastic weather generator (WG) to produce synthetic, long-term weather data for a given location based on observed statistical weather patterns. It also features a hydrology module based on a modified Green-Ampt infiltration equation and kinematic wave equations to describe runoff. Additional components include daily water balance, plant growth, and residue management (Laflen et al., 1997; Nearing et al., 2005).

WEPP is best suited for fields with a maximum size of around 256 ha, depending on the complexity of the landscape. This erosion modeling method works well for agricultural land with small, temporary erosion features like rills or shallow gullies that do not significantly interfere with farming operations. However, it is not suitable for areas with large, deep, and permanent erosion features such as classical gullies or stream channels. In rangeland and forest settings, the model can handle gullies that are about 1–2 m wide and around 1 m deep.

16.4.4 Chemical Runoff and Erosion from Agricultural Management Systems

Chemical Runoff and Erosion from Agricultural Management Systems (CREAMS) is a physically based, continuous simulation model developed by the Agricultural Research Service (ARS) of the U.S. Department of Agriculture (USDA) to estimate erosion, runoff, sediment transport, plant nutrient losses, and pesticide yield from field-sized areas around the 1980s (Knisel, 1980). The model comprises three main components: hydrology, sediment yield, and chemical transport. In the hydrologic component, surface runoff is modeled using two datasets—daily rainfall data and hourly or breakpoint rainfall data—corresponding to the SCS curve number model and an infiltration-based model, respectively. The erosion and sediment yield component estimates soil erosion and sediment transport by comparing soil detachment and transport capacity in overland flow. It also calculates sediment and chemical losses at the bottom edge of the field based on soil type, slope, crop, and management factors. The chemical transport component includes two parts: plant nutrients and pesticides. The plant nutrient section is further divided into nitrogen and phosphorus sub-models. The nitrogen model accounts for nitrogen losses through surface runoff, mineralization, denitrification, plant uptake, and nitrate (NO_3^-) leaching from the root zone, while the phosphorus model focuses on phosphorus loss via surface runoff. The pesticide section models can simulate pesticide loss in surface runoff through foliar wash-off, adsorption, and degradation, which are complex and variable processes (Manrique, 1993).

A major limitation of the CREAMS model is that it is not suitable for large areas. It requires the field being studied to be an independent unit, free from the influence of surrounding fields or external factors such as neighboring runoff, shared irrigation systems, or mixed land uses. The model also assumes that the field is homogeneous; however, in reality, no field is truly homogeneous. CREAMS relies on the assumption of a uniform field surface, relatively consistent soil properties, uniform rainfall, and a single management practice. Additionally, the model does not account for several important factors such as soil compaction, irrigation, salinization, waterlogging, reservoir silting, drainage, wind erosion, and the use of organic fertilizers—all of which are significant in real-world agricultural modeling (Svetlosanov, 1982).

16.5 ET MODELING

ET includes two distinct processes: evaporation from surfaces such as lakes, rivers, soils, and wet vegetation, which is primarily driven by solar radiation and air temperature; and transpiration, which occurs through plant stomata and also includes direct temperature, air humidity, crop characteristics, agricultural practices, and wind speed. Modeling ET is essential to determine how much water is lost through these two processes (Ghiat et al., 2021; Granata, 2019). A flowchart depicting various methods used in the estimation of ET is given in Figure 16.2.

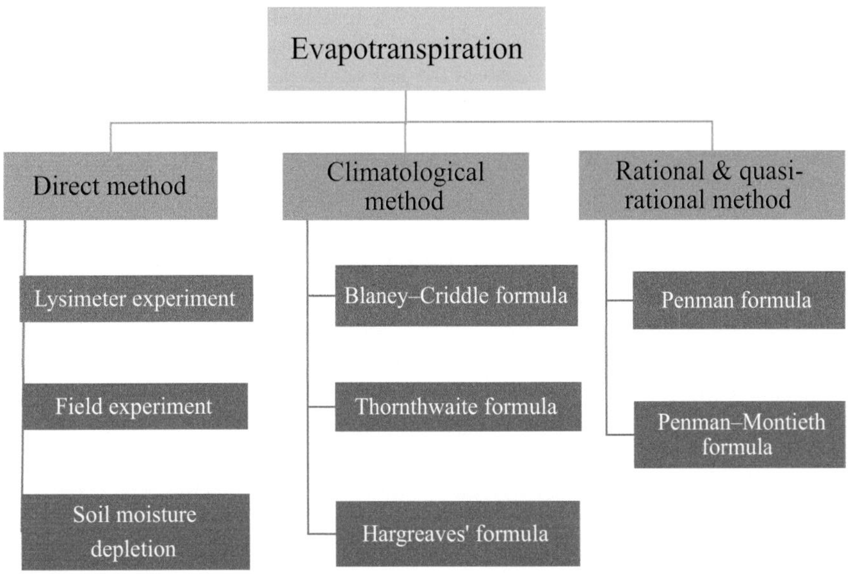

FIGURE 16.2 Methods to estimate ET losses.

16.5.1 Direct Methods

1. *Lysimeter Experiment:* A lysimeter is a device used to measure water loss through ET and deep percolation (Yang et al., 2000; Xu et al., 1998). It consists of a soil container with planted vegetation, buried within the surrounding soil. Typically employed for short-term studies, lysimeters are designed to replicate natural environmental conditions. Two main types exist: non-weighing lysimeters and weighing lysimeters (Bergström, 1990; Payero and Irmak, 2008; Howell, 2004). Non-weighing and weighing lysimeters employ opposite measurement methods to determine vegetation water use and/or soil–water evaporation. While capable of providing precise measurements, their use is limited in reproducing physical conditions such as temperature, water table, soil texture, structure, and density within the lysimeter (Michael, 2009; Holmes, 1984).
2. *Field Experiments Plots Method:* This method involves measuring how much water is supplied to plots in the field and how much change occurs in the soil moisture content of these field plots after the supply. It depends on the seasonal water requirement of crops, which is the summation of irrigation water, effective rainfall during the season, and the contribution of soil moisture, as expressed in Equation 16.6 (Michael, 2009):

$$WR = IR + ER + \sum_{i=1}^{n} \frac{M1}{M2} A.D \tag{16.6}$$

where WR denotes seasonal water requirement (mm); IR is total irrigation water applied (mm); M_1 is moisture percentage at the beginning of the season; M_2 is moisture percentage at the end of the season; A is apparent specific gravity of the soil; D is depth of the soil within the root zone (mm), and ER is effective rainfall.

This approach is appropriate for evaluating water needs over an entire growing season. However, it is not well suited for examining immediate water consumption, intermediate soil moisture, studying the soil profile, quantifying water loss through deep percolation, and determining the maximum rate at which the crop utilizes water (Michael, 2009).

3. *Soil Moisture Depletion Method:* Soil moisture measurements are utilized in homogeneous soil to determine the consumptive use of irrigated field crops when the groundwater table's depth does not impact soil moisture fluctuations within the root zone. Fundamentally, this involves taking soil moisture readings at different depths and at numerous points throughout the crop growth period (Blaney and Criddle, 1962). The accuracy of the measurement increases with a greater number of measurements. The total water consumed by plants (Cu/ET) can be calculated from changes in soil water content between successive samplings, according to this relationship shown in Equation 16.7 (Michael, 2009):

$$Cu = \sum_{i=1}^{n} \frac{M_{1i} - M_{2i}}{100} \times A_i \times D_i \tag{16.7}$$

where Cu is water use from the root zone for successive sampling period or within one irrigation cycle (mm); n is number of soil layers sampled in the root zone depth D; M_{1i} is soil moisture percentage at the time of the first sampling, the ith layer; M_{2i} is soil moisture percentage at the time of the second sampling, the ith layer; A_i is apparent specific gravity of the ith layer of the soil; D_i is depth of the ith layer of the soil (mm).

16.5.2 Climatological Methods

1. *Blaney–Criddle Formula:* The Blaney–Criddle approach is employed to approximate the Potential Evapotranspiration (PET) or Reference Crop Evapotranspiration (ET_o) by taking into account average monthly temperatures, the proportion of daily or monthly sunlight hours, and the duration of the plant's growth period. PET or ET_o can be expressed as Equation 16.8 (Michael, 2009):

$$PET = TP / 100 \tag{16.8}$$

   where PET is monthly potential evapotranspiration; P is monthly percentage of total annual day time; T is mean monthly temperature (average of daily maximum and minimum values).

   This method's accuracy is primarily suited for estimating seasonal consumptive water use, based on the assumption that a crop's water consumption solely depends on temperature and day length. This assumption is not entirely accurate; water consumption also varies with humidity and wind conditions. Consequently, it is often assumed that the Blaney–Criddle method is most applicable in areas experiencing minimal variations in relative humidity and wind velocity. Furthermore, the Blaney–Criddle method is inappropriate for areas with stable temperatures but fluctuating weather conditions, regions where air temperature is affected by nearby sea temperatures, locations where daytime solar energy is largely unrelated to nighttime temperatures, and climates experiencing significant variations in sunlight duration during the change of seasons.
2. *Thornthwaite Formula:* The Thornthwaite equation serves as a tool to categorize the diverse global climate types in an organized and logical manner (Michael, 2009). It establishes an empirical and exponential relationship between mean monthly temperature and mean monthly PE. This equation relies on temperature as a key variable (Michael, 2009; Kumar et al., 1987). The mathematical expression developed by Thornthwaite is presented in Equation 16.9 (Li et al., 2022; Pereira and Pruitt, 2004):

$$PE = 1.6\left(10t / I\right)^{a} \tag{16.9}$$

where PE is unadjusted PET (cm per month) (month of 30 days each and 12 hours day time); t is mean air temperature (°C); I is annual or seasonal heat index; a denotes an empirical component.

A key drawback of this formula is its origin in data primarily from the central and eastern United States, leading to reduced accuracy in other parts of the world. It fails to incorporate the influence of wind speed, which can be crucial in determining ET rates in certain areas (Kumar et al., 1987).

3. *Hargreaves' Formula:* This formula was developed for measuring ET using a grass lysimeter, based on average daily temperature and global solar radiation. The equation for Reference Evapotranspiration PET (ET_0) (Michael, 2009) is given in Equation 16.10:

$$PET = 0.0135\left(t + 17.78\right)R_s \tag{16.10}$$

where PET is reference crop potential evapotranspiration, mm day^{-1}; t is temperature, °C; R_s is incident solar radiation, mm day^{-1}.

A notable drawback of this formula is its failure to consider wind speed and relative humidity, both of which significantly influence ET. In the updated version of Hargreaves' formula, while Evapotranspiration (ET_0) and extraterrestrial radiation can be estimated even without relative humidity, solar radiation, and wind speed

16.5.3 Rational and Quasi-Rational Methods

1. *Penman's Method:* The Penman equation was developed to estimate the amount of evaporation that occurs from different surfaces specifically open water, grass-covered land, and bare soil (Hussein, 2022; Subramanya, 2021]). It is based on a combination of the energy balance approach, which considers heat availability, and the mass transfer approach, which accounts for water movement. Equation 16.11 shows the mathematical formula for computing evaporation through Perman method:

$$E_{PEN} = \frac{\Delta R_a + \gamma E_a}{\Delta + \gamma} \tag{16.11}$$

where E_{PEN} is evaporation in mm day^{-1}; Δ is slope of the saturation vapour pressure temperature curve at mean air temp, hP_d/k; γ is psychrometric constant, hP_d/K; R_a is net long-wave and short-wave radiation, mm day^{-1}; E_a is aerodynamic vapor transport, mm day^{-1}. The Penman method only estimates evaporation from a free water surface (Michael, 2009) and does not calculate transpiration. However, obtaining this data can be challenging in rural or underdeveloped areas (Valiantzas, 2006).

2. *Penman–Montieth Formula:* The Penman–Monteith method is revised version of the Penman equation. It considers all the factors that govern energy exchange and corresponding latent heat flux. This method is designed for plant areas that are large enough and have a good distribution of leaves, allowing it to be treated as a single "big leaf" (Sumner and Jacobs, 2005; FAO, Food and Agriculture Organization). Basically, unlike the Penman method which only estimates evaporation, the Penman–Monteith method is used to estimate ET. It does this by including resistance factors, specifically surface resistance and aerodynamic resistance (Michael, 2009) and expressed as Equation 16.12:

$$\lambda ET_o = \frac{\Delta(R_n - G) + \rho C_p (e_a - e_d) / r_a}{\Delta + \gamma(1 + r_c / r_a)} \quad (16.12)$$

where ET_o is latent heat flux of evaporation, kJ m^{-2} s^{-1}; R_n is net radiation flux at surface, kJ m^{-2} s^{-1}; G is soil heat flux, kJ m^{-2} s^{-1}; ρ is atmospheric density, kg m^{-3}; C_p is specific heat of moist air, kJ kg^{-1} °C; $(e_a - e_d)$ is vapor pressure deficit, kPa; r_c is crop canopy resistance, s / m; r_a is aerodynamic resistance, s / m; Δ is slope of vapor pressure curve, kPa $°C^{-1}$; γ is psychrometric constant, kPa $°C^{-1}$; λ is latent heat of vaporization, MJ / kg.

16.6 CONCLUSION

The present chapter highlighted the various limitation involved in the hydrological models which can be proved of value in the sustainable development of land and water resources. This chapter delved into the various models covering almost the entire hydrological processes occurring into the environment. For land resource management, models covered in the chapter are erosion models and watershed models, whereas for the water resource managements such as rainfall prediction models, surface runoff models, and ET models are being covered in this chapter.

REFERENCES

Abbenante, S. E., Althoff, I., and Valdes-Abellan, J. (2025). Improving hydrological modelling with Topkapi through irrigation and open-channel flow in semi-arid agricultural catchments. *Hydrological Processes*, 39(4), e70140.

Aloui, S., Mazzoni, A., Elomri, A., Aouissi, J., Boufekane, A., and Zghibi, A. (2023). A review of Soil and Water Assessment Tool (SWAT) studies of Mediterranean catchments: Applications, feasibility, and future directions. *Journal of Environmental Management*, 326, 116799.

Arekhi, S., Shabani, A., and Rostamizad, G. (2012). Application of the modified universal soil loss equation (MUSLE) in prediction of sediment yield (Case study: Kengir Watershed, Iran). *Arabian Journal of Geosciences*, 5, 1259–1267.

Arnold, J. G., Allen, P. M., and Bernhardt, G. (1993). A comprehensive surface-groundwater flow model. *Journal of Hydrology*, 142(1-4), 47–69.

Baalousha, H. (2009). Fundamentals of groundwater modelling. In L. F. Konig and J. L. Weiss (Eds.), *Groundwater: Modelling, Management and Contamination* (pp. 149–166).

Batista, P. V., Davies, J., Silva, M. L., and Quinton, J. N. (2019). On the evaluation of soil erosion models: Are we doing enough? *Earth-Science Reviews*, 197, 102898.

Ben Simon, R., Bernard, S., Meurville, C., and Rebour, V. (2015). Flow-through stream modeling with MODFLOW and MT3D: Certainties and limitations. *Groundwater*, 53(6), 967–971.

Benson, D. A., Wheatcraft, S. W., and Meerschaert, M. M. (2000). Application of a fractional advection-dispersion equation. *Water Resources Research*, 36(6), 1403–1412.

Bergström, L. (1990). Use of lysimeters to estimate leaching of pesticides in agricultural soils. *Environmental Pollution*, 67(4), 325–347.

Blaney, H. F., and Criddle, W. D. (1962). Determining consumptive use and irrigation water requirements (No. 1275). US Department of Agriculture.

Borrelli, P., Alewell, C., Alvarez, P., Anache, J. A. A., Baartman, J., Ballabio, C., and Panagos, P. (2021). Soil erosion modelling: A global review and statistical analysis. *Science of the Total Environment*, 780, 146494.

Brunner, P., Simmons, C. T., Cook, P. G., and Therrien, R. (2010). Modeling surface water-groundwater interaction with MODFLOW: Some considerations. *Groundwater*, 48(2), 174–180.

Chavira, L. G., Villanueva-Rosales, N., Heyman, J., Pennington, D. D., and Salas, K. (2022). Supporting regional water sustainability decision-making through integrated modeling. In *2022 IEEE International Smart Cities Conference* (ISC2) (pp. 1–7). IEEE.

Chu, X., and Steinman, A. (2009). Event and continuous hydrologic modeling with HEC-HMS. *Journal of Irrigation and Drainage Engineering*, 135(1), 119–124.

Devatha, C. P., Deshpande, V., and Renukaprasad, M. S. (2015). Estimation of soil loss using USLE model for Kulhan Watershed, Chattisgarh-A case study. *Aquatic Procedia*, 4, 1429–1436.

Dhakal, N., Fang, X., Cleveland, T. G., and Thompson, D. B. (2011). Revisiting modified rational method. In *World Environmental and Water Resources Congress 2011: Bearing Knowledge for Sustainability* (pp. 751–762).

Diersch, H. J. G. (2013). FEFLOW: Finite element modeling of flow, mass and heat transport in porous and fractured media. Springer Science & Business Media.

Gebre, S. L. (2015). Application of the HEC-HMS model for runoff simulation of Upper Blue Nile River Basin. *Hydrology: Current Research*, 6(2), 1.

Geetha, K., Mishra, S. K., Eldho, T. I., Rastogi, A. K., and Pandey, R. P. (2008). SCS-CN-based continuous simulation model for hydrologic forecasting. *Water Resources Management*, 22, 165–190.

Ghiat, I., Mackey, H. R., and Al-Ansari, T. (2021). A review of evapotranspiration measurement models, techniques and methods for open and closed agricultural field applications. *Water*, 13(18), 2523.

Granata, F. (2019). Evapotranspiration evaluation models based on machine learning algorithms—A comparative study. *Agricultural Water Management*, 217, 303–315.

Guo, Z., Fogg, G. E., Brusseau, M. L., LaBolle, E. M., and Lopez, J. (2019). Modeling groundwater contaminant transport in the presence of large heterogeneity: a case study comparing MT3D and RWhet. *Hydrogeology Journal*, 27(4), 1363.

Gwapedza, D., Nyamela, N., Hughes, D. A., Slaughter, A. R., Mantel, S. K., and van der Waal, B. (2021). Prediction of sediment yield of the Inxu River catchment (South Africa) using the MUSLE. *International Soil and Water Conservation Research*, 9(1), 37–48.

Gwapedza, D., Slaughter, A., Hughes, D., & Mantel, S. (2018). Regionalising MUSLE factors for application to a data-scarce catchment. *Proceedings of the International Association of Hydrological Sciences*, 377, 19–24.

Harbaugh, A. W., Banta, E. R., Hill, M. C., and McDonald, M. G. (2000). MODFLOW-2000, the US Geological Survey modular ground-water model: User guide to modularization concepts and the ground-water flow process. Open-File Report 00-92, U.S. Geological Survey.

Hattermann, F. F., Wattenbach, M., Krysanova, V., and Wechsung, F. (2005). Runoff simulations on the macroscale with the ecohydrological model SWIM in the Elbe catchment–validation and uncertainty analysis. *Hydrological Processes: An International Journal*, 19(3), 693–714.

Holmes, J. W. (1984). Measuring evapotranspiration by hydrological methods. *Agricultural Water Management*, 8(1-3), 29–40.

Howell, T. A. (2004). *Lysimetry. Encyclopedia of Soils in the Environment* (pp. 379–386). Oxford, UK: Elsevier Press.

Hussein, M. M. (2022). Comparison between simplified and standardized Penman evaporation equation: Egyptian Mediterranean coast case study. *Egyptian Journal of Aquatic Research*, 48(4), 315–323.

Kateb, Z., Bouchelkia, H., Benmansour, A., and Belarbi, F. (2020). Sediment transport modeling by the SWAT model using two scenarios in the watershed of Beni Haroun dam in Algeria. *Arabian Journal of Geosciences*, 13, 1–17.

Kinzelbach, W. (1986). *Groundwater Modelling: An Introduction with Sample Programs in BASIC* (Vol. 25). Elsevier.

Knisel, W. G. (1980). CREAMS: A field scale model for chemicals, runoff, and erosion from agricultural management systems (No. 26). Department of Agriculture, Science and Education Administration, Science & Education Administration, United States Department of Agriculture.

Koskinen, L., Laitinen, M., Lofman, J., Meling, K., and Meszaros, F. (2025). FEFLOW: A finite element code for simulating groundwater flow, heat transfer and solute transport. *WIT Transactions on Ecology and the Environment*, 10, 287–296.

Krysanova, V., Wechsung, F., Arnold, J., Srinivasan, R., and Williams, J. (2000). SWIM (soil and water integrated model), Number 69, User Manual, Potsdam Institute for Climate Impact Research.

Kumar, C. P. (2012). Groundwater modelling software–capabilities and limitations. *IOSR Journal of Environmental Science, Toxicology and Food Technology*, 1(2), 46–57.

Kumar, K. K., Kumar, K. R., and Rakhecha, P. R. (1987). Comparison of Penman and Thornthwaite methods of estimating potential evapotranspiration for Indian conditions. *Theoretical and Applied Climatology*, 38, 140–146.

Laflen, J. M., Elliot, W. J., Flanagan, D. C., Meyer, C. R., and Nearing, M. A. (1997). WEPP-predicting water erosion using a process-based model. *Journal of Soil and Water Conservation*, 52(2), 96–102.

Li, H., Jiang, C., Choy, S., Wang, X., Zhang, K., and Zhu, D. (2022). A comprehensive study on factors affecting the calibration of potential evapotranspiration derived from the Thornthwaite Model. *Remote Sensing*, 14(18), 4644.

Liu, Z., and Todini, E. (2002). Towards a comprehensive physically-based rainfall-runoff model. *Hydrology and Earth System Sciences*, 6(5), 859–881.

Luo, Y., Yang, S., Liu, X., Liu, C., Zhang, Y., Zhou, Q., ... and Dong, G. (2016). Suitability of revision to MUSLE for estimating sediment yield in the Loess Plateau of China. *Stochastic Environmental Research and Risk Assessment*, 30, 379–394.

Lyu, P., Song, J., Yin, Z., Wu, J., and Wu, J. (2024). Integrated SEAWAT model and GALDIT method for dynamic vulnerability assessment of coastal aquifer to seawater intrusion. *Science of the Total Environment*, 925, 171740.

Majhi, A., Shaw, R., Mallick, K., and Patel, P. P. (2021). Towards improved USLE-based soil erosion modelling in India: A review of prevalent pitfalls and implementation of exemplar methods. *Earth-Science Reviews*, 221, 103786.

Manrique, L. A. (1993). Technology for soil erosion assessment in the tropics: A review. *Communications in Soil Science and Plant Analysis*, 24(9-10), 1033–1064.

Merritt, W. S., Letcher, R. A., and Jakeman, A. J. (2003). A review of erosion and sediment transport models. *Environmental Modelling and Software*, 18(8-9), 761–799.

Michael, A. M. (2009). *Irrigation Theory and Practice-2nd Edition: Theory and Practice*. Vikas publishing house.

Mohtashami, A., Al-Ghafri, A., and Akbarpour, A. (2025). Qanats' assistance in reviving groundwater resources using numerical groundwater model. *Paddy and Water Environment*, 23, 443–454.

Morway, E. D., Feinstein, D. T., Hunt, R. J., and Healy, R. W. (2023). New capabilities in MT3D-USGS for simulating unsaturated-zone heat transport. *Groundwater*, 61(3), 330–345.

Narayanamurthi, V., and Ramasamy, A. (2023). Comparison of three groundwater models with finite element methods for groundwater head simulation. In *International Conference on Civil Engineering Innovative Development in Engineering Advances* (pp. 469–476). Singapore: Springer Nature Singapore.

Nearing, M. A., Jetten, V., Baffaut, C., Cerdan, O., Couturier, A., Hernandez, M., and Van Oost, K. (2005). Modeling response of soil erosion and runoff to changes in precipitation and cover. *Catena*, 61(2-3), 131–154.

Niswonger, R. G., Panday, S., and Ibaraki, M. (2011). MODFLOW-NWT, a Newton formulation for MODFLOW-2005. *US Geological Survey Techniques and Methods*, 6(A37), 44.

Nyeko, M. (2015). Hydrologic modelling of data scarce basin with SWAT model: Capabilities and limitations. *Water Resources Management*, 29, 81–94.

Payero, J. O., and Irmak, S. (2008). Construction, installation, and performance of two repacked weighing lysimeters. *Irrigation Science*, 26, 191–202.

Pereira, A. R., and Pruitt, W. O. (2004). Adaptation of the Thornthwaite scheme for estimating daily reference evapotranspiration. *Agricultural Water Management*, 66(3), 251–257.

Provost, A. M., and Voss, C. I. (2019). SUTRA, a model for saturated-unsaturated, variable-density groundwater flow with solute or energy transport—Documentation of generalized boundary conditions, a modified implementation of specified pressures and concentrations or temperatures, and the lake capability (No. 6-A52). US Geological Survey.

Ramos, M. C., and Martínez-Casasnovas, J. A. (2015). Soil water content, runoff and soil loss prediction in a small ungauged agricultural basin in the Mediterranean region using the Soil and Water Assessment Tool. *The Journal of Agricultural Science*, 153(3), 481–496.

Rushton, K. R. (2004). *Groundwater Hydrology: Conceptual and Computational Models*. Wiley.

Sahu, M. K., Shwetha, H. R., and Dwarakish, G. S. (2023). State-of-the-art hydrological models and application of the HEC-HMS model: A review. *Modeling Earth Systems and Environment*, 9(3), 3029–3051.

Singh, A. (2013). Groundwater modelling for the assessment of water management alternatives. *Journal of Hydrology*, 481, 220–229.

Singh, V. P., and Frevert, D. K. (2003). Watershed modeling. In *World Water & Environmental Resources Congress 2003* (pp. 1–37).

Sisay, B. M., Nedaw, D., Birhanu, B., and Gigar, A. G. (2023). Application of SWAT and MODFLOW models for characterization of surface–groundwater interaction in the Modjo River catchment, central Ethiopia. *Environmental Earth Sciences*, 82(13), 341.

Soulis, K. X., and Valiantzas, J. D. (2012). SCS-CN parameter determination using rainfall-runoff data in heterogeneous watersheds–the two-CN system approach. *Hydrology and Earth System Sciences*, 16(3), 1001–1015.

Subramanya, K. (2021). *Engineering Hydrology* (pp. 155–159). Tata McGraw-Hill Publication Company Limited, New Delhi.

Sumner, D. M., and Jacobs, J. M. (2005). Utility of Penman–Monteith, Priestley–Taylor, reference evapotranspiration, and pan evaporation methods to estimate pasture evapotranspiration. *Journal of Hydrology*, 308(1-4), 81–104.

Svetlosanov, V. A. (1982). Capabilities and Limitations of the Creams Model (Methodological Aspects), International Institute For Applied Systems Analysis, A-2361 Laxenburg, Austria.

Todini, E., and Ciarapica, L. (2001). The TOPKAPI model. In *Mathematical Models of Large Watershed Hydrology*, Water Resources Publications, LLC, Colorado, USA, 471–506.

Valiantzas, J. D. (2006). Simplified versions for the Penman evaporation equation using routine weather data. *Journal of Hydrology*, 331(3-4), 690–702.

Wischmeier, W. H., and Smith, D. D. (1978). Predicting rainfall erosion losses: a guide to conservation planning (No. 537). Department of Agriculture, Science and Education Administration.

Xu, X., Zhang, R., Xue, X., and Zhao, M. (1998). Determination of evapotranspiration in the desert area using lysimeters. *Communications in Soil Science and Plant Analysis*, 29(1-2), 1–13.

Yang, J., Li, B., and Shiping, L. (2000). A large weighing lysimeter for evapotranspiration and soil-water–groundwater exchange studies. *Hydrological Processes*, 14(10), 1887–1897.

Young, C. B., McEnroe, B. M., and Rome, A. C. (2009). Empirical determination of rational method runoff coefficients. *Journal of Hydrologic Engineering*, 14(12), 1283–1289.

17 Glacial Lakes of Garhwal Himalaya

An Inventory and Change Detection Analysis Using Multi-Decadal Satellite Imagery

Vikas Rawat, Shekhar Singh, and Mahabir Singh Negi

17.1 INTRODUCTION

The Garhwal Himalaya is home to over 1,500 glaciers, covering an area of approximately 4,000 km^2 (Shukla et al., 2018). These glacial systems receive accumulation through winter precipitation and undergo ablation in the summer season. The melting of these glaciers results in the development of glacial lakes (GLs), which are classified based on their origin and morphology (Gardell et al., 2011; Bolch et al., 2012). These GLs, formed as a consequence of glacier recession, play a pivotal role in the local environment and are of significant interest due to their potentiality to cause glacial lake outburst floods (GLOFs) (Fujita et al., 2013; Veh et al. 2019). In the Garhwal Himalaya, proglacial lakes that develop at the base or terminus of a glacier are considered the most hazardous among all lake types (Dubey and Goyal, 2020; Rawat et al., 2024). These lakes are typically shallow and have a high sediment load due to the erosion and deposition of glacial debris. The current retreat of glaciers due to rising temperatures has significant consequences on the genesis and enlargement of GL, leading to their formation and expansion, respectively, and posing potential hazards to the surrounding communities (Khadka et al., 2021; Singh et al., 2024). The size and volume of GL in the Garhwal Himalaya vary widely. Some lakes are small and temporary, while others are large and persistent. The largest lake in the region is Gangotri Tal, located at the foot of the Gangotri Glacier (Shukla et al., 2018). This lake has a surface extent of approximately 36 hectares and a maximum depth of 45 m. Other notable lakes include Kedartal, Roopkund Tal, and Bedni Bugyal Tal. The potential for GLOFs from GLs in the Garhwal Himalaya has gained increasing consideration in recent years due to the rapid melting of glaciers and the resulting increase in lake volume (Bhambri et al., 2017). GLOFs can have catastrophic consequences on downstream communities, causing loss of life, destruction to infrastructure, and disruption to agriculture and livelihoods. Therefore, understanding the characteristics

DOI: 10.1201/9781003596080-17

and behavior of GL is crucial for managing GLOF risks. One key factor that influences GLOF potential is lake morphology. Shallow lakes with steep slopes are more prone to outburst floods than deep lakes with gentle slopes (Bhambri et al., 2017). In addition, lakes with high sediment loads are more likely to experience landslides and debris flows that can trigger outburst floods. Therefore, monitoring lake morphology is an important component of GLOF risk management. Another important factor is lake volume. As lake volume increases due to melting glaciers or heavy precipitation events, the likelihood of outburst floods also increases (Bhambri et al., 2017). Therefore, monitoring lake volume is crucial for early warning systems and disaster preparedness. In addition, understanding how lake volume changes over time can provide insights into long-term trends in GLOF risk. Remote sensing techniques have been instrumental in inventorying and monitoring GL in the Himalaya, enabling the identification of dangerous lakes and the assessment of their spatiotemporal patterns. The imprint of climate alteration on glacial settings has been suggested as a cause for the accelerated growth of GL in the Garhwal Himalaya, emphasizing the dynamic nature of these water bodies (Shukla S. et al., 2018). Overall, the GL of the Garhwal Himalaya is a subject of significant scientific interest due to its environmental, geological, and societal implications. This book chapter's objective is to provide a comprehensive analysis of the GL in the Garhwal Himalaya, incorporating insights from remote sensing and hazard assessment.

17.2 STUDY AREA

Uttarakhand comprises two primary administrative divisions, with a total geographical area of approximately 53,485 km^2 (Singh and Kumar, 2021; Singh et al., 2023). Garhwal is one of them, encompassing seven districts within the state (Figure 17.1). It spans a geospatial extent of 32,383 km^2, encompassing altitudes ranging from 246 m in Roorkee to 7,816 m at Nanda Devi above mean sea level. Situated within the latitudinal coordinates of 29°34′50.93″ to 31°27′39.69″ N and longitudinal coordinates of 77°34′33.47″ to 80°05′45.35″ E, Garhwal is characterized by a varied and dynamic topographical profile. Alpine meadows, dense forests, and high-altitude ecosystems offer habitats for various plant and animal species, some of which are endemic to the region (Bahuguna and Pandey, 2017). It is renowned for its formidable geophysical features, delineated by precipitous gradients, profound ravines, and elevated mountain summits. Important peaks encompass Nanda Devi, Kamet, Trisul, and Badrinath (Maikhuri et al., 2000). Garhwal is home to numerous glaciers, including the Gangotri Glacier, one of the primary sources of the Ganges River. Glacial meltwater feeds into several major rivers, shaping the hydrology of the region. The Bhagirathi, Alaknanda, and Yamuna rivers originate from the Garhwal Himalaya, contributing to the vast network of the Ganges basin (Negi and Samal, 2012; Sati, 2015; Dobhal et al., 2013).

17.3 DATA USED

The present catalogue of GL in the Garhwal Himalaya employs recent multispectral imagery obtained from the Landsat 9 and Resourcesat-1 satellites (Yao et al., 2010;

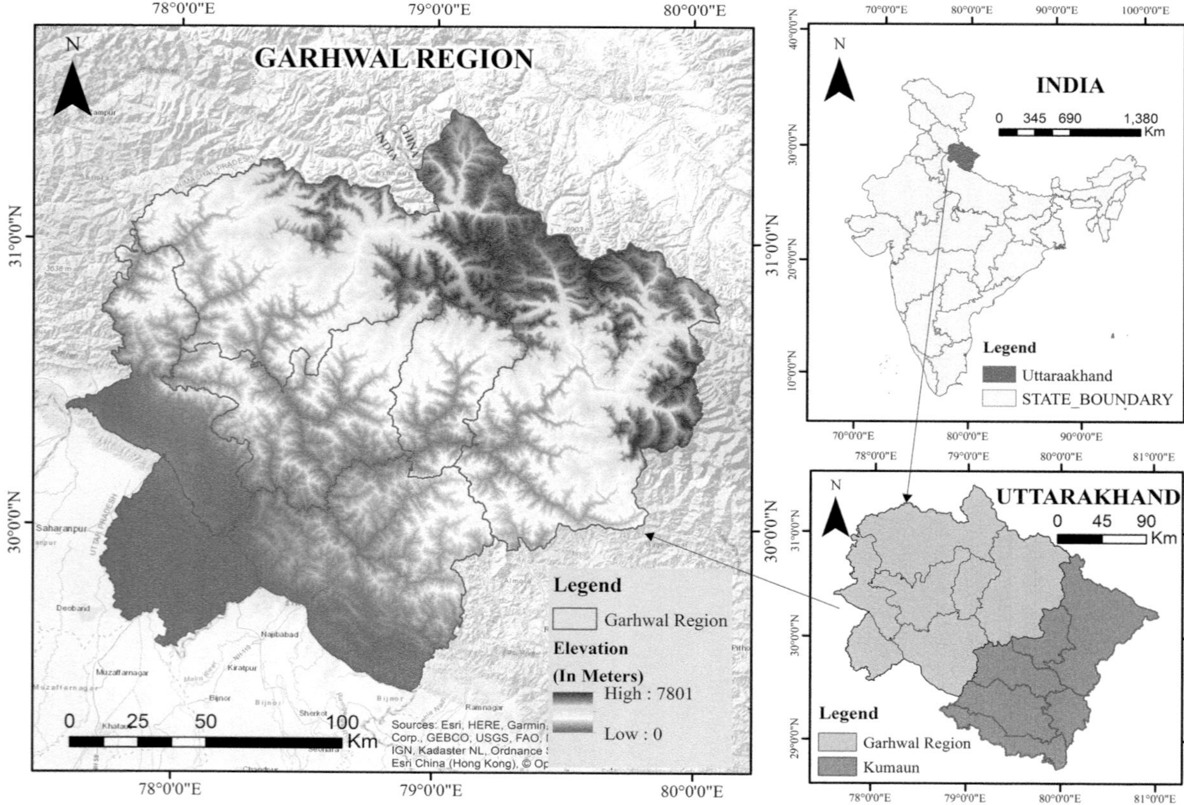

FIGURE 17.1 Map showcasing the physical extent of the study area, which pertains to the Garhwal Region located in Uttarakhand, encompasses an approximate surface area of 32,383 km^2 aligned with the Digital Elevation Model.

Zhang et al., 2023). The study area is delimited by three Landsat 9 satellite imagery scenes and 10 Resourcesat acquired images. The Landsat images were procured from the United States Geological Survey (USGS) platform, while the Sentinel imagery was downloaded from the Copernicus program. The Earth observation satellite Copernicus Sentinel-2A, which is operated by the European Union, is outfitted with a multispectral instrument (MSI) that acquires multispectral imagery of the Earth's surface at distinct spatial resolutions with a repeat cycle of 10 days (Wang et al., 2015). The European Copernicus Sentinel-2A satellite incorporates an MSI with a total of 13 spectral bands (Wang et al., 2020). Four of these bands offer a spatial resolution of 10 m, six provide a resolution of 20 m, and the remaining three bands offer a resolution of 60 m. The Resourcesat-1 satellite is equipped with a Linear Imaging Self-Scanning Sensor (LISS-III) that captures high-resolution images in four spectral bands.

In the course of this analysis, digital elevation data (DEM) attained from the Shuttle Radar Topography Mission (SRTM) during the years 2010–2012, procured from the USGS, were utilized. To identify the increasing number and expansion of GL, satellite images from the Corona and Hexagon programs, dating back to 1968, were also employed in this study. The inception of the Corona satellite program transpired in 1959, initiated by the Central Intelligence Agency (CIA) as a countermeasure to the Soviet Union's deployment of Sputnik, marking the advent of artificial satellite technology (Dashora et al., 2007). The Hexagon satellite program, initiated by the National Reconnaissance Office (NRO) in 1971 as a successor to Corona, aimed to provide improved imagery capabilities using larger film capsules and more advanced camera systems. Both programs offer high-resolution panchromatic images, and their contributions to the development of advanced image processing techniques are significant in remote sensing. Sentinel-2A employs 13 distinct spectral bands, 4 of which have a spatial resolution of 10 m, 6 have 20 m, and the remaining 3 have 60 m of resolution. To accomplish this investigation, we employed satellite images that were free of clouds and had the least amount of snow for optimal results.

17.4 RESEARCH METHODOLOGY

17.4.1 Inventory of GLs

The preeminent technique employed for delineating GLs involves the utilization of the Normalized Difference Water Index (NDWI) presented in Equation 17.1 (Huggel et al., 2002). These indices are applied to ascertain individual GLs or glacier extents, based on the spectral bands characterized by the highest reflectivity (Sarp and Ozcelik, 2017):

$$NDWI = \frac{Band_{Green} - Band_{NIR}}{Band_{Green} + Band_{NIR}} \tag{17.1}$$

The methodology for delimiting lakes from Landsat and Resourcesat LISS III images entails a semi-automated mapping process utilizing the NDWI, with a specified threshold value of 0.15 to discriminate between surface water features. Positive values are allotted to water bodies, while values in negative values denote non-water features, as elaborated upon subsequently. To enhance precision, Google Earth Pro and satellite images are employed in conjunction with a false-color composite (FCC) comprising bands of 432 for LISS III and 532 for LS (RGB) (Jain and Mir, 2019). These techniques facilitate meticulous manual digitization of GL through visual assessment. Subsequently, emphasis is placed on prioritizing lakes with the maximum coverage area.

17.4.2 GL Change

The methodology employed to discern changes in GL within the study area involved a comprehensive analysis utilizing satellite imagery spanning multiple decades. Initially, historical images such as Corona or Hexagon from 1968 were utilized to establish a baseline, providing insights into the initial configuration of GLs. Subsequently, Landsat imagery from 1994 and Resourcesat imagery from 2008 were incorporated to facilitate a spatiotemporal analysis, capturing the evolving dynamics of the GLs over the intervening years. The Landsat 9 and Sentinel-2 image from 2023 was used as the most recent dataset
to assess the contemporary state of GLs. Through a systematic application of supervised image classification techniques and comparative analyses across these diverse temporal datasets, the study aimed to delineate and quantify changes in GL extents, providing an inclusive understanding of the temporal evolution of these crucial environmental features within the specified study area (Jain and Mir, 2019).

17.5 RESULTS AND DISCUSSION

17.5.1 Inventory of GLs

In this chapter, it has been determined that around 323 lakes were detected through automated mapping techniques employing the Landsat 8–9 image from 2023. The methodology involved semi-automated mapping utilizing Landsat images, Google Earth Pro, and Resourcesat satellite data, enhancing precision through false-color composites (FCC) and visual assessment. This meticulous process prioritized lakes based on coverage area, and their boundaries were accurately delineated. The most well-known lakes of the region are Roopkund, Satopanth, Hemkund, Vauki Tal, Kedar Tal, etc., and the suffix Tal is employed in the Himalayan region of Uttarakhand to indicate a body of water. For the purpose of this chapter, only lakes with a surface area surpassing one hectare were evaluated. The total count of such lakes, exceeding 1 hectare in size, was identified as 67. In the Garhwal region, the majority of lakes are relatively small, with 256 lakes, constituting nearly 79.25% of the overall total, having an area of less than 1 hectare. Moreover, there are 35 lakes with an intermediate area ranging from 1 to 3 hectares. Furthermore, among the 323 lakes in the region, 14 are proglacial, posing a potential threat for GLOFs. As Figure 17.2 shows, Chamoli district in Uttarakhand has had the highest concentration of lakes since 1968.

TABLE 17.1
List of Satellite Data and Their Properties

Satellite	Image ID	Date of Acquisition	Spatial Resolution (in mtr.)	Spectral Resolution	Number of Bands
Corona/ hexagon	DS1048-1134DF109_109_a	27 Sep 1968	4	Panchromatic	1
	DS1048-1134DF109_109_b				
	DS1048-1134DF109_109_c				
	DS1048-1134DF109_109_d				
	DS1048-1134DF109_109GF				
	DZB1207-500047L023001_a	24 Nov1973	7	Panchromatic	1
	DZB1207-500047L023001_b				
Landsat 5	LT05_L2SP_145039_19941025_20200913_02_T1	25 Oct 1994	79	Multispectral	7
	LT05_L2SP_145039_19941025_20200913_02_T1				
	LT05_L2SP_145038_19941009_20200913_02_T2	09 Oct 1994			
	LT05_L2SP_145039_19941009_20200913_02_T2				
Resourcesat 1	l3h44a0723oct08	23 Oct 2008	24	Multispectral	4
	l3h44a0823oct08				
	l3h44a1223oct08				
	l3h44g0523oct08				
	l3h44h1309oct08				
	l3h44h1509oct08				
	l3h44i0109oct08				
	l3h44i0209oct08				
	l3h44i0309oct08				
	l3h44a1123oct09	23 Oct 2009			
	l3h44a1523oct09				
	l3h44a1623oct09				
	l3h44b0423oct09				
	l3h44g0923oct09				
	l3h44g1323oct09				

(*continued*)

TABLE 17.1 (Continued)
List of Satellite Data and Their Properties

Satellite	Image ID	Date of Acquisition	Spatial Resolution (in mtr.)	Spectral Resolution	Number of Bands
	13h44h0123oct09				
Landsat 8–9	LC08_L2SP_145039_20231009_20231017_02_T1 LC09_L2SP_146038_20231008_20231009_02_T1 LC09_L2SP_146039_20231008_20231009_02_T1	09 Oct 2023	30	Multispectral	7

17.5.2 GL Change

For assessing changes in GLs, a multi-decadal analysis was conducted using historical images from Corona and Hexagon, Landsat, Resourcesat, and the most recent Landsat 9 and Sentinel-2 imagery. The systematic application of supervised image classification techniques facilitated the quantification of changes in GL extents over time. This comprehensive approach provided insights into the temporal evolution of GLs, contributing to a holistic understanding of their dynamics within the specified study area. Owing to the impact of global warming, a discernible trend has been observed wherein a significant number of glaciers are undergoing retreat at varying rates and manners. Consequently, there has been a substantial increase in the formation of new lakes and the expansion of previously existing ones (Mayewski and Jeschke, 1979; Kulkarni et al., 2007). After assessing the NDWI and analyzing the error reduction of the Landsat 5 satellite image dated 1994, it was determined that in the Garhwal region, there were 177 lakes initially. However, by the year 2023, this figure had risen to a total of 323 lakes. The numerical variation signifies an 80% increase in the quantity of GLs, exclusively encompassing the genesis of newly formed lakes.

Moreover, it is imperative to highlight the heightened trajectory of relative changes in the lake area during the temporal span from 1994 to 2023. Figure 17.3 delineates the dynamics pertaining to both the numerical count and surface dimensions of lakes within the upper Alaknanda River basin. As shown in the above images, the progression elucidates the incremental rise in lake numbers across successive decades. In the initial phases, these lakes exhibited a limited count, but over subsequent decades, their proliferation and amalgamation with newly formed lakes became evident. Notably, the cumulative surface area of GLs, quantified at 1.08 km^2 in 1994, underwent a subsequent augmentation, reaching 2.16 km^2 by the termination of 2023.

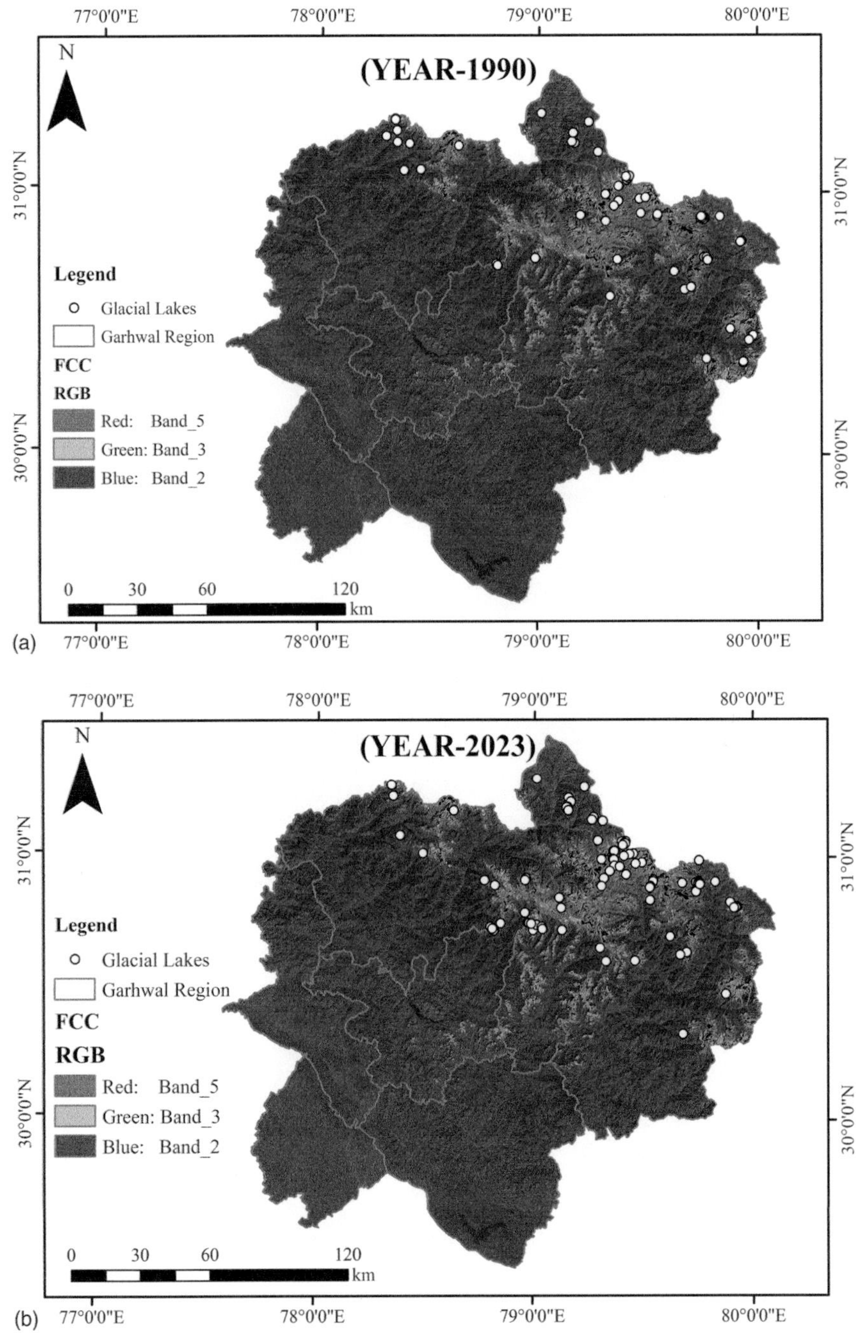

FIGURE 17.2 Spatial distribution of GLs in the Garhwal Region as delineated in the contemporary research inventory, illustrating (a) the progressive rise in lake numbers using a false-color composite derived from Landsat-9 imagery from the year 2023, and (b) the initial count of GLs in 1994 aligned with the Digital Elevation Model.

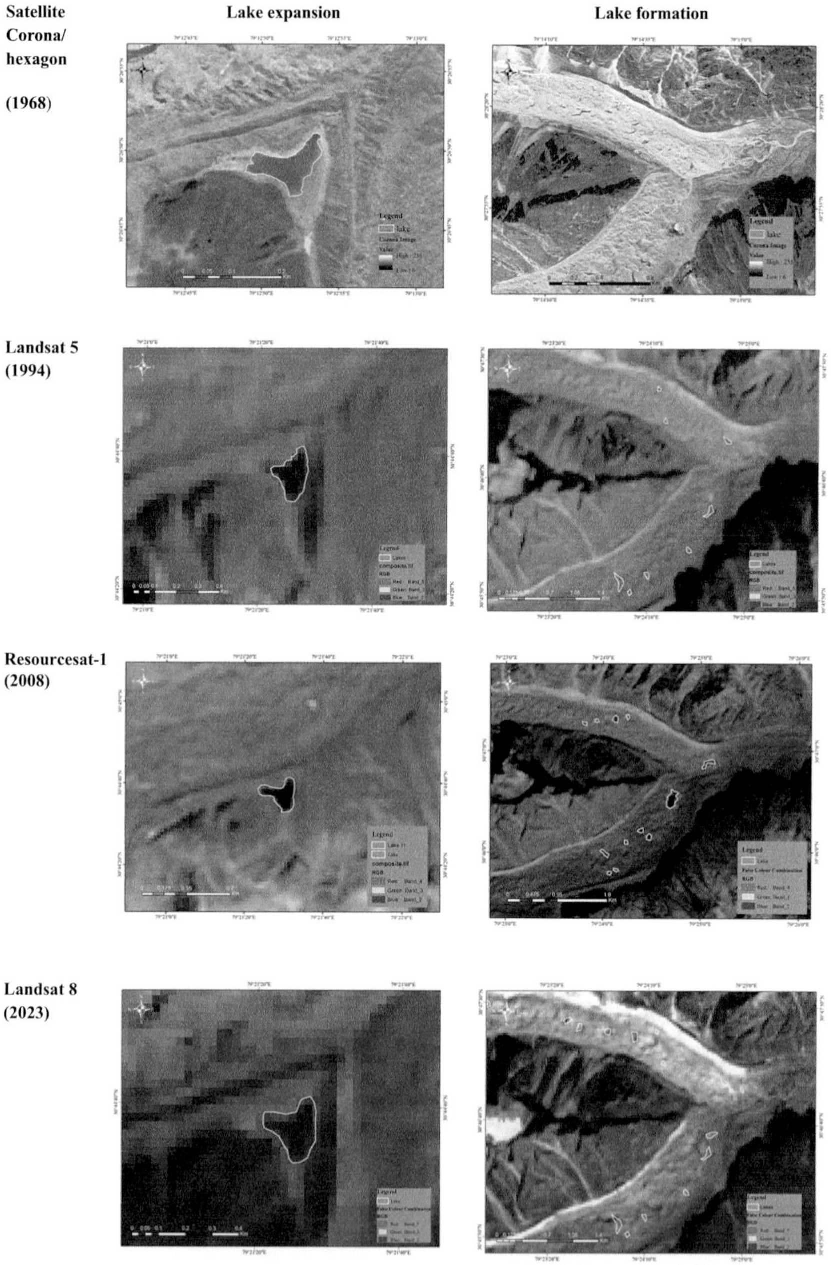

FIGURE 17.3 GL change detection through different satellite images.

17.6 CONCLUSION

The research carried out in the Garhwal region of Uttarakhand, India, aimed to provide an inclusive assessment of the lakes in the area, incorporating insights from remote sensing and hazard assessment. The study utilized recent multispectral imagery obtained from various satellite sources, including Landsat 9, Sentinel-2A, and Resourcesat-2. The study also employed historical images from Corona and Hexagon dating back to 1968 to establish a baseline and facilitate a spatiotemporal analysis. The study found that around 323 lakes were identified through automated mapping techniques using the Landsat 8–9 image from 2023. Most of the lakes in the region are relatively small, with 256 lakes (nearly 79.25%) having an area of less than 1 hectare. However, among the 323 lakes in the region, 14 are proglacial, posing a potential threat for GLOFs. The research also discovered a notable increase in the creation of new lakes and the enlargement of preexisting ones, attributed to the effects of global warming. By 2023, there were 323 lakes in the region, compared to just 177 lakes in 1994, signifying an 80% increase in the quantity of lakes, exclusively encompassing the genesis of newly formed lakes. The study also found that there has been a substantial increase in the surface extent of GLs over time. In 1994, the combined surface area of GL was measured at 1.08 km^2, and it later increased to 2.16 km^2 by the end of 2023. This highlights the heightened trajectory of relative changes in the lake area during the temporal span from 1994 to 2023. The study emphasized that it is imperative to monitor these changes closely as they pose a potential threat to GLOFs. This study delivers valuable insights into the dynamics of GL in the Garhwal region of Uttarakhand, India, which can aid in developing effective strategies for hazard assessment and mitigation measures. Additional investigation is warranted to understand the enduring consequences of these alterations on the environment and the nearby communities residing in this area.

REFERENCES

Bahuguna, V. K., and Pandey, A. (2017). Socio-cultural and economic impact of tourism in Garhwal Himalaya, India. *Tourism Management Perspectives*, 23, 10–21.

Bhambri, R., Bolch, T., and Shepherd, A. (2017). Glacier change in high mountain Asia: current status and future scenarios. *Journal of Glaciology & Geography*, 6(3), 1–14.

Bolch, T., Peters, J., Yegorov, A., Pradhan, B., Buchroithner, M., and Blagoveshchensky, V. (2012). Identification of potentially dangerous glacial lakes in the Northern Tien Shan. *Natural Hazards*, 59, 1691–1714.

Dashora, A., Lohani, B., and Malik, J. N. (2007). A repository of earth resource information-CORONA satellite programme. *Current Science*, 92(7), 926–932.

Dobhal, D. P., Mehta, M., Bhambri, R., and Kumar, S. (2013). Influence of debris cover on terminus retreat and mass changes of Chorabari Glacier, Garhwal region, central Himalaya. *Journal of Glaciology, 59*(214), 961–971.

Dubey, S., & Goyal, M. (2020). Glacial lake outburst flood hazard, downstream impact, and risk over the Indian Himalayas. *Water Resources Research*, 56(4), e2019WR026533.

Fujita, K., Sakai, A., Takenaka, S., Nuimura, T., Surazakov, A., Sawagaki, T., and Yamanokuchi, T. (2013). Potential flood volume of Himalayan glacial lakes. *Natural Hazards and Earth System Science*, 13(7), 1827–1839.

Gardell, J., Arnaud, Y., and Berthier, É. (2011). Contrasted evolution of glacial lakes along the Hindu Kush Himalaya mountain range between 1990 and 2009. *Global and Planetary Change*, 75(1-2), 47–55.

Huggel C, Kaab A, Haeberli W, Teysseire P, Paul F (2002) Remote sensing based assessment of hazards from glacier lake outbursts: a case study in the Swiss Alps. *Canadian Geotechnical Journal*, 39, 316–330

Jain, S. K., and Mir, R. A. (2019). Glacier and glacial lake classification for change detection studies using satellite data: a case study from Baspa basin, western Himalaya. *Geocarto International*, *34*(4), 391–414.

Khadka, N., Chen, X., Nie, Y., Thakuri, S., Zheng, G., and Zhang, G. (2021). Evaluation of glacial lake outburst flood susceptibility using multi-criteria assessment framework in Mahalangur Himalaya. *Frontiers in Earth Science*, 8, 601288.

Kulkarni, A. V., Bahuguna, I. M., Rathore, B. P., Singh, S. K., Randhawa, S. S., Sood, R. K., and Dhar, S. (2007). Glacial retreat in Himalaya using Indian Remote Sensing satellite data. *Current Science*, 92(1), 69–74.

Maikhuri, R. K., Nautiyal, S., Rao, K. S., and Saxena, K. G. (2000). Role of medicinal plants in the traditional healthcare system of the Garhwal Himalaya, India. *Conservation Biology*, 14(5), 1183–1192.

Mayewski, P. A., and Jeschke, P. A. (1979). Himalayan and TransHimalayan glacier fluctuations since AD 1812. *Arctic and Alpine Research*, 11(3), 267–287.

Rawat, V., Singh, S., and Negi, M. S. (2024). Assessing GLOF susceptibility and risk mapping using optical remote sensing data: A case chapter of upper Alakananda River Basin. In *New Advancements in Geomorphological Research: Issues and Challenges in Quantitative Spatial Science* (pp. 33–42). Cham: Springer Nature Switzerland.

Sarp, G., and Ozcelik, M. (2017). Water body extraction and change detection using time series: A case study of Lake Burdur, Turkey. *Journal of Taibah University for Science*, 11(3), 381–391.

Sati, V. P. (2015). Climate change and socio-ecological transformation in high mountains: an empirical study of Garhwal Himalaya. *Change and Adaptation in Socio-Ecological Systems*, 2(1), 45–56.

Shukla, S., Kumar, S., & Singhvi, A. K. (2018). Glacial lakes in Garhwal Himalaya: An overview. *International Journal of Geosciences*, 10(4), 488–496.

Singh, S., and Kumar, D. (2021). Investigation of rainfall variability of the southern part of Uttarakhand using entropy theory. *Indian Journal of Soil Conservation*, 49(2), 83–88.

Singh, S., Kumar, D., Kumar, A., & Kuriqi, A. (2023). Entropy-based assessment of climate dynamics with varying elevations for hilly areas of Uttarakhand, India. *Sustainable Water Resources Management*, 9(4), 130.

Singh, S., Kumar, D., Vishwakarma, D. K., Kumar, R., and Kushwaha, N. K. (2024). Seasonal rainfall pattern using coupled neural network-wavelet technique of southern Uttarakhand, India. *Theoretical and Applied Climatology*, 155, 5185–5201.

Veh, G., Korup, O., von Specht, S., Roessner, S., and Walz, A. (2019). Unchanged frequency of moraine-dammed glacial lake outburst floods in the Himalaya. *Nature Climate Change*, 9(5), 379–383.

Wang, S., Che, Y., and Xinggang, M. (2020). Integrated risk assessment of glacier lake outburst flood (GLOF) disaster over the Qinghai–Tibetan Plateau (QTP). *Landslides*, 17(12), 2849–2863.

Yao, T., Li, Z., Yang, W., Guo, X., Zhu, L., Kang, S., and Yu, W. (2010). Glacial distribution and mass balance in the Yarlung Zangbo River and its influence on lakes. *Chinese Science Bulletin*, 55, 2072–2078.

Zhang, D., Zhou, G., Li, W., Han, L., Zhang, S., Yao, X., and Duan, H. (2023). A robust glacial lake outburst susceptibility assessment approach validated by GLOF event in 2020 in the Nidu Zangbo Basin, Tibetan Plateau. *Catena*, 220, 106734.

18 Handling Big Data
Remote Sensing in Land and Water Resource Management

Shubhi Khare, Deepak Kumar, and Victor Nyabuti Ong'era

18.1 INTRODUCTION

Remote sensing is one of the techniques that has now become a necessity in every field of research. Remote sensing is basically the technique of acquiring and measuring information about the properties of an object, place, or phenomenon without being in physical contact with them, with the help of a recording instrument. This information is collected in the form of electromagnetic radiation emitted by the object, and because the information is gathered by the sensors mounted on the satellites or aircraft, there is an opportunity for the scientists or researchers to capture large geographical areas in a single frame or scene and unveil the information about them. The objects are differentiated based on varied levels of reflected electromagnetic radiation emitted by them (Camphell and Randolph, 2011; Gibson, 2013). All this information is stored in the form of data, and the sources of this data are satellites, aerial photography, global positioning system (GPS), light detection and ranging (LiDAR), ground-based sensors, radar systems, crowdsourcing, social media platforms, and historical records (van Wesemael et al., 2023; Yahia et al., 2021; Muhadi et al., 2020; Abdulraheem et al., 2023). As there is continuous technical evolution in remote sensing technology, the size of the data is also expanding simultaneously, and for this exploding data, there arises the concept of big data.

Big data can be defined as the data of a significantly large volume and cannot be handled by a single software for data capture, curation, management, and processing data within a tolerable elapsed time. The concept of big data started to professionalize around the 1970s. It originates from database management tactics and highly relies on the storage, extraction, and optimization techniques applied by the data stored in a relational database management system (RDBMS) (Gantz and Reinsel, 2012; Chen et al., 2012; Chen et al., 2021). The key concepts associated with big data were initially the three Vs, that is, volume, variety, and velocity, in which veracity and value were added later. Combining all five Vs constitutes the quality possessed by big data (Cappa et al., 2021). Big data finds its application in every field since the time the internet and corresponding web applications started to generate tremendous amounts

DOI: 10.1201/9781003596080-18

of data, to the current phase of evolution of big data, where more than 10 million devices connected to the internet are generating more than 44 zettabytes (44 trillion gigabytes) till 2020 (Hilbert and López, 2011).

In remote sensing technology, big data plays a crucial role as tremendous amounts of data are generated by the instruments used for collecting information. For instance, the data streaming from a mission of NASA mission could approximately be 1.73 GB (Ma et al., 2015). In a similar fashion, there are numerous satellites wandering in space, capturing data. According to Singh (2022), after the launch of Sputnik in 1957, more than 7,900 satellites were launched in space, including probes, rockets, and other such devices, as reported by the United Nations Office for Outer Space Affairs. The increasing number of satellites raises the problem of generating of vast volume of big data, which needs to be efficiently handled considering its significant importance in the research and development field.

18.2 SENSORS USED FOR THE ACQUISITION OF REMOTE SENSING DATA

Sensors are basically the instruments mounted on a satellite or aircraft to capture the radiation emitted by the object. Sensors can be active or passive depending on the availability of an energy source. If the sensors have their own source of energy, they emit their radiation toward the targeted objects and detect and measure the radiation emitted by them. Active remote sensors do not rely on time, day, or season and can be used to take measurements anytime, as in the cases of radioactive detection and ranging (RADAR). To capture a better illumination of the object, the energy requirement in the case of active sensors is high. The examples of active sensors include synthetic aperture radar (SAR), LASER, and European Remote Sensing Satellite (ERS) (Masud and Bastiaanssen, 2017; Roy et al., 2017). Another type of sensor is a passive sensors, which rely on an external source of energy to emit the radiation on the targeted objects and capture the electromagnetic radiation emitted by them. The sun is primarily the largest source of energy for these satellites, and thus they are bound to rely on the time of day or season for the acquisition of data (Kuenzer and Dech, 2013).

In the field of land and water management, sensors play a chief role due to their ability to access a vast area for study. Spatial, temporal, and spectral resolutions are the significant factors driving the selection of sensors for the acquisition of data for various land use/land change patterns, hydrological phenomena, hydrometeorological phenomena, spatiotemporal changes, and other phenomena related to land and water management. In this regard, the sensors used for acquiring such data can be classified into spatial, temporal, radiometric, and spectral parameters.

18.2.1 Coarse Resolution Sensors

1. *Advanced Very High-Resolution Radiometer (AVHRR):* The advanced very high-resolution radiometer, beginning with TRIOS-N, is a broadband four- or five-channel scanner dealing with the visible, near-infrared, and thermal

infrared portions of the spectrum. The carrier of this sensor is the Polar Orbiting Environmental Satellites (POES) of the National Oceanic and Atmospheric Administration (NOAA). The central application of the AVHRR sensor lies in monitoring and processing land use management by allowing access to analyze a vast area in an instantaneous view (Holben 1986; Cracknell 1997). The data acquired by AVHRR can be in three formats: High-resolution picture transmission data which are high-resolution data transmitted to the ground station simultaneously with the collection, local area coverage (LAC) are full resolution data that are recorded on an on broad tape for subsequent transmission during a station overpass, and global area coverage (GAC) data that are derived by taking the average of sample data of AVHRR. Four out of every five samples are taken, and an average value is calculated out of them. Also, the data of every third scan line is processed to prepare GAC. The AVHRR sensor provides global data with a 2,399 km wide swath by orbiting the earth 14 times a day from 833 km above its surface (McClain 1985).

2. *MODIS:* MODIS is one of the key sensors included in the Surface Imaging and Sounding Package (SISP) provided by the Earth Observing System (EOS) operated by the National Oceanic and Atmospheric Administration (NOAA) and the National Aeronautics and Space Administration (NASA). Sensors such as the Landsat Multispectral Scanner (MSS) and Thematic Mapper (TM), the NOAA Advanced Very High-Resolution Radiometer (AVHRR), High-Resolution Infrared Sounder (HIRS), and the Nimbus Coastal Zone Color Scanner (CZCS) are the sensors used by MODIS to acquire the information about Earth. The entire earth surface is surveyed by MODIS in a span of every 1–2 days, acquiring data in 36 spectral bands between 0.405 and 14.385 mm with a swath width of 2,330 km, having 250, 500, and 1,000 m spatial resolution. The requirement to avoid sun-glint and also provide continuous nadir viewing made it necessary to split MODIS into two separate instruments called MODIS-T (Tilt) and MODIS-N (Nadir). The specifications of both MODIS-T and MODIS-N are provided in Tables 18.1 and 18.2, respectively (Salomonson et al., 1989; Gracia-Mora et al., 2012; Kumar and Deoli, 2020).

TABLE 18.1
Characteristics of MODIS-T

Specification	MODIS-T
Bands	64
Spectral range	400–1,040 nm
Altitude	824 km
Swath width	64 km along track
Coverage area	1,780 km
Spatial resolution	10 nm
Coverage time	2 days
Weight	110 lbs

TABLE 18.2
Characteristics of MODIS-N

Bands	Spatial Resolution	Band Width	Use
5–13	1,000 m	662–1,250	To obtain set of ocean color
1–4 and 20–23	500 m	620–565	To observe land surface and cloud characteristics.
26–40	1 km	1.36–14.385	To provide tropospheric sounding and surface temperature measurements.
14–16	1 km	673–877	To observe polarization effects over large regions.
17–19	1 km	890–965	To measure the amount of atmospheric water vapor.

18.2.2 Multispectral Sensors

1. *Landsat 1, 2, 3*: Landsat 1 is the first Earth Resources Technology Satellite (ERTS-I), designed and launched by NASA in July 1972. The satellite was launched at a height of 920 km in near-polar, sun-synchronous orbit with an orbital period of 103 minutes, covering the entire Earth in an 18-day repeat cycle. The sensors used in Landsat 1 are the return beam vidicon (RBV) multispectral scanner (MSS). The RBV, a television camera that was initially installed as the primary part but was later overshadowed by MSS, was used for various cartographic purposes. The multispectral scanner (MSS) was used to capture four spectral bands: Visible green, visible red, and two near-infrared bands. The satellite Landsat 1 was powered by solar panels and was decommissioned in January 1978. Enhanced Thematic Mapper Plus (ETM+) and Operational Land Imager (OLI) are two important sensors mounted on LANDSAT, which are briefly discussed below.

   ETM+ sensors, the advanced version of thematic mapper used in Landsat 4 and 5, are used in Landsat 7, which was successfully launched from the Vanderburg Air Force Base on April 15, 1999. The primary features of ETM+ sensors include full aperture solar calibration, a panchromatic band with 15 m spatial resolution, and a thermal infrared channel with a fourfold improvement in spatial resolution. The carrier satellite of ETM+, that is, Landsat 7, has an inclination of about 98.2 degrees that orbits the Earth in a 705 km sun-synchronous polar orbit. The satellite has a 16-day repeat cycle circling the Earth every 99 minutes. Also, the pixel size of ground sampling interval (GSI) is 30 m reflective and 60 m thermal, delivering the 8-bit image with 256 gray levels. The band details of Landsat 7 are as follows:

   - *Band 1:* Blue, having a wavelength of 0.45–0.52 μm and a spatial resolution of 30 m.
   - *Band 2:* Green having a wavelength of 0.52–0.60 μm and a spatial resolution of 30 m.

- *Band 3:* Red, having a wavelength of 0.63–0.69 μm and spatial resolution of 30 m.
- *Band 4:* Near-infrared, having a wavelength of 0.77–0.90 μm and a spatial resolution of 30 m.
- *Band 5:* Short-wave infrared having a wavelength of 1.55–1.75 μm and a spatial resolution of 30 m.
- *Band 6:* Thermal, having a wavelength of 10.40–12.50 μm and spatial resolution of 60 m, low gain/high gain.
- *Band 7:* Mid-infrared, having a wavelength of 2.08–2.35 μm and a spatial resolution of 30 m.
- *Band 8:* Panchromatic (PAN) having a wavelength of 0.52–0.90 μm and a spatial resolution of 15 m.

The application of Landsat 7 images lies in various sectors including land cover mapping, climate change detection, ecological monitoring, as well as natural resource management (Deoli et al., 2022; Roy et al., 2016).

Landsat 8, formerly known as Landsat Data Continuity Mission (LDCM), is the satellite that carries an OLI sensor along with the thermal infrared sensor (TIRS) was launched by Vanderberg Air Force Base on February 11, 2013 on an Atlas-V rocket The sensor OLI was built by Ball Aerospace and Technologies Corporation and measures in the visible, near-infrared, and shortwave regions of the spectrum. The spatial resolution of the sensor is 15 m panchromatic and 30 m multispectral. The sensor has a swath width of 185 km. The satellite has a repeat cycle of 16 days due to its near-polar orbit at an altitude of 705 km and completes one Earth revolution in 99 minutes. The spatial resolution of the TIRS sensor is 100 m. The application of Landsat 8 imagery includes land use planning and monitoring, disaster response and evaluation, water management (Kumari et al., 2019; Gupta et al., 2020), agriculture efficiency, coastal management, ecological forecasting, geological and mineral exploration, aquatic science, and surface water mapping (Deoli and Kumar, 2019; Deoli et al., 2021). The satellite imagery includes 9 bands whose detailed information is provided below:

- *Band 1:* Coastal aerosol having a wavelength of 0.43–0.45 μm and a spatial resolution of 30 m.
- *Band 3:* Green having a wavelength of 0.53–0.59 μm and a spatial resolution of 30 m.
- *Band 4:* Red, having a wavelength of 0.64–0.67 μm and a spatial resolution of 30 m.
- *Band 5:* Near-infrared having a wavelength of 0.85–0.88 μm and a spatial resolution of 30 m.
- *Band 6:* SWIR 1 having a wavelength of 1.57–1.65 μm and a spatial resolution of 30 m.
- *Band 7:* SWIR 2, having a wavelength of 2.11–2.29 μm and spatial resolution of 30 m.
- *Band 8:* Panchromatic (PAN) having a wavelength of 0.50–0.68 μm and a spatial resolution of 15 m.
- *Band 9:* Cirrus having a wavelength of 1.36–1.38 μm and spatial resolution of 30 m.

Two more bands for the TIRS sensor are also included in the satellite imagery, namely,

- *Band 10:* TIRS 1 having a wavelength of 10.6–11.19 μm and a spatial resolution of 100 m.
- *Band 11:* TIRS 2 having a wavelength of 11.5–12.51 μm and a spatial resolution of 100 m.

2. *ASTER:* The Advanced Spaceborne Thermal Emission and Reflection Radiometer (ASTER) is an instrument provided by the Ministry of International Trade and Industry (MITI), Tokyo, Japan, to be launched on NASA's Earth Observing System morning (EOS-AM1) platform in 1998. The primary objective of the ASTER mission was to cover the phenomenon taking place near or on the surface of the Earth, extending to the lower atmosphere of the Earth, including surface-atmosphere interactions. The instrument consists of three optical divisions, namely the visible and near-infrared (VNIR) radiometer, shortwave-infrared (SWIR) radiometer, and thermal infrared (TIR) radiometer. The basic characteristics of all three divisions are listed in Table 18.3.
3. *SPOT-1:* It was the first satellite of the SPOT family that was launched in February 1986 for a mission that was supposed to be completed in 3 years, but took 17 years to complete. Being a heavy satellite carrying a weight of two tonnes, it was stabilized by three axes to facilitate its mission of monitoring the entire Earth. The position of the satellite in the orbit, as well as changes in the orbit, was monitored and controlled by the Altitude and Orbit Control System (AOCS) to control any perturbations in its performance as well as to maintain the orbital characteristics (Baraldi et al., 2009). The Earth and sun sensors were used by AOCS to determine the satellite's orientation, and the rate gyroscopes to measure spin rates. The satellite was placed into an orbit whose perigee was 580 km and whose apogee was 801 km, where it could have possibly stayed for approximately 200 years, but due to certain problems raised after the damage to the solar array of the satellite in 2001.

TABLE 18.3
Characteristics of the Three Optical Divisions of ASTER

	Spectral Features	Radiometric Performance	Geometric Performance
VNR	Band in RGB 3, 2, 1	More than ±4% high-level input radiance	Spatial resolution 15 m
SWIR	Band in RGB 8, 6, 4	More than ±4% high-level input radiance	Spatial resolution 30 m
TIR	Band in RGB 14, 12, 10	Less than or equal to ±3 K	Spatial resolution 90 m

Thus, to avoid any future mishappening, the satellite was then deorbited and passivated after 17 years by following the international recommendations for space debris mitigation (Alby, 2005).

4. *IRS 1C:* The Indian Remote Sensing Satellite (IRS) 1C mission is a second-generation satellite launched in 1995 and is an improved version of its predecessors, IRS 1A/1B. The sensors included in the satellite were PAN, Linear Image Self-Scanning Sensors (LISS-III), and a wide field sensor (WiFS). The spatial resolution of PAN is 5.8 m, and radiometric resolution is 6-bit, having a spectral band range of 0.50–0.75 µm and a swath of 70 km covering the entire earth phenomenon on a repeat cycle of every 24 days. In the case of LISS-III, the spatial and radiometric resolutions are 23.5 m and 7 bits, respectively, for green, red, and NIR bands and 70.5 m and 7 bits for SWIR band with a swath of 141 km and spectral band range of 0.52–0.59 µm for green band, 0.62–0.68 µm for red band, and 0.77–0.86 µm for NIR band along with 1.55–0.70 µm for SWIR band with the same repetition time as PAN (Armenakis and Savopol, 1998; Kumar et al., 2021). The spatial and radiometric resolutions for the third sensor of IRS 1C, that is, WiFS, are 188 m and 7- bits, along with two spectral bands having a range of 0.62–0.68 for the red band and 0.77–0.86 for the NIR band, covering a swath of 810 km in the same repetition time as the other two sensors. The three-tier unique capability of IRS-1C to capture high-resolution imagery data as well as generation of digital elevation map (DEM) facilitates the extraction of information at varied scales of mapping using visual interpretation techniques or conventional to advanced classifiers, and opened up new vistas for diverse applications, including urban and regional studies (Reddy et al., 1996).
5. *IKONOS:* IKONOS, previously known as commercial remote sensing satellite (CRSS), was launched on September 24, 1999 to acquire global high-resolution imagery for individuals, organizations, and governmental institutes for development, mapping, and monitoring. The data from space imaging was provided to NASA under the scientific data purchase (SDP) contract to utilize the commercial imagery for research purposes. There are two sensors equipped in IKONOS that are a PAN sensor with an array width of 13,816 pixels and a multispectral sensor having 3,454-pixel array width (Helder et al., 2003). The ground sample distance of the PAN sensor and multispectral sensor constituted in IKONOS to provide high-resolution, intelligence quality imagery are 0.82 m and 3.28 m at nadir, respectively. The multispectral sensor collects data in blue, green, red, and NIR bands with the bandpass of 445–516 nm, 506–595 nm, 632–698 nm, and 757–853 nm, respectively. The revisit time of the IKONOS satellite is 3 days for 60° elevation, 11 days at 72° elevation, and 141 days for 89° elevation (Dial et al., 2003). The IKONOS satellite is highly efficient as it can pitch and roll to acquire data from a large strip, thereby collecting numerous small images as well as stereo images of the orbital pass area. During a single orbital pass, IKONOS can image a contiguous area of 4,700 km^2 at a 1-m GSD or 10,000 km^2 with a GSD.

6. *QUICKBIRD:* QUICKBIRD is a satellite very similar to IKONOS, besides the fact that IKONOS was launched in 1999, and it was launched in 2001 by Digital Globe. QUICKBIRD also has similar sensors to those of IKONOS, that is, PAN and multispectral sensors. QUICKBIRD has one of the highest resolution imaging capacities, which is 61 cm in nadir for the PAN sensor. The radiometric resolution of this satellite is 11 bit, which implies that it can be suitably used in urban terrain areas. The spectral range of panchromatic sensor of QUICKBIRD is 450–900 nm, whereas the multispectral sensor of QUICKBIRD captures images in four bands which are blue with spectral range 450–520 nm, green with spectral range 520–600 nm, red with spectral range 630–690 nm, and NIR with 760–900 nm spectral range (Jacobsen, 2003; Alexandrov et al., 2004). The image provided by QUICKBIRD varies in three different types depending on the processing of the raw image. The three types of imagery products are basic, standard, and orthorectified imagery.
7. *RapidEye AG:* RapidEye AG is a geospatial information provider constellation consisting of a set of five satellites providing valuable information in the field of agriculture, forestry, infrastructure, security, and emergency monitoring applications. It was launched in the year 2009 with the objective of capturing high-resolution multispectral imagery with the help of five satellites equally spaced in their orbits with a spatial distance of about 19 m to each other. The weight of the satellite is approximately 150 kg, placed in a sun-synchronous orbit of 97.9^0 inclination and 630 km height. The swath width of the satellites is 77 km, with the ground sample distance of approximately 6.5 m. The image is captured in various regions of the multispectral band, and the regions are red, red edge, green, blue, and near-infrared region. The headquarters of RapidEye is located in the city of Brandenburg a.d. Havel in Germany, with the station Stub Series Terminated Logic (SSTL) as the backup for taking responsibility for maintenance and emergency (Stoll et al., 2012). The satellite's constellation is beneficial for providing the continuous high-resolution data with a daily revisit option to acquire image of up to 4 million km^2 per day, repetitive remote sensing of large area and generating temporal stack to assess changes easily as well as stitching of single images to create a large scene allows the generation of a mosaic for the entire country or similar large areas, and with the capability of the constellation to acquire data twice a day with the daily revisit interval, the constellation is capable of capturing the data of an incidental site by latest one day time period and that will also depend on the time of the day incident occurred as well as Area of Interest (AOIs) (Sang et al., 2015).

18.2.3 Hyperspectral Sensors

1. *Hyperion:* Hyperion, the first spaceborne imaging spectrometer having identical orbital characteristics as the Landsat ETM+ multispectral sensor, is the spaceborne hyperspectral sensor launched in 2000 onboard the Earth Observer-1 (EO-1) satellite under the New Millennium Program of NASA.

The spatial resolution of Hyperion is 30 m with a swath width of 7.6 km across track and approximately 80.4 km along track. The Hyperion sensor has two spectrometers, one in the visible and near-infrared (VNIR) (bands 8–57, region 427–925 nm) and the other in the shortwave infrared (SWIR) region (bands 77–224, region 912–2,395 nm) (Folkman et al., 2001). Hyperion imagery for land use/land cover acquires a significant potential and was studied by various investigators like Goodenough et al. (2003), Walsh et al. (2008), Christian and Krishnayya (2009), and Pignatti et al. (2009).

18.3 TECHNIQUES USED IN PROCESSING REMOTE SENSING DATA

The big data in remote sensing and GIS needs extraordinarily smart technology to process the large volumes of satellite data in a flash of time. Although there are various software opted for this specified purpose like Walmart that combines machine learning and statistical techniques to extract patterns from the large volume of transaction data, Taobao, a Chinese company like eBay, uses large stream data mining techniques on users' browse data recorded on its website, and exploits a good deal of valuable information to facilitate their decision-making, etc. Still, the core techniques for all these software include a vast number of disciplines involving statistics, data mining, machine learning, neural networks, social network analysis, signal processing, pattern recognition, optimization methods, and visualization approaches. These are some significant techniques playing a crucial role in processing the remote sensing big data and overlap with each other frequently for an effective processing outcome.

18.3.1 Optimization Technique

The strategies used for enhancing the effectiveness and efficiency of the procedure used in the processing of remote sensing big data are generally referred to as optimization techniques. The basic objectives behind using the optimization techniques can be: reducing the time and power used in the processing procedure, minimizing as much space utilized by the data as possible to make the management of data easier, enhancing the quality of the data by fixing most of the distortion in the data, as well as effectively handling large sets of big data files and managing them (Sun et al., 2019). The basic processes opted within this technique can be listed as:

- *Compression:* This process is used to reduce the size of the data with reference to maintaining the integrity of data, as in the case of PNG format for images, FLAC for audio files, as well as discarding the less critical information by compressing the file to a greater extent, as in the case of JPEG for images and MP3 for audio files. Wavelet compression is also a very common optimization technique used for effectively processing big data in remote sensing (Zhang et al., 2024).

- *Dimensional Reduction:* Operations such as principal component analysis, as well as feature selection, are used to select the relevant band and simplify the data in accordance with data having higher variance.
- *Noise Filtration:* To preserve the details of the data, median filtering or a wavelet denoising process is used by the optimization techniques.

18.3.2 Data Mining

With reference to the big data of remote sensing, data mining techniques are used to extract the useful information from sets of large data by creating patterns or trends using a number of techniques such as the classification techniques that are employed to categorize the data in predefined categories as well as for image recognition with the help of various algorithms such as neural networks, support vector machines (SVMs), Naive Bayes, k-Nearest Neighbor (k-NN), etc. Other techniques include clustering, used for arranging similar data into a cluster with the help of algorithms such as k-Means, DBSCAN, hierarchical clustering, clustering large applications(CLARA) algorithm, clustering large applications based upon randomized search(CLARANS), balanced iterative reducing using cluster hierarchies (BIRCH) algorithm; regression which is used to predict a value on the basis of a variable with the help of algorithms such as linear regression, polynomial regression, support vector regression.

18.3.3 Machine Learning

Machine learning is one of the most important tools across all fields in the current scenario, as it enables the proper synchronization of discovering knowledge and handling it intelligently. The role of machine learning in handling big data is paramount as the increasingly large volumes of structured and unstructured data are troublesome to handle with traditional software and applications (Kumar et al., 2022). Various machine learning algorithms along with their features are listed in Table 18.4.

18.4 SOFTWARE USED FOR HANDLING BIG DATA

To handle big data of remote sensing, scientists have developed a wide variety of extensive and efficient new technologies to acquire, curate, analyze, and store big data. Still, the results are far from the needs and expectations. These technologies extend to various fields such as medical, computer science, mathematics, academic research, remote sensing, and several others. There is a need to adopt a multidisciplinary approach for the development of new technologies more efficient and effective than the present ones. In this study, the current technologies and software being used for handling big data are presented.

18.4.1 Cloud-Based Platforms

1. *Google Earth Engine (GEE):* It is one of the major platforms launched by Google in 2010, gaining significant attention nowadays with respect to the processing of remote sensing big data. GEE is an open-source software that

TABLE 18.4
Details of Various ML Algorithms along with Their Citations

S. No.	ML Algorithm	Type of Algorithm	Use of ML Algorithm	References
1	Naive Bayes Classifier	Supervised machine learning	Classification of big data	Bayes (1968), Murphy (2006), Yang (2018)
2	k-Nearest Neighbor	Unsupervised machine learning	Regression as well as classification of big data	Islam et al. (2007), Steinbach and Tan (2009)
3	Support Vector Machine	Supervised machine learning	Classification of big data as well as anomaly detection	Abdullah and Abdulazeez (2021), Chandra and Bedi (2021)
4	Random Forest	Ensemble machine learning	Regression, time series analysis as well as classification of big data	Pal (2005), Rodriguez et al. (2012), Parmar et al. (2019), Kumar et al. (2022)
5	Artificial Neural Network	Unsupervised machine learning	Regression, time series analysis, anomaly detection as well as classification of big data	Wong and Hsu (2006), Azarmdel et al. (2020), Anami and Elemmi (2022), Singh et al. (2024)
6	Multilayer Perceptrons	Supervised machine learning		Murtagh (1991), Kruse et al. (2022)
7	Dimensionality Reduction	Unsupervised machine learning	Anomaly detection	Velliangiri and Alagumuthukrishnan (2019), Reddy et al. (2020)

processes real-time satellite data and provides it globally using the Google Cloud. It hosts petabytes of data including temporal and geospatial data from more than 40 years. GEE also provides a set of state-of-the-art used for the classification of images on the basis of pixels, so that it can be used for crop mapping (Tamiminia et al., 2020; Deoli et al., 2023). In addition, GEE users do not need to download the data or any other application to process it; instead, they get the privilege to enjoy the packages provided by the GEE software, or they can process their own database on GEE. GEE can be accessed via an internet-based application programming interface (API) and a web-based interactive development environment. Accumulation of various built-in algorithms to analyze data at a planetary scale and also helps scientists to develop their own algorithms with less effort than before makes the platform more popular among users (Amani et al., 2020).

2. *Amazon Web Services:* Amazon Web Services (AWS) is a cloud-based platform launched by Amazon in 2006. AWS is a pay-as-you-go platform where

you need to pay the charges according to the hours of usage. AWS also provides a dedicated Earth Observation cloud, also known as "Earth AWS" as a part of its Public Dataset program (Ferreira et al., 2020). AWS processes data from various high-resolution satellites such as Landsat-8, Sentinel-1, Sentinel-2, China–Brazil Earth Resources Satellite program, National Oceanographic and Atmospheric Administration Advanced (NOAA) image datasets, as well as global model outputs. AWS also hosts open data supplied by DigitalGlobe with its SpaceNet challenges as well, and AWS offers the largest suite of machine learning services (Xu et al., 2022; Yusoff et al., 2022).

3. *Distributed File System (DFS):* It is another type of cloud-based platform used for handling a more complex and comprehensive interface. The files stored in DFS can only be accessed through mounted hosts. It includes software such as Colossum, which is the name of Google files; and Hadoop, also known as Apache Hadoop, is a software used for handling large data (Li et al., 2016; Pekturk and Unal, 2018). The Apache Hadoop software library is a framework that allows for the distributed processing of large datasets across clusters of computers using simple programming models. It is designed to scale up from single servers to thousands of machines, each offering local computation and storage. Instead of relying on hardware for delivering high availability, the library is designed such that it self-manages the problem caused by the application.
4. *Relational Database Management System (RDBMS):* It is a cloud-based platform that allows users to incorporate changes in the database without putting restrictions on the format of the database. It is a widely used software and is generally used for transactional operations of a large database. The RDBMS works on the principle of atomicity, consistency, isolation, and durability (ACID) model to ensure the consistency and integrity of the data. RDBMS stores data in the form of tables using structured query language (SQL). The RDBMS stores data in a row-like structure that connects to each other, and it also enables functions to maintain the integrity, accuracy, and consistency of data, which makes it different from DBMS (Powell et al., 2011; Ifimov et al., 2017).
5. *NoSQL:* Not only is structured query language a platform derived from RDBMS and SQL. The name is NoSQL because it neither inherited the properties of RDBMS nor completely of SQL. The NoSQL neither follows the ACID model nor focuses on transactional operations; instead, it emphasizes the principles of consistency, availability, and partition tolerance (CAP). The NoSQL technologies are generally classified into four types, including wide-column, key-value, document, and search engine, with each type having its individual characteristics, making it suitable for use in different scenarios (Zafar et al., 2016; Hein and Blankenbach, 2021).

18.4.3 Desktop-based Remote Sensing Big Data Software

1. *ArcGIS Pro:* ArcGIS, the latest version of ESRI's ArcGIS, is a premium desktop GIS application that provides users some advanced features such as generation 2D, 3D, and 4D visualization and analyzes them using tools unparalleled by

any other desktop GIS software. It also provides users with seamless sharing of ArcGIS software to foster GIS solutions to speed up the processing and sharing of data (Rasoul, 2024). Some of the key features of ArcGIS Pro are:

Cartography: The state-of-the-art tools such as smart mapping, multi-attribute symbology, and flexible labelling and annotation options provided by the ArcGIS software, make the maps more significant and effective, as well as increase their credibility on a global scale.

2D/3D Mapping: ArcGIS Pro offers users a workspace where they can work or create projects by incorporating 2D as well as 3D maps.

Real-Time Visualization: With the use of ArcGIS Pro software, one can build creative yet impactful maps, tables, and charts in informative and powerful dashboards (Barsai, 2018).

2. *ERDAS IMAGINE:* It is the fourfold advanced version of the first ERDAS software, known as ERDAS 4, launched in 1978. ERDAS IMAGINE was demonstrated in October 1991 and was launched in 1992 as ERDAS 8. With the integration of ARC/INFO and ERDAS IMAGINE, ERDAS launched a new software in 1993 known as ERDAS IMAGINE 8.2, incorporating vector data processing and licensed by ESRI. Another updated version, known as ERDAS IMAGINE 9.3, was released in September 2008. Basically, ERDAS IMAGINE is a raster data editor with primary emphasis on processing the raster data and allowing users to prepare and enhance the imagery for mapping in geographical information systems (GIS) or computer-aided design (CADD) software (Imagine E.R.D.A.S, 2008; Nelson and Khorram, 2018).
3. *ENVI:* ENVI is one of the most popular software for image processing in the current scenario due to its robust capabilities in the field of multispectral and hyperspectral imagery. It is based on a user-oriented approach and provides an interface that can be dealt with easily by any beginner or field expert. This image processing software provides users with a suite of well-equipped tools including calibration, preprocessing, atmospheric correction, pan sharpening, orthorectification, and image registration (Li et al., 2020). ENVI also has an automated workflow for analysis and processing of images to ease the users and allow them to access information from all sorts of geospatial data. The tools provided by the ENVI set a standard for the industry. For instance, ENVI spectral tools can detect targets, calculate vegetation and forest health, map materials of interest, and even they are used to measure marine debris and pollution, analyze wildlife habitats, map oil slicks, evaluate water quality, mitigate wildfires, detect methane leaks, identify minerals, map vegetation health, and assist with many Défense and Intelligence applications (Barrett 2013; Chuvieco 2020).

18.5 OPEN-SOURCE TOOLS FOR HANDLING REMOTE SENSING BIG DATA

Quantum GIS was founded by Gary Sherman of Alaska in 2002 and was launched by the Open-Source Geospatial Foundation in 2009. It is a free, open-source GIS software for desktop that facilitates users to analyze and edit spatial information. It also provides the feature of composing and exporting geographical maps by using both raster as well as vector data. The platform-independent software employs operating systems like Linux, Unix, Mac OSX, Windows, and Android with the help of various development platforms like C++ or Python (Khan and Mohiuddin, 2018). The software integrates with various free and open-source GIS software such as Geospatial Data Abstraction Library, Geographic Resources Analysis Support System, Sextante, System for Automated Geo-scientific Analysis, Orfeo Tool Box, along with R software, to support numerous GIS formats and databases (Congedo, 2021). The GIS formats supported by QGISA as raster data include TIFF, JPEG, GIF, Erdas Imagine, ECW, MisterSID, NetCDF, ESRI ArchInfo Grid, and the supported vector formats are ESRI shapefile, DXF, SpatiaLite, Mapinfo, GML, KML, Interlis (Switzerland) (Denis, 2021).

The Geographic Resources Analysis Support System (GRASS) was developed by the U.S. Army Corps of Engineers as a military assistance tool in 1982. It uses both raster and vector data to integrate image processing along with the data visualization through the involvement of numerous modules within the interface of the software. GNU/Linux, Mac OSX, and MS Windows operating systems are some operating systems on which GRASS is operational. The latest version of GRASS, that is, GRASS 6.4, introduces a number of improvements including support for vector network analysis as well as a 2D/3D vector engine. The DBF file and the SQL-based DBMS are used for managing attributes in the software by integrating with GDAL/OGR libraries to support a set of data formats. The software uses an NVIZ tool for visualizing the vector data as well as 2D/3D raster data (Neteler et al., 2008).

The GRASS software allows multiple users to work on the same project with the help of its LOCATION/MAPSET directory and can import data in numerous formats such as PNG, TIF, and PDF, with the use of OGC web services. GRASS supports a wide array of map projections and geodetic datums and includes georeferencing and reprojection tools. GRASS integration with software like Quantum GIS and R provides users with the facility to statistically analyze and support databases like PostgreSQL, MySQL, and SQLite. LiDAR processing, hydrological modelling, terrain analysis, and more constitute some of the diverse functions of the GRASS software (Ciolli and Zatelli, 2005).

18.5 CONCLUSION

The data generated by various sources, such as Internet of Things, social media platforms, and satellites, is of tremendously large volume, having significant distinction in variety, indicating forms of data obtained to be raw, structured, unstructured, or semi-structured, and velocity referring to the speed of capturing and collecting data for processing. This bulk data which is extremely hard to analyze, process, and store, is known as big data and plays a crucial role in today's advanced pacing technology. The role of big data intensifies even more when it is referred to in the field of remote sensing, and with the advancement in technologies, various software are now in the market providing reliable results, as per the case with handling big data in remote sensing is concerned. One of such software, which according to Google is the most searched software for big data, is Hadoop. As far as processing big data is concerned, machine learning has gained new heights for improved handling of big data by merging two or more machine learning algorithms. Even though the techniques are there to manage the big data, there still stands a need to refine the techniques and dive into the insight of the upcoming big data load and problems associated with it, to adapt to the continuous bulging pressure of the big data management system in remote sensing.

REFERENCES

Abdullah, D. M., and Abdulazeez, A. M. (2021). Machine learning applications based on SVM classification a review. *Qubahan Academic Journal*, 1(2), 81–90.

Abdulraheem, M. I., Zhang, W., Li, S., Moshayedi, A. J., Farooque, A. A., and Hu, J. (2023). Advancement of remote sensing for soil measurements and applications: A comprehensive review. *Sustainability*, 15(21), 15444.

Alby, F. (2005). Spot 1 end of life disposition manoeuvres. *Advances in Space Research*, 35(7), 1335–1342.

Alexandrov, A., Hristova, T., Ivanova, K., Koeva, M., Madzharova, T., and Petrova, V. (2004, July). Application of QuickBird satellite imagery for updating cadastral information. In *XX Congress of ISPRS* (pp. 386–391). Istanbul.

Amani, M., Ghorbanian, A., Ahmadi, S. A., Kakooei, M., Moghimi, A., Mirmazloumi, S. M., and Brisco, B. (2020). Google earth engine cloud computing platform for remote sensing big data applications: A comprehensive review. *IEEE Journal of Selected Topics in Applied Earth Observations and Remote Sensing*, 13, 5326–5350.

Anami, B. S., and Elemmi, M. C. (2022). Comparative analysis of SVM and ANN classifiers for defective and non-defective fabric images classification. *The Journal of the Textile Institute*, 113(6), 1072–1082.

Armenakis, C., and Savopol, F. (1998). Mapping potential of the IRS-1C PAN satellite imagery. *International Archives of Photogrammetry and Remote Sensing*, 32, 23–26.

Azarmdel, H., Jahanbakhshi, A., Mohtasebi, S. S., and Muñoz, A. R. (2020). Evaluation of image processing technique as an expert system in mulberry fruit grading based on ripeness level using artificial neural networks (ANNs) and support vector machine (SVM). *Postharvest Biology and Technology*, 166, 111201.

Baraldi, A., Durieux, L., Simonetti, D., Conchedda, G., Holecz, F., and Blonda, P. (2009). Automatic spectral-rule-based preliminary classification of radiometrically calibrated SPOT-4/-5/IRS, AVHRR/MSG, AATSR, IKONOS/QuickBird/OrbView/GeoEye, and DMC/SPOT-1/-2 imagery—Part I: System design and implementation. *IEEE Transactions on Geoscience and Remote Sensing*, 48(3), 1299–1325.

Barrett, E. C. (2013). *Introduction to Environmental Remote Sensing*. Routledge.

Barsai, G. (2018). Getting to know ArcGIS PRO. *Photogrammetric Engineering & Remote Sensing*, 84(4), 181–182.

Bayes, T. (1968). Naive bayes classifier. Article Sources and Contributors, 1–9.

Campbell, James B., and Wynne, Randolph H. (2011). *Introduction to remote sensing*. Guilford press.

Cappa, F., Oriani, R., Peruffo, E., and McCarthy, I. (2021). Big data for creating and capturing value in the digitalized environment: Unpacking the effects of volume, variety, and veracity on firm performance. *Journal of Product Innovation Management*, 38(1), 49–67.

Chandra, M. A., and Bedi, S. S. (2021). Survey on SVM and their application in image classification. *International Journal of Information Technology*, 13(5), 1–11.

Chen, H., Chiang, R. H., and Storey, V. C. (2012). Business intelligence and analytics: From big data to big impact. MIS *Q*uarterly, 1165–1188.

Chen, S. Y., Liu, C. Y., Huang, C. Y., Hsu, S. C., Li, H. W., Lin, P. H., Cheng, J. P., and Huang, C. Y. (2021). An analysis study of FORMOSAT-7/COSMIC-2 radio occultation data in the troposphere. *Remote Sensing*, 13(4), 717.

Christian, B., and Krishnayya, N. S. R. (2009). Classification of tropical trees growing in a sanctuary using Hyperion (EO-1) and SAM algorithm. *Current Science*, 96(12), 1601–1607.

Chuvieco, E. (2020). *Fundamentals of Satellite Remote Sensing: An Environmental Approach*. CRC press.

Ciolli, M., and Zatelli, P. (2005). GRASS GIS an overview of the applications in 2005'. *Geomatics Workbooks*, 6, 1–39.

Congedo, L. (2021). Semi-Automatic Classification Plugin: A Python tool for the download and processing of remote sensing images in QGIS. *Journal of Open Source Software*, 6(64), 3172.

Cracknell, A. P. (1997). *Advanced Very High Resolution Radiometer AVHRR*. Crc Press.

Denis, A. (2021). Introduction to QGIS with QGIS 3.10. 0-Geographic Information System (GIS) *practical work*, agris.fao.org.

Deoli, V., and Kumar, D. (2019). Remote sensing and GIS approach for spatiotemporal mapping of Ramganga reservoir. *International Journal of Current Microbiology and Applied Sciences*, 8(5), 775–783.

Deoli, V., Kumar, D., Kumar, M., Kuriqi, A., and Elbeltagi, A. (2021). Water spread mapping of multiple lakes using remote sensing and satellite data. *Arabian Journal of Geosciences*, 14, 1–15.

Deoli, V., Kumar, D., and Kuriqi, A. (2022). Detection of water spread area changes in eutrophic lake using landsat data. *Sensors*, 22(18), 6827.

Deoli, V., Kumar, D., Shikha, D., Saw, S., and Patel, R. (2023). Mapping and trend analysis in water spread area of upper and lower lakes of Bhopal, using remote sensing technique. In *Surface and Groundwater Resources Development and Management in Semi-arid Region: Strategies and Solutions for Sustainable Water Management,* (pp. 177–189). Cham: Springer International Publishing..

Dial, G., Bowen, H., Gerlach, F., Grodecki, J., and Oleszczuk, R. (2003). IKONOS satellite, imagery, and products. *Remote Sensing of Environment*, 88(1-2), 23–36.

Ferreira, K. R., Queiroz, G. R., Camara, G., Souza, R. C. M., Vinhas, L., Marujo, R. F. B., Simoes, R. E. O., Noronha, C.A.F., Costa, R. W., Arcanjo, J. S., Gomes, V. C. F., and Zaglia, M. C. (2020, March). Using remote sensing images and cloud services on AWS to improve land use and cover monitoring. In *2020 IEEE Latin American GRSS & ISPRS Remote Sensing Conference (LAGIRS)* (pp. 558–562), IEEE.

Folkman, M., J., Pearlman, L. Liao, P., and Jarecke, (2001). EO-1/Hyperion hyperspectral imager design, development, characterization, and calibration. *In: Hyperspectral Remote Sensing of the Land and Atmosphere, Proceedings of SPIE* (Vol. 4151 pp. 40–51), Sendai, Japan.

Gantz, J., and Reinsel, D. (2012). The digital universe in 2020: Big data, bigger digital shadows, and biggest growth in the far East. *IDC iView: IDC Analyze the Future*, 2007(2012), 1–16.

García-Mora, T. J., Mas, J. F., and Hinkley, E. A. (2012). Land cover mapping applications with MODIS: A literature review. *International Journal of Digital Earth*, 5(1), 63–87.

Gibson, P. (2013). *Introductory Remote Sensing Principles and Concepts*. Routledge.

Goodenough, D. G., Dyk, A., Niemann, K. O., Pearlman, J. S., Chen, H., Han, T., ... and West, C. (2003). Processing Hyperion and ALI for forest classification. *IEEE Transactions on Geoscience and Remote sensing*, 41(6), 1321–1331.

Gupta, P. K., Kumari, B., Gupta, S. K., and Kumar, D. (2020). Nitrate-leaching and groundwater vulnerability mapping in North Bihar, India. *Sustainable Water Resources Management*, 6(3), 48.

Hein, N., and Blankenbach, J. (2021). Evaluation of a NoSQL Database for Storing Big Geospatial Raster Data. *GI_Forum*. Munster, Issue 1, 76–84.

Helder, D., Coan, M., Patrick, K., and Gaska, P. (2003). IKONOS geometric characterization. *Remote Sensing of Environment*, 88(1-2), 69–79.

Hilbert, M., and López, P. (2011). The world's technological capacity to store, communicate, and compute information. *Science*, 332(6025), 60–65.

Holben, B. N. (1986). Characteristics of maximum-value composite images from temporal AVHRR data. *International Journal of Remote Sensing*, 7(11), 1417–1434.

Ifimov, G., Pigeau, G., Arroyo-Mora, J. P., Soffer, R., and Leblanc, G. (2017, October). A geospatial database model for the management of remote sensing datasets at multiple spectral, spatial, and temporal scales. In *Earth Resources and Environmental Remote Sensing/GIS Applications VIII, SPIE*, (Vol. 10428, pp. 560–570), Warsaw, Poland.

Islam, M. J., Wu, Q. J., Ahmadi, M., and Sid-Ahmed, M. A. (2007, November). Investigating the performance of naive-bayes classifiers and k-nearest neighbor classifiers. In *2007 International Conference on Convergence Information Technology (ICCIT 2007)* (pp. 1541–1546). IEEE, Gwangju, Korea (South).

Jacobsen, K. A. R. S. T. E. N. (2003). Geometric potential of IKONOS-and QuickBird-images. *Photogrammetric Weeks '03*, 101–110.

Khan, S., and Mohiuddin, K. (2018). Evaluating the parameters of ArcGIS and QGIS for GIS Applications. *International Journal of Advanced Research in Science, Engineering and Technology*, 7, 582–594.

Kruse, R., Mostaghim, S., Borgelt, C., Braune, C., and Steinbrecher, M. (2022). Multi-layer perceptrons. In *Computational Intelligence: A Methodological Introduction* (pp. 53–124).Cham: Springer International Publishing.

Kuenzer, C., and Dech, S. (2013). Thermal infrared remote sensing: Sensors, methods, applications. *Photogrammetric Engineering and Remote Sensing*, 81, 359–360.

Kumar, D., and Deoli, M. V. (2020). Study of land surface temperature using Landsat-8 imagery: A case study of Udhamsingh nagar, Uttarakhand. The Institution of Engineers (India), 88.

Kumar, D., Soni, A., and Kumar, M. (2022). Retrieval of land surface temperature from landsat-8 thermal infrared sensor data. *Journal of Human, Earth, and Future*, 3(2), 159–168.

Kumar, M., Elbeltagi, A., Pande, C. B., Ahmed, A. N., Chow, M. F., Pham, Q. B., Kumari, A., and Kumar, D. (2022). Applications of data-driven models for daily discharge estimation based on different input combinations. *Water Resources Management*, 36(7), 2201–2221.

Kumar, P., Rao, K. V., Ravindranath, S., Maithani, S., Siddiqui, A., Bharath, B. D., and Bhavani, S. V. L. (2021). A review of IRS-1C applications in urban and regional studies, and infrastructure planning. *Journal of the Indian Society of Remote Sensing*, 49(1), 161–177.

Kumari, B., Gupta, P. K., and Kumar, D. (2019). In-situ observation and nitrate-N load assessment in Madhubani District, Bihar, India. *Journal of the Geological Society of India*, 93, 113–118.

Li, X. C., Liu, L. L., and Huang, L. K. (2020). Comparison of several remote sensing image classification methods based on envi. *The International Archives of the Photogrammetry, Remote Sensing and Spatial Information Sciences*, 42, 605–611.

Li, Z., Li, X., Liu, T., and Xie, J. (2016, March). Towards building a distributed file system for remote sensing image process. In *2016 IEEE International Conference on Big Data Analysis (ICBDA)*, (pp. 1–4).

Ma, Y., Wu, H., Wang, L., Huang, B., Ranjan, R., Zomaya, A., and Jie, W. (2015). Remote sensing big data computing: Challenges and opportunities. *Future Generation Computer Systems*, 51, 47–60.

Masud, M. J., and Bastiaanssen, W. G. (2017). Remote sensing and GIS applications in water resources management. *Water Resources Management*, 351–373.

Mcclain, E. P. (1985). Advanced Very High Resolution Radiometer (AVHRR) sensor comparison characteristics. JPL Satellite-Derived Sea Surface Temperature.

Muhadi, N. A., Abdullah, A. F., Bejo, S. K., Mahadi, M. R., and Mijic, A. (2020). The use of LiDAR-derived DEM in flood applications: A review. *Remote Sensing*, 12(14), 2308.

Murphy, K. P. (2006). Naive bayes classifiers. University of British Columbia, 18(60), 1–8.

Murtagh, F. (1991). Multilayer perceptrons for classification and regression. *Neurocomputing*, 2(5-6), 183–197.

Nelson, S. A., and Khorram, S. (2018). *Image Processing and Data Analysis with ERDAS IMAGINE®*. CRC Press.

Neteler, M., Beaudette, D. E., Cavallini, P., Lami, L., and Cepicky, J. (2008). Grass gis. In *Open Source Approaches in Spatial Data Handling* (pp. 171–199).Berlin, Heidelberg: Springer Berlin Heidelberg.

Pal, M. (2005). Random forest classifier for remote sensing classification. *International Journal of Remote Sensing*, 26(1), 217–222.

Parmar, A., Katariya, R., and Patel, V. (2019). A review on random forest: An ensemble classifier. In *International Conference on Intelligent Data Communication Technologies and Internet of Things (ICICI) 2018* (pp. 758–763). Springer International Publishing.

Pektürk, M. K., and Ünal, M. (2018, August). Performance-aware high-performance computing for remote sensing big data analytics. In *Data Mining BoD–Books on Deman* (Vol. 69).

Pignatti, S., Cavalli, R. M., Cuomo, V., Fusilli, L., Pascucci, S., Poscolieri, M., and Santini, F. (2009). Evaluating Hyperion capability for land cover mapping in a fragmented ecosystem: Pollino National Park, Italy. *Remote Sensing of Environment*, 113(3), 622–634.

Powell, J., Erukulla, G., Buhisi, M., and Velauthapillai, B. (2011). A relational database management system for atmospheric compensation research. *Computers & Geosciences*, 37(4), 588–597.

Rasoul, K. (2024). The introduction of ArcGIS Pro Software. *The Scientific-Cultural Journal of Ecosphere*, 7(3), 71–76.

Reddy, G. T., Reddy, M. P. K., Lakshmanna, K., Kaluri, R., Rajput, D. S., Srivastava, G., and Baker, T. (2020). Analysis of dimensionality reduction techniques on big data. *IEEE Access*, 8, 54776–54788.

Reddy, P. R., Kumar, K. V., and Seshadri, K. (1996). Use of IRS-1C data in groundwater studies. *Current Science*, 600–605.

Rodriguez-Galiano, V. F., Ghimire, B., Rogan, J., Chica-Olmo, M., and Rigol-Sanchez, J. P. (2012). An assessment of the effectiveness of a random forest classifier for land-cover classification. *ISPRS Journal of Photogrammetry and Remote Sensing*, 67, 93–104.

Roy, D. P., Kovalskyy, V., Zhang, H. K., Vermote, E. F., Yan, L., Kumar, S. S., and Egorov, A. (2016). Characterization of landsat-7 to landsat-8 reflective wavelength and normalized difference vegetation index continuity. *Remote Sensing of Environment*, 185, 57–70.

Roy, P. S., Behera, M. D., and Srivastav, S. K. (2017). Satellite remote sensing: sensors, applications and techniques. *Proceedings of the National Academy of Sciences, India Section A: Physical Sciences*, 87, 465–472.

Salomonson, V. V., Barnes, W. L., Maymon, P. W., Montgomery, H. E., and Ostrow, H. (1989). MODIS: Advanced facility instrument for studies of the Earth as a system. *IEEE Transactions on Geoscience and Remote Sensing*, 27(2), 145–153.

Sang, H., Zhang, J., Zhai, L., Xie, W., and Sun, X. (2015, December). Analysis of RapidEye imagery for agricultural land mapping. In *International Conference on Intelligent Earth Observing and Applications 2015, SPIE* (Vol. 9808, pp. 561–569).

Singh, R. (2022). Introduction: Re-envisioning Advances in Remote Sensing. In *Re-envisioning Advances in Remote Sensing* (pp. 1–22). CRC Press.

Singh, S., Kumar, D., Vishwakarma, D. K., Kumar, R., and Kushwaha, N. L. (2024). Seasonal rainfall pattern using coupled neural network-wavelet technique of southern Uttarakhand, India. *Theoretical and Applied Climatology*, 155(6), 5185–5201.

Steinbach, M., and Tan, P. N. (2009). kNN: K-nearest neighbors. In *The Top Ten Algorithms in Data Mining* (pp. 165–176). Chapman and Hall/CRC.

Stoll, E., Konstanski, H., Anderson, C., Douglass, K., and Oxfort, M. (2012, March). The RapidEye constellation and its data products. In *2012 IEEE Aerospace Conference* (pp. 1–9).

Sun, J., Zhang, Y., Wu, Z., Zhu, Y., Yin, X., Ding, Z., ... and Plaza, A. (2019). An efficient and scalable framework for processing remotely sensed big data in cloud computing environments. *IEEE Transactions on Geoscience and Remote Sensing*, 57(7), 4294–4308.

Tamiminia, H., Salehi, B., Mahdianpari, M., Quackenbush, L., Adeli, S., and Brisco, B. (2020). Google Earth Engine for geo-big data applications: A meta-analysis and systematic review. *ISPRS Journal of Photogrammetry and Remote Sensing*, 164, 152–170.

van Wesemael, B., Chabrillat, S., Sanz Dias, A., Berger, M., and Szantoi, Z. (2023). Remote sensing for soil organic carbon mapping and monitoring. *Remote Sensing*, 15(14), 3464.

Velliangiri, S., and Alagumuthukrishnan, S. J. P. C. S. (2019). A review of dimensionality reduction techniques for efficient computation. *Procedia Computer Science*, 165, 104–111.

Walsh, S. J., McCleary, A. L., Mena, C. F., Shao, Y., Tuttle, J. P., González, A., and Atkinson, R. (2008). QuickBird and Hyperion data analysis of an invasive plant species in the Galapagos Islands of Ecuador: Implications for control and land use management. *Remote Sensing of Environment*, 112(5), 1927–1941.

Wong, W. T., and Hsu, S. H. (2006). Application of SVM and ANN for image retrieval. *European Journal of Operational Research*, 173(3), 938–950.

Xu, C., Du, X., Fan, X., Giuliani, G., Hu, Z., Wang, W., ... and Guo, H. (2022). Cloud-based storage and computing for remote sensing big data: A technical review. *International Journal of Digital Earth*, 15(1), 1417–1445.

Yahia, O., Guida, R., and Iervolino, P. (2021). Novel weight-based approach for soil moisture content estimation via synthetic aperture radar, multispectral and thermal infrared data fusion. *Sensors*, 21(10), 3457.

Yang, F. J. (2018, December). An implementation of naive bayes classifier. In *2018 IEEE International Conference on Computational Ccience and Computational Intelligence (CSCI)* (pp. 301–306).

Yusoff, F. H., Kamarudin, S. N. K., and Harun, N. (2022). Big data-based image handling—A review of implementation using amazon web services. *Multimedia Technologies in the Internet of Things Environment*, 3, 95–106.

Zafar, R., Yafi, E., Zuhairi, M. F., and Dao, H. (2016, May). Big data: The NoSQL and RDBMS review. In *2016 IEEE International Conference on Information and Communication Technology (ICICTM)* (pp. 120–126).

Zhang, S., Xue, Y., Zhang, H., Zhou, X., Li, K., and Liu, R. (2024). Improved Hungarian algorithm–based task scheduling optimization strategy for remote sensing big data processing. *Geo-Spatial Information Science*, 27(4), 1141–1154.

Index

Note: Figures are indicated by *italics*. Tables are indicated by **bold**.

For Product Safety Concerns and Information please contact our EU representative GPSR@taylorandfrancis.com Taylor & Francis Verlag GmbH, Kaufingerstraße 24, 80331 München, Germany